Materials Modelling using Density Functional Theory

Materials Modelling using Density Functional Theory

Properties and Predictions

Feliciano Giustino

Department of Materials, University of Oxford

OXFORD
UNIVERSITY PRESS

OXFORD
UNIVERSITY PRESS

Great Clarendon Street, Oxford, OX2 6DP,
United Kingdom

Oxford University Press is a department of the University of Oxford.
It furthers the University's objective of excellence in research, scholarship,
and education by publishing worldwide. Oxford is a registered trade mark of
Oxford University Press in the UK and in certain other countries

First Edition published in 2014

Published in the United States of America by Oxford University Press
198 Madison Avenue, New York, NY 10016, United States of America

British Library Cataloguing in Publication Data

Data available

Library of Congress Control Number: 2013953616

ISBN 978–0–19–966243–2 (hbk.)
ISBN 978–0–19–966244–9 (pbk.)

Printed and bound by
CPI Group (UK) Ltd, Croydon, CR0 4YY

To Nicola and Anna

'For whatever it's worth, I'm here to tell you that it is possible. It is possible.'

— V. A. Freeman

Preface

This book is intended to be an introduction to the modelling of materials starting from the first principles of quantum mechanics. For the reader who is not familiar with the notion of 'first-principles calculations', one could simply say that this book is about the art of using the periodic table, quantum mechanics and computers in order to understand and possibly predict certain properties of materials.

The primary audience for this book are senior undergraduate and first-year graduate students in Materials Science, Physics, Chemistry and Engineering. Experienced researchers and engineers who are approaching the quantum theory of materials for the first time may also find this book useful as a light-touch introduction.

As advanced materials make their way into every aspect of modern life and society, from electronics to construction, transport and energy, the use of quantum mechanics in the quantitative non-empirical modelling of materials is experiencing a rapid growth in university departments, research institutes and industrial laboratories alike.

Following this trend, in the Department of Materials at the University of Oxford we felt the need to complement the undergraduate curriculum in Materials Science with introductory courses on materials modelling using quantum mechanics. This book was inspired by a series of lectures delivered at Oxford by the author, for two undergraduate modules in Materials Science: *Electronic structure of materials* (second year) and *Prediction of material properties using density functional theory* (third year). The latter course is also attended by doctoral students in their first year. The present book was written to make the learning of atomistic materials modelling as easy as possible, in anticipation that this subject will become a core discipline in Materials Science education.

The key idea of this book is to introduce the reader to advanced concepts in quantum mechanics and materials science by assuming only elementary prior knowledge of classical mechanics, electrodynamics and quantum mechanics, at the level expected for students in Materials Science or Engineering. Students reading Physics or Chemistry will be already familiar with many of the concepts presented in this book, and will mostly benefit from the concise and pragmatic introduction to density functional theory and its uses.

At the time of writing, the many textbooks available on materials modelling broadly fall into two categories. On the one hand we find books devoted to the theory of molecules and solids, including several classics which shaped entire generations of scientists, such as Kittel (1976) and Ashcroft and Mermin (1976) for solids. These books generally pre-dated the rise of density functional theory; hence they inevitably miss those aspects of materials modelling relating to 'first-principles calculations'. On the other hand we have presentations devoted to density functional theory and its applications, which are widely adopted and have become definitive references for doctoral students and experienced researchers in this field, such as those by Parr and Yang (1989) and Martin (2004). Textbooks in this second category tend to address a relatively specialized audience, and are slightly too advanced for undergraduate students. The present textbook is meant to fill the gap between these two categories, by presenting density functional theory and first-principles materials modelling in a way which should be accessible to undergraduates. In this sense, the author's hope is to bridge between elementary notions of quantum mechanics, the theory of materials and first-principles calculations, starting from a minimal set of prerequisites and without requiring a formal training in advanced theoretical physics.

The presentation in this book revolves around the idea that it is possible to formulate a theory of materials whereby apparently unrelated properties can be rationalized and quantified within a single mathematical model, namely the Schrödinger equation. For this reason emphasis is placed on the unifying concepts, and simple heuristic arguments are provided whenever possible. Formal derivations are also given in many cases, in order to convince the reader that there is no need to formulate a new theory for each material property, and that in many cases we can find answers in the common root represented by the Schrödinger equation.

The inspiring principle of this book is borrowed from one of the slogans of the Perl programming language, 'Easy things should be easy and hard things should be possible' (L. Wall). In the present context, whenever a property can be understood using simple intuitive arguments we will make no attempts at a rigorous mathematical justification. Conversely, whenever it seems useful to dissect the mathematical formalism, we will go through the derivations in order to convince ourselves that there is nothing to be afraid of.

The presentation style is somewhat cross-disciplinary, insofar as an attempt is made to seamlessly combine materials science, quantum mechanics, electrodynamics and numerical analysis, without using a compartmentalized approach. This choice truly reflects the spirit of first-principles materials modelling, which finds its place at the intersection between traditional disciplines such as materials science, physics, chemistry and modern high-performance computing and software engineering.

Throughout the book emphasis is placed on numerical values. One of the greatest achievements of the theory addressed in this book is to make it possible, in many cases, to accurately predict the outcomes of actual measurements. This feature is perhaps the most distinctive and unique strength of density functional theory, as it is precisely the evolutionary step from *qualitative* theories of materials to *quantitative* predictions of their properties that marked the beginning of first-principles materials modelling. This book tries to reflect the importance that we attach to numerical values by relentlessly calculating, comparing and discussing numbers. Numbers allow us to develop meaningful approximations to the complex equations of quantum mechanics, and numbers are at the origin of more abstract conceptual developments. A theory of materials is in the first place a theory about quantities that can be measured; hence it is important to develop a sense of the numerical values corresponding to the properties discussed in each chapter.

An effort was made throughout the book to provide references to the original literature whenever possible, as opposed to referring the reader to other books. This choice was motivated by the fact that, owing to the digitization and online archiving of many scientific journals, we now have almost complete and direct access to original papers dating back to the works of Bohr on the structure of the atom in 1913. All of the journal references cited (over 350) are accessible within a standard university subscription. All the citations to the original sources are hyper-referenced, in such a way that the readers of the e-book version will be just one click away from the sources.

The choice of referring to the original works was meant to show how the topics discussed in this book do not form a static piece of knowledge, but are made up of a myriad of contributions spanning over a century of science. It was felt that even only glancing at some of the references provided in this book, taking a look at the language, the equations and the presentation style, would be helpful for understanding how certain aspects of the theory of materials were developed, and how complex and non-linear is the way to achieve a formal theory which is both effective and aesthetically satisfactory.

An attempt was also made to avoid as much as possible sweeping the more difficult concepts 'under the carpet'. While this is somewhat standard practice in undergraduate education owing to obvious time constraints, it usually generates significant confusion, and may give the wrong impression that the theory of materials is a set of rules lacking a unifying principle. In those cases where certain notions require too advanced a discussion to fit in this book, references to more comprehensive presentations are always provided.

Similarly to the references, also all the equations and sections are hyper-referenced. This should make it easier for the e-book readers to hop back and forth between equations while following some of the derivations.

Exercises are provided within each chapter, and serve the dual purpose of making the study of the subject more interactive, and exploring some more advanced aspects which are not explicitly addressed in the main text. Most exercises are designed in such a way as to guide the reader through each step, and important intermediate and final expressions are always provided for cross-checking. The exercises are scattered within the chapters, instead of being collected at the end as in most textbooks. This should motivate the reader to actually put some effort into solving the exercises, without being discouraged by the more traditional lists of questions at the end of chapters. The assignments in each exercise are clearly identified by a symbol ($\blacktriangleright$) in order to distinguish them from those parts which form complements to the theory in the main text ($\square$). Some exercises are intended to give a flavour of the key steps involved in actual first-principles calculations of material properties. Some others are included for acquiring familiarity with techniques of calculus and numerical methods which may not have been introduced in previous undergraduate courses. During a first reading of this book it may be advantageous to skip the exercises altogether, and then go back and try them during a second reading.

It goes without saying that some of the topics discussed in this book could be presented more rigorously and elegantly using advanced conceptual tools, such as perturbation theory and second quantization. However, such an approach would defy the very purpose of this book, which is to bring first-principles materials modelling into the undergraduate curriculum, without requiring preparatory modules other than elementary electrodynamics and quantum mechanics.

It is assumed that the reader is already familiar with multivariate and vector calculus, linear algebra, Fourier analysis, Taylor series and differential equations. The topics covered require also some background in elementary quantum mechanics and electrodynamics. In particular, it is assumed that the reader has a good understanding of the Schrödinger equation for the hydrogen atom, of electrostatics in dielectric media and of the electromagnetic field. This text is not meant to serve as an introduction to solid-state physics, and accordingly the reader is expected to be familiar with the notions of crystal lattices and Brillouin zones. For completeness some of these aspects are briefly summarized in the appendices. The presentation is reasonably self-contained and concise, and could be used for a taught module covering one academic semester.

As computational techniques in this area are in constant and rapid evolution, a choice was made not to focus on the more technical aspects required to perform actual calculations, and place instead the emphasis on those general concepts which are expected to hold unchanged even in a not-so-distant future. For example, despite its historical and practical importance, the theory of pseudopotentials appears a disproportionate complication for readers approaching this subject for the first time. Accordingly this topic is discussed in a dedicated appendix. In the same way, in order to avoid narrowing down the target audience to specialist communities, the representation of electron wavefunctions using specific basis sets is discussed in the appendices rather than in the main text.

The choice of keeping technical details to a minimum partly reflects current trends whereby reliable materials modelling software is widely available, and its usage as a 'black-box' is becoming increasingly popular. In this context it seems more useful to place the emphasis on the most general aspects of the calculations, and less critical to discuss the details of specific computational implementations, some of which may become obsolete in the Darwinian evolution of first-principles calculations.

The book is organized as follows. The first chapter is a qualitative introduction to first-principles materials modelling, and defines the scope and limits of the methods discussed throughout the book. Chapter 2 establishes the link between elementary quantum mechanics and many-body electronic structure theory using simple intuitive arguments. In the third chapter density functional theory is introduced on a formal ground. Chapters 4 and 5 describe how to use density functional theory in order to determine the equilibrium structures of materials. From the equilibrium structures we move on to discuss the deformations of materials (Chapter 6) and their vibrations (Chapter 7). The quantum theory of vibrations is developed in Chapter 8, and its implications in vibrational spectroscopy and phase stability are discussed. Chapters 9 and 10 address electronic and optical properties of materials. In particular the calculation of band structures and their connection with photoelectron spectroscopy is discussed in Chapter 9, while Chapter 10 is devoted to the theory of optical excitations. The discussion of spin is deferred to the final chapter of this book, where a short introduction to the relativistic theory of the electron and a glimpse into calculations for magnetic materials are provided. The first two appendices are devoted to formal derivations of important equations which do not fit into the introductory presentations of Chapters 2 and 3. The remaining three appendices cover the numerical representation of electron wavefunctions, the notion of the reciprocal lattice and Brillouin zone in solids, and the pseudopotential method. In every chapter an effort is made to connect the theory with experiments, in the hope that the reader will be able to develop a sense of what is the level of accuracy that can be expected from the methods presented here.

It is hoped that this book will make the reader enthusiastic about the quantum theory of materials and first-principles materials modelling. To the author it still seems inconceivable that a single differential equation is capable of describing so many properties of materials with such impressive accuracy.

F. Giustino
Oxford, April 2014

Acknowledgments

The origin of this book is connected with the visionary idea that materials modelling using density functional theory should become a core subject in undergraduate education in Materials Science. I am indebted to D. G. Pettifor for being fully supportive of this idea since its early inception in 2009. I also owe a great deal to D. G. Pettifor and to J. R. Yates for their thorough reading of the manuscript and for providing me with invaluable comments and suggestions. The contents and the presentation style reflect what I have learned from great mentors and colleagues, especially A. Pasquarello, who introduced me to this field in Lausanne, and S. G. Louie and M. L. Cohen, who gave me the unique opportunity to join their team in Berkeley as a postdoctoral researcher. During my years in Oxford I had the privilege to interact with very inspiring colleagues at home and internationally, including in alphabetical order: S. C. Benjamin, G. A. D. Briggs, A. C. Ferrari, X. Gonze, L. M. Herz, K. Kern, A. Marini, N. Marzari, J. J. L. Morton, J. B. Neaton, P. Ordejón, M. C. Payne, D. G. Prendergast, P. Radaelli, G.-M. Rignanese, P. Rinke, A. Rubio, J. M. Smith, H. J. Snaith, P. Umari, J. H. Warner, A. A. R. Watt, and many others. I am grateful to all of them for contributing, either directly or indirectly, to shaping the ideas and vision upon which this book rests. I am also grateful to C. R. M. Grovenor for always being supportive of all my initiatives, including this book. The presentation is also the result of many stimulating discussions with current and former members of my research group, in alphabetical order: A. Buccheri, K. Cao, M. R. Filip, H. I. Fisher, H. Hübener, R. E. Margine, K. Noori, C. E. Patrick, M.-A. Perez-Osorio, and with the undergraduate students in the Department of Materials at Oxford who attended my lectures and classes. To all of them goes my warmest gratitude for creating a very pleasant research and teaching environment. In particular, I am indebted to K. Noori for patiently preparing the cover art using data from his Ph.D. work. I would also like to thank D. Alfè, H. J. Choi, G. Csányi, G. Galli, X. Gonze, T. Iitaka, A. Marini, J. Moriarty, J. Nørskov, A. Pasquarello, J.-Y. Raty, C. Umrigar, for agreeing to have some of their work reproduced in this book, T. Yokoyama for sharing original STM images, and T. Yokoya for providing original ARPES spectra. Finally my deepest gratitude goes to D. A. I. Mavridou for bearing with me during the most intense moments of this activity, and for supporting me all the way in this project.

Most of the example calculations and exercises presented in this book were performed using the 'Quantum ESPRESSO' (Giannozzi et al., 2009) and the 'fhi98pp' (Fuchs and Scheffler, 1999) software packages. The atomistic models were rendered using 'XCrysden' (Kokalj, 2003), 'VESTA' (Momma and Izumi, 2008) and 'Rasmol' (Sayle and Milner-White, 1995); the cover art was rendered using 'VMD' (Humphrey et al., 1996). All other graphics were generated using 'Gnuplot' and 'GIMP'. I wish to express my gratitude to the developers of these software packages for making them available to the community.

Contents

Notation

Units of measurement

m   metre
kg   kilogram
K   kelvin
s   second
A   ampere

J	joule	$1\,\mathrm{J}$	$= 1\,\mathrm{m\,kg\,s^{-2}}$
C	coulomb	$1\,\mathrm{C}$	$= 1\,\mathrm{A\,s}$
V	volt	$1\,\mathrm{V}$	$= 1\,\mathrm{m^2\,kg\,s^{-3}\,A^{-1}}$
F	farad	$1\,\mathrm{F}$	$= 1\,\mathrm{C\,V^{-1}}$
Pa	pascal	$1\,\mathrm{Pa}$	$= 1\,\mathrm{m^{-1}\,kg\,s^{-2}}$
Hz	hertz	$1\,\mathrm{Hz}$	$= 1\,\mathrm{s^{-1}}$
T	tesla	$1\,\mathrm{T}$	$= 1\,\mathrm{kg\,s^{-2}\,A^{-1}}$
eV	electronvolt	$1\,\mathrm{eV}$	$= 1.60218{\cdot}10^{-19}\,\mathrm{J}$
Ha	hartree	$1\,\mathrm{Ha}$	$= 27.2114\,\mathrm{eV}$
Å	ångström	$1\,\mathrm{Å}$	$= 10^{-10}\,\mathrm{m}$
bohr		$1\,\mathrm{bohr}$	$= 0.529177\,\mathrm{Å}$

Metric prefixes

G	giga	10^{9}
M	mega	10^{6}
k	kilo	10^{3}
m	milli	10^{-3}
μ	micro	10^{-6}
n	nano	10^{-9}
p	pico	10^{-12}
f	femto	10^{-15}

Universal constants

e	electron charge	$e = 1.60218 \cdot 10^{-19}\,\mathrm{C}$
m_e	electron mass	$m_e = 9.10938 \cdot 10^{-31}\,\mathrm{kg}$
m_p	proton mass	$m_p = 1.67262 \cdot 10^{-27}\,\mathrm{kg} = 1836.15\,m_e$
c	speed of light	$c = 2.99792 \cdot 10^{8}\,\mathrm{m\,s^{-1}}$
$\hbar$	reduced Planck constant	$\hbar = 1.05457 \cdot 10^{-34}\,\mathrm{J\,s}$
ϵ_0	dielectric permittivity of vacuum	$\epsilon_0 = 8.85419 \cdot 10^{-12}\,\mathrm{F\,m^{-1}}$
a_0	Bohr radius	$a_0 = 4\pi\epsilon_0\hbar^2/m_e e^2 = 0.529177\,\mathrm{Å}$
μ_B	Bohr magneton	$\mu_\mathrm{B} = e\hbar/2m_e = 5.78838 \cdot 10^{-5}\,\mathrm{eV\,T^{-1}}$
k_B	Boltzmann constant	$k_\mathrm{B} = 1.38065 \cdot 10^{-23}\,\mathrm{m^2 kg\,s^{-2}K^{-1}}$
		$k_\mathrm{B}T = 25.8\,\mathrm{meV}$ for $T = 300\,\mathrm{K}$
$\hbar c$		$\hbar c = 197.327\,\mathrm{MeV\,fm}$
$m_e c^2$		$m_e c^2 = 0.510999\,\mathrm{MeV}$

1
Computational materials modelling from first principles

'Materials modelling' is the development and use of mathematical models for describing and predicting certain properties of materials at a quantitative level. Mathematical models of materials constitute a key component of research and development in materials science, and span a very broad range of techniques and applications, from finite elements methods in structural mechanics to atomic-scale simulations in nanoelectronics.

When we specify that the model is 'from first principles', or equivalently that we are using an *ab initio* model, we refer to the choice of a bottom-up modelling strategy, as opposed to a top-down approach. For example, a top-down approach could be that of extracting empirical laws by fitting experimental measurements. Conversely, in a bottom-up strategy we would like to understand and predict certain properties without relying on empirical parameters. This is possible if we can identify a fundamental theory of materials which is at once general, flexible and reliable.

Luckily enough, such a theory does exist and is called *quantum mechanics*. During introductory courses we learn that quantum mechanics is a strange theory, that it involves counterintuitive concepts such as entanglement and non-locality, and that it carries many interesting philosophical implications. However, there is an aspect of quantum mechanics which goes probably unnoticed in undergraduate introductions to the subject. That is, quantum mechanics is in many ways a spectacular 'engineering tool'. In fact, as we will learn throughout this book, quantum mechanics is a very general and powerful mathematical model of reality. As such, it is extremely useful for understanding and predicting many properties of materials.

As we will see in Chapter 2, the use of quantum mechanics in the study of material properties goes through the solution of one complicated Schrödinger equation. The complexity of the mathematical problem is so formidable that not only can we practically forget closed-form solutions for any system except the hydrogen atom, but also standard personal computers are not enough for studying any but the simplest systems. In this context the word 'computational' attached to 'materials modelling from first principles' indicates that most calculations require the use of high-performance computers or *supercomputers*. For example, while a simple band structure calculation on a crystal with a small unit cell (Chapter 9) can take a fraction of a minute on a laptop computer, the study of an organic/inorganic interface (e.g. polymer/semiconductor) can require weeks of computing time on a parallel supercomputing architecture consisting of hundreds of CPUs.

The use of supercomputers is probably the most distinctive trait of materials modelling from first principles, and places this discipline at the boundary between traditional materials science, physics and chemistry on the one side, and applied mathematics, numerical analysis and software engineering on the other side.

Perhaps the second most characteristic feature of materials modelling from first principles is a healthy obsession with the systematic comparison between calculations and experimental data. Most scientists and engineers working in this field are aiming for models which can predict experimental measurements, for example the dielectric constant of an insulator, with an accuracy of a few percent. The pursuit of reliable and accurate predictions, and the relentless efforts to validate the models against experiment, are possibly the key to the success of this emerging discipline.

1.1 Density functional theory

Modern computational modelling of materials from first principles relies on a variety of theoretical and computational techniques. The common denominator of such techniques is that they are based upon, or are related to, so-called *density functional theory*. The precise meaning of density functional theory, traditionally shortened as 'DFT', will become clear in the following two chapters. For the time being we only anticipate that, from a practical standpoint, DFT is a very effective technique for studying molecules, nanostructures, solids, surfaces and interfaces, by directly solving approximate versions of the Schrödinger equation.

Historically the birth date of DFT coincides with a manuscript published by Hohenberg and Kohn in 1964 in the journal *Physical Review*, entitled 'Inhomogeneous electron gas'. Clearly, anyone new to this field would find it hard to believe that gases of electrons and materials science might have anything in common. However, as it turned out, understanding the properties of a gas of electrons was the key to the development of reliable *ab initio* models of materials.

In order to appreciate the importance of DFT in materials modelling, it is useful to examine some standard metrics of scientific impact. When in a scientific publication one uses a methodology or an observation from a previous work, it is good practice to acknowledge the influence of that work by a citation. It is therefore reasonable to expect that the number of citations to a paper provides some kind of measure of its impact on the field.

Figure 1.1 shows the citations to Hohenberg and Kohn's article and a few other seminal contributions to this field. We can regard the number of citations during a given year as an estimate of the number of researchers using DFT during that period. The trends in the figure are self-explanatory: this number has been increasing steadily at a very fast pace since the mid-1960s. For example, during 2012 a new scientific study using DFT was published on average every 2 hours.

The trends in Figure 1.1 are useful to form an idea of the size of the community working in this area, and provide a rough measure of the impact of some pioneering contributions. However, we should not be misled into thinking that DFT-based materials modelling is one man's achievement. Probably it is fair to say that materials modelling using DFT is so popular today thanks to a concerted community effort and a myriad of contributions from at least two generations of scientists worldwide.

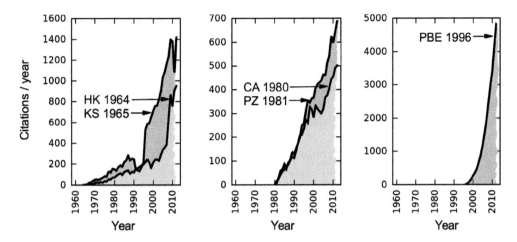

Fig. 1.1 Number of citations per year to the works where density functional theory (DFT) was introduced, HK 1964 (Hohenberg and Kohn, 1964) and KS 1965 (Kohn and Sham, 1965), as well as some of the works which established DFT as a predictive computational method for *ab initio* materials modelling, CA 1980 (Ceperley and Alder, 1980), PZ 1981 (Perdew and Zunger, 1981), PBE 1996 (Perdew *et al.*, 1996). The citation statistics are from the Thomson Reuters Web of Science database (http://apps.webofknowledge.com).

In fact, the main strength of DFT is that it is truly an *enabling technology* for materials modelling. In this sense, when we say 'DFT modelling' we are really talking about an ensemble of methods which, starting from the original ideas of Hohenberg and Kohn, have been and still are experiencing a steady 'Darwinian' evolution towards better accuracy and shorter computing times.

1.2 Examples of materials modelling from first principles

In this section we briefly review a few examples of materials modelling studies based on density functional theory. As 'one picture is worth ten thousand words' we refrain from going through the technical details for each example, and we discuss only some general ideas, with the help of the figures. The presentation in the following examples is very qualitative, and should be understood as a taster of some of the applications and uses of DFT.

It goes without saying that the following selection is by no means exhaustive, and partly reflects the interests of the author. This selection is in the first place an opportunity for touching upon topics which are slightly too advanced for an introductory presentation, and would not otherwise find a place in this book. Topics which will be covered explicitly in the following chapters (e.g. band structures, phonons, optical properties and magnetism) are not presented here, in order to avoid unnecessary duplication. The reader interested in a comprehensive review of the scope and potential of DFT-based materials modelling will find an authoritative and up-to-date presentation in the book by Martin (2004).

1.2.1   Exploring structures at the nanoscale: detonation diamonds

Starting from the mid-1980s there has been a growing interest in nanoscale diamond, i.e. carbon nanoparticles exhibiting the same sp^3 tetrahedral bonding arrangement as in diamond crystals. Initially these particles were discovered in meteoric fragments and were mostly of interest within an astrophysical context. Subsequently it has become possible to produce these particles in the laboratory using explosives (hence the name 'detonation diamonds'), and more recently by means of chemical vapour deposition. Small molecules with the diamond structure, called diamondoids, were also discovered in crude oil. From a technological point of view nanodiamond is attractive since it is biocompatible and ultrahard; it can be used, for example, to deliver drugs into the body or as a nanolubricant. In addition, chemically modified nanodiamonds might be used as electron guns for field-emission displays, or as fluorescent markers for biological labelling.

When a new material is discovered, the first question that one typically asks is what is its underlying structure. In order to answer this question one can perform a variety of characterization tests, e.g. UV-Vis and X-ray spectroscopy, Raman spectroscopy,

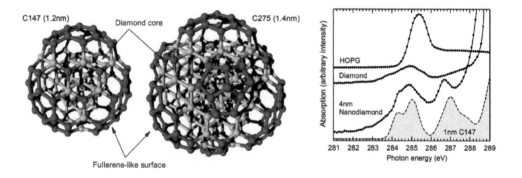

Fig. 1.2   Raty *et al.* (2003) performed atomic-scale calculations based on DFT in order to elucidate the structure of nanoscale diamond particles. By simulating the temperature evolution of nanometre-sized diamonds using 'first-principles molecular dynamics', they observed that the surface of the nanoparticles rearranged in an sp^2-bonded configuration typical of graphite. The resulting stable structures consisted of 'bucky diamonds', i.e. nanoparticles with a diamond core and a graphitic shell, as shown in the ball-and-stick models on the left. The DFT calculation by Raty *et al.* of the X-ray absorption spectra of these nanostructures (yellow spectrum on the right) reproduced the experimental peaks at 285 eV and 287 eV (curve labelled '4nm Nanodiamond'). The match between these theoretical predictions and the experiment established the presence of bucky diamonds in the film. The label HOPG in the figure stands for 'highly oriented pyrolitic graphite'. Additional information on nanodiamonds can be found in Raty and Galli (2003) and Mochalin *et al.* (2011). Adapted by permission from the American Physical Society: *Physical Review Letters* (Raty *et al.*, 2003), copyright 2003.

nuclear magnetic resonance, etc. The problem with nanoscale materials is that, owing to the small length scales and the structural complexity, the interpretation of measured spectra can be rather challenging. In these cases the use of *first-principles materials modelling* techniques can go a long way in assisting the structural assignment.

In a study of synthetic nanodiamond films by Raty *et al.* (2003), while the electron diffraction analysis indicated a high content of crystalline diamond, the X-ray absorption spectra exhibited unknown features not observed in bulk diamond. This suggested that the structure of nanodiamonds was not exactly the same as that of diamond. In order to clarify this, Raty *et al.* used atomic-scale simulations based on the *ab initio* molecular dynamics method of Car and Parrinello (1985). In short the procedure consisted of preparing atomistic models of nanometre-sized diamonds, and then simulating the time evolution of these structures at low temperature using DFT. In these simulations one follows the motion of each atom as the quantum-mechanical electron wavefunctions rearrange around the evolving structure. Importantly this procedure does not involve *any* empirical parameters.

Raty *et al.* observed that, during the course of the simulations, the surface of the nanoparticle underwent a transformation from the sp^3 bonding structure typical of diamond to an sp^2 bonding pattern typical of graphite. The final stable structures consisted of a diamond-like core surrounded by a graphite-like shell (Figure 1.2). These diamond nanoparticles were denoted 'bucky diamonds', owing to their similarity with fullerenes, i.e. C_{60} molecules resembling Buckminster Fuller's geodesic domes.

When one proposes a new structure based on simulations, it is important to validate it by comparing the calculations with experimental spectroscopy. Figure 1.2 shows that the X-ray absorption spectrum calculated for the bucky diamonds agrees with experiment, thereby confirming the presence of these particles in the film.

This example speaks for the usefulness of DFT calculations for identifying unambiguously newly discovered materials.

1.2.2 Superconducting materials: the heat capacity of magnesium diboride

One of the most stunning macroscopic manifestations of quantum mechanics is the phenomenon of superconductivity. The key property of superconducting materials is that, below a temperature called the *critical temperature*, T_c, the electrical resistivity vanishes completely. As a result, the electrons in a superconductor can travel without energy dissipation. Superconductors are useful in many applications, including magnets for nuclear magnetic resonance and magnetic resonance imaging, particle accelerators, distribution grids and magnetic energy storage.

The critical temperatures of all known superconducting materials are rather small, the largest observed value at ambient pressure being $T_c = 133$ K for the copper oxide $HgBa_2Ca_2Cu_3O_{8+x}$, shortened as Hg-1223 (Schilling *et al.*, 1993).

Traditionally one distinguishes between low-T_c and high-T_c superconductors, depending on whether the critical temperature falls below or above the boiling point of liquid nitrogen (77 K). This distinction is important in applications, as refrigeration by liquid nitrogen is considerably cheaper than by liquid helium (boiling point 4 K).

The microscopic mechanism of superconductivity is understood for a class of low-T_c superconductors which are referred to as *conventional superconductors*. Elemental metals such as Pb fall within this category. For this class of materials the theory of Bardeen, Cooper and Schrieffer (1957), in short the 'BCS theory', established that the onset of superconductivity is due to the interaction between the electrons and the vibrations of the lattice.

The BCS theory was developed in order to explain the general principles of superconducting phenomena and rationalize the experimental observations. However, this theory does not tell us how to calculate the superconducting properties of materials from first principles. For example, it does not tell us how to calculate the superconducting critical temperature of Pb using its atomic number $Z = 82$ as the only parameter.

Luckily enough it is possible to extend the BCS theory in order to address precisely this type of question. The formalism is due to Migdal (1958) and Eliashberg (1960), and is reviewed extensively by Allen and Mitrovich (1982). The Migdal–Eliashberg theory of superconductors relates the critical temperature, T_c, to atomic-scale properties of materials, in particular the electronic band structure, the phonon dispersion relations, and their interrelation. Since these properties can now be calculated with good accuracy using DFT, as we will see in Chapters 8 and 9, we are in a good position to study conventional superconductors *quantitatively* starting from the first principles of quantum mechanics.

As an example of first-principles calculations for superconductors, we consider the case of magnesium diboride. MgB_2 was found to superconduct below $T_c = 39$ K by Nagamatsu *et al.* (2001), and currently holds the record for highest T_c among conventional superconductors. The discovery by Nagamatsu *et al.* attracted immense attention in the scientific community, and was followed by many studies of MgB_2 and its superconducting properties in rapid succession. It would be impossible to do justice to all the important contributions to this field in the limited space of an example, therefore we only point the interested reader to the very first contributions to the theory of superconductivity in MgB_2: Kortus *et al.*, An and Pickett and Kong *et al.*, all published their theories within a few months of the discovery.

A standard signature of the superconducting state is a sudden jump of the heat capacity vs. temperature plot at the superconducting critical temperature. Figure 1.3 shows the comparison between measurements of the temperature-dependent heat capacity of MgB_2 and *ab initio* calculations. The calculations were performed by Choi *et al.* (2002) by combining DFT and the Migdal–Eliashberg theory. We can see that the calculations (red line) are in good agreement with experiment. In particular, the discontinuity at the critical temperature, 39 K, which marks the phase transition from the normal to the superconducting state, is well reproduced. Furthermore, we can observe that an additional feature in the measured curves, which appears as a knee around 7 K, is also correctly captured by the calculations. A simpler BCS model of MgB_2 (represented by the dashed black line) would not be able to account for this feature. The critical temperature calculated by Choi *et al.* was 39.4 K, in excellent agreement with the experimental value. This example should give an idea of the accuracy and predictive power of DFT calculations when combined with sophisticated

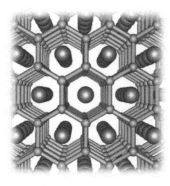
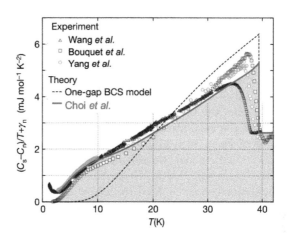

Fig. 1.3 The left panel shows a ball-and-stick representation of the superconductor MgB_2. Magnesium diboride consists of sheets of B atoms (green) arranged in a hexagonal lattice, intercalated by layers of Mg atoms (orange). The right panel shows a comparison between the heat capacity of MgB_2 measured by Wang *et al.*, Yang *et al.* and Bouquet *et al.* (2001, open symbols), and the first-principles calculations of Choi *et al.* (2002; red line, highlighted by the yellow shading). The calculations were performed by combining DFT and the Migdal–Eliashberg theory of superconductivity, and are in very good agreement with experiment. A simpler model based on the BCS theory is shown as a dashed black line. The quantity reported in the plot, $(C_s - C_n)/T + \gamma_n$, is obtained as the difference between the heat capacity in the superconducting state, C_s, and that in the normal state, $C_n = \gamma_n T$ with $\gamma_n = 2.62$ mJ·mol^{-1}K^{-2}. Right panel adapted by permission from Macmillan Publishers Ltd: *Nature* (Choi *et al.*, 2002), copyright 2002.

methodologies such as the Migdal–Eliashberg theory of superconductivity.

As a matter of fairness we should point out that the calculations in Figure 1.3 involve one semi-empirical approximation. This approximation has to do with the description of the Coulomb interaction between the electrons. Since this matter is rather advanced we will not discuss it, but simply mention that it is now possible to remove this approximation and perform calculations entirely from first principles, as shown by Moon *et al.* (2004). An alternative strategy would be to use an extension of density functional theory to superconducting materials (Lüders *et al.*, 2005; Marques *et al.*, 2005). As shown by Floris *et al.* (2005), this alternative first-principles method gives results similar to those in Figure 1.3.

1.2.3 Phase diagrams from first principles: iron in the Earth's core

In some cases it happens that measurements of material properties are so difficult that different laboratories may produce results in substantial disagreement with one another. In such situations first-principles DFT calculations may help to resolve the discrepancy by providing independent and complementary information.

A classic example of this situation is the study of the temperature at the centre of the Earth. A direct measurement of this quantity is not possible, and the temperature has to be inferred from the thermodynamic properties of the materials found in the Earth's core. The analysis of seismic data indicates that the core is mostly made of Fe, and that a solid nucleus called the 'inner core' is surrounded by a liquid shell, the 'outer core'. Since the solid and the liquid coexist at the boundary between the inner and outer core, the temperature at their boundary must correspond to the melting point of Fe at the local pressure, which is around 350 GPa. The measurement of the melting point of Fe at such high pressures requires complicated experiments using diamond anvil cells or shock waves, and the temperatures obtained in different laboratories differ by up to 2,000 K. Given this uncertainty, it makes sense to ask whether DFT calculations could be used to narrow down the error bar.

At pressures above 10–15 GPa, iron crystallizes in a hexagonal close-packed structure called the ϵ-Fe phase (Figure 1.4). In order to determine the temperature of the inner core, Alfè *et al.* (1999) set to study the melting curve of ϵ-Fe using a combination of *ab initio calculations* and *thermodynamics* techniques.

The determination of melting curves relies on the observation that, when a solid and its liquid are in thermodynamic equilibrium at a given temperature, T_m, then the corresponding Gibbs energies are equal: $G_s(p, T) = G_l(p, T)$ when $T = T_m$ for a given pressure p (the Gibbs energy will be discussed in Chapter 8). This means that one can determine the melting curve T_m vs. p by calculating the Gibbs energies of solid and liquid for a range of pressures and temperatures, and determine where the two energy surfaces intersect. The calculation of the Gibbs energy for the solid phase requires the

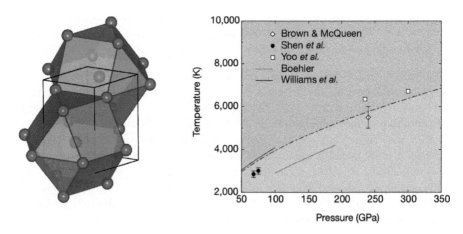

Fig. 1.4 Above 10–15 GPa the stable phase of iron is the hexagonal close-packed structure denoted ϵ-Fe. The ball-and-stick model on the left shows the structure of ϵ-Fe and its hexagonal unit cell. The panel on the right shows the high-pressure phase diagram of ϵ-Fe calculated by Alfè *et al.* (1999). The yellow region corresponds to the solid, the blue region to the liquid. Experimental data are from Brown and McQueen (1986), Williams *et al.* (1987), Boehler (1993), Yoo *et al.* (1993) and Shen *et al.* (1998). Right panel adapted by permission from Macmillan Publishers Ltd: *Nature* (Alfè *et al.*, 1999), copyright 1999.

study of lattice vibrations, which we will discuss in Chapter 8. The calculation of the energy for the liquid is based instead on the simulation of the dynamics of each atom at fixed temperature and density, using a method due to Car and Parrinello (1985). The interested reader will find details on the calculations of phase diagrams using this and other techniques in Gillan *et al.* (2006) and Laio *et al.* (2000).

The high-pressure phase diagram of ϵ-Fe calculated by Alfè *et al.* is shown in Figure 1.4 together with a selection of experimental data. We can see that the calculations are in good agreement with most of the experiments. From these calculations it can be concluded that the temperature, T_m, in the core is at least 6,670 K.

It is important to appreciate that this DFT calculation of the melting curve was performed by Alfè *et al.* entirely from first principles, without the use of any empirical parameters. This is especially important here, as it guarantees that the results are not biased by existing measurements, and provide instead independent information.

1.2.4 Multiscale simulations: understanding brittle fracture in silicon

In the previous examples we discussed situations where DFT calculations were performed for systems containing a few to a few hundreds of atoms. However, many properties of materials involve length scales corresponding to many hundreds of thousands of atoms. In these cases it is not possible yet, or at least is very difficult, to perform direct DFT simulations. In order to circumvent this difficulty one strategy is to acknowledge that the details of the quantum-mechanical electron wavefunctions may not be strictly necessary everywhere. Following this reasoning one can try and separate the system under consideration into one region where quantum mechanics is important, the *quantum mechanics region* (QM), and the remainder of the system where a classical description using Newton's equations is sufficient, the *molecular mechanics region* (MM). The resulting hybrid quantum/classical modelling method, denoted 'QM/MM', can be useful for describing systems with many thousands of atoms by retaining the accuracy of quantum mechanics only where the action takes place. These methods originated from the study of chemical reactions in biomolecular systems (Warshel and Levitt, 1976), and a review of the current state-of-the-art can be found in Senn and Thiel (2009).

In materials science an example of the above situation is the phenomenon of crack propagation in brittle solids. Here the regions of the material far from the crack can be described using linear elasticity theory. However, as one approaches the crack tip, it becomes necessary to describe atomistic processes such as interatomic bond breaking and bond rotations. In this case the use of parametrized interatomic forces is not sufficient, since the bonding environment (e.g. the number of neighbours and the bond lengths) is continuously and rapidly changing as the crack front advances. Near the crack tip it is important to describe the region surrounding the tip using *ab initio* quantum-mechanical calculations.

Csányi *et al.* (2004) proposed an efficient QM/MM strategy to address this class of 'multiscale' problems. In their method, which they called 'learn on the fly', all the atoms in the systems evolve according to classical Newton's equations. The forces between the atoms far from the crack tip are calculated using standard parametrized models. The forces acting on the atoms near the crack tip are obtained from quantum

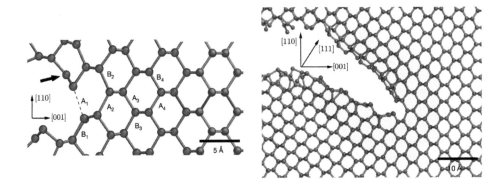

Fig. 1.5   The ball-and-stick models represent two snapshots of the multiscale simulations of brittle fracture in silicon by Kermode *et al.* (2008). In both panels the red spheres represent Si atoms for which forces were calculated using quantum mechanics, while blue spheres are the Si atoms described using parametrized (non-quantum-mechanical) forces. The simulated system consists of 174,752 atoms. The panel on the left illustrates a slow propagation of a crack through the (110) plane along the [001] crystallographic direction. The plane is perpendicular to the arrow indicating the [110] direction. Here the Si–Si bonds labelled 'A' at the crack tip have a nearest-neighbour Si atom which is under-coordinated (indicated by the arrow). This atom draws some electron charge away from the A-bond. As a result, this bond is weaker than 'B' bonds and tends to break. In this condition the crack propagates along the [001] direction indicated by the arrow. The panel on the right shows a situation where the crack initially propagates at high speed along the [001] direction. In this case the kinetic energy available near the crack tip allows the exposed surfaces to reconstruct. As a result, A-bonds and B-bonds are now equally strong, and the breaking of B-bonds leads to a deflection of the crack to the (111) plane, i.e. the plane perpendicular to the arrow indicating the [111] direction. Adapted by permission from Macmillan Publishers Ltd: *Nature* (Kermode *et al.*, 2008), copyright 2008.

mechanics, by 'carving out' a small region around the crack and performing separate DFT calculations for this subsystem.

Figure 1.5 shows a simulation of crack propagation in a silicon single crystal performed by Kermode *et al.* (2008) using this method. The authors set out to explain the deflection of the crack from the (110) crystallographic plane to the (111) plane which is observed in experiments when the crack propagates along the [001] direction. They found that, when the crack propagates very slowly, bond breaking takes place preferentially for those bonds labelled by 'A' in the figure. Such a selective breaking enables the crack to travel along the initial [001] direction. The breaking of A-bonds is connected with the presence of neighbouring Si atoms which are left under-coordinated by the passage of the crack tip. In fact, these under-coordinated atoms draw some electron charge away from the A-bonds, thereby making them weaker than the B-bonds in the figure.

When the applied stress is increased, and hence the crack front propagates at higher speed, the bond breaking can cause the exposed atoms to snap back and initiate a reconstruction of the surface. As this leads to the removal of under-coordinated Si atoms, A-bonds and B-bonds become equivalent and the crack can proceed through the breaking of B-bonds. Under these circumstances the crack is deflected away from the initial propagation direction, and settles on to the (111) plane.

These observations based on multiscale simulations were confirmed by direct measurements of crack propagation in silicon crystals (Kermode *et al.*, 2008).

1.2.5 Discovering new materials from first principles: catalysts for hydrogen production

In all the examples discussed so far, DFT calculations were used for *understanding* the properties of materials which already exist. Another class of applications of DFT methods consists of *predicting* the properties of materials which have not yet been made.

The concept of predicting material properties is meaningful because computational methods based on the first principles of quantum mechanics are entirely general, and can be applied to any combinations of atoms. This concept can be especially useful in the optimization and design of new materials.

The use of DFT for 'materials discovery' has been gaining popularity during the past decade (Kang *et al.*, 2006; Greeley *et al.*, 2006), and can be expected to become more widespread as the calculations become more accurate and less time-consuming.

As an example of 'materials design from first principles' we discuss briefly the optimization of catalysts for the production of hydrogen. H_2 represents an important alternative to fossil fuels, as it burns cleanly by transforming into water ($2\,H_2 + O_2 \rightarrow 2\,H_2O$). One means of producing hydrogen gas is to exploit the 'hydrogen evolution reaction'. In this reaction protons are transferred from an electrolyte solution to a metal electrode which acts as a catalyst. The protons are first chemisorbed on the catalyst, and subsequently they are chemically reduced to yield H_2. The best catalyst in terms of H_2 yield is platinum; however, Pt is also among the rarest elements on Earth and hence very expensive. As one might expect, many laboratories around the world are working on the development of effective alternatives to Pt as a catalyst.

Greeley *et al.* (2006) addressed this problem by searching for potential Pt replacements using DFT calculations. The first challenge in this kind of search is to understand which material properties identify a good catalyst, and how to calculate those properties using DFT. In the case of the hydrogen evolution reaction, it turns out that the performance of the catalyst correlates strongly with the Gibbs energy of H adsorption on the catalyst surface, denoted as ΔG_H. In fact Parsons (1958) observed that the catalytic activity is maximized when $|\Delta G_H|$ is as close to zero as possible. For example, in the case of Pt we have $\Delta G_H = -0.1$ eV. In the specialized literature it is common to say that ΔG_H is a good *descriptor* of the catalytic activity.

The energy of adsorption can be evaluated within DFT by calculating the change in total energy upon the chemisorption of H on the catalyst (Nørskov *et al.*, 2005). As this is a relatively straightforward procedure, Greeley *et al.* decided to search for a new catalyst by calculating ΔG_H for a very large number of binary alloys. They performed calculations for 736 binary alloys, considering the solid solutions of

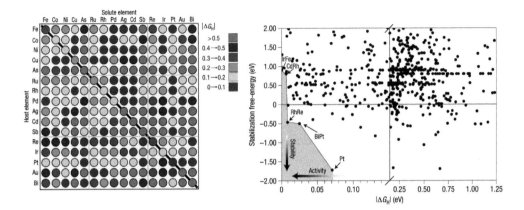

Fig. 1.6 The left panel shows the energy of H adsorption $|\Delta G_H|$ calculated by Greeley *et al.* (2006) for some of the possible binary alloys of the transition metals indicated on the axes. The solute/host ratio at the surface of the catalyst is 1:3 in all cases, and the values in the legend are in eV. The blue discs indicate alloys whose energy of adsorption gives the highest catalytic activity. In the right panel the information on $|\Delta G_H|$ is plotted against the worst possible figure of merit in the stability analysis for each alloy. The closer the data-point to the bottom left corner, the higher the expected 'quality' of the alloy in terms of efficiency and stability. Apart from pure Pt, we can see that BiPt is one of the most promising candidates. Adapted by permission from Macmillan Publishers Ltd: *Nature Materials* (Greeley *et al.*, 2006), copyright 2006.

16 transition metals. It is important to stress that most of these alloys were purely 'hypothetical' materials.

Figure 1.6 shows the values of $|\Delta G_H|$ for a subset of alloys calculated by Greeley *et al.* Already in this subset we can see that 49 combinations are promising, as they give $|\Delta G_H| \leq 0.1$ eV. At this stage the search space is still very wide; therefore it is necessary to narrow down the range using more stringent optimization criteria.

Without going into the details, we simply point out that another important performance parameter for a catalyst is its stability in an electrochemical environment. A catalyst alloy can degrade, for example by solute segregation, oxidation and corrosion. Greeley *et al.* calculated Gibbs energies corresponding to the most common degradation phenomena, and combined those with the information on ΔG_H in order to identify the alloys which could be, at once, efficient and stable. The panel on the right of Figure 1.6 shows that, out of the many potentially efficient alloys, only a few are expected to be stable in an electrochemical environment. This further selection reduces the number of candidates to a handful of possibilities.

The final step in this procedure was to go ahead and synthesize the materials identified through this computational screening. Greeley *et al.* succeeded in synthesizing one of the most promising alloys, BiPt. As predicted from the DFT calculations, the measured catalytic activity of this new compound turned out to be superior to that of pure Pt. What we should retain from this example is that DFT calculations have reached a point where they can be used for screening potential new materials (possibly hypothetical), before going ahead with the synthesis in the laboratory.

The reader interested in a review of recent developments in the computational screening of materials and materials design using DFT can consult, among others, Hafner *et al.* (2006) and Ceder (2010).

1.3   Timeline of DFT calculations in materials modelling

Now that we are a little more familiar with the notion of 'materials modelling from first principles', it is meaningful to discuss briefly how this discipline developed over time. The field is so vast today that an exhaustive listing of contributions would probably resemble a phone book; therefore we limit ourselves to highlighting some of the milestones, and we refer to the comprehensive review by Martin (2004) for a more thorough overview.

The time frame is purposely restricted to the period going from about 1965 to 2000. Earlier contributions definitely helped shape the key ideas underlying DFT and its applications, but they belong more to the 'prehistory of DFT' and will not be discussed. Contributions post-2000 are also omitted, since the field is evolving at a very fast pace, and it seems too early to place very recent developments in a historical perspective.

Most of the terminology used below is rather specialized. Instead of cluttering the presentation with detailed explanations in each case, we will proceed with a *semi-qualitative* description, confident that the meaning of each new term will become clearer throughout the course of this book.

We distinguish milestones in two areas, the *foundations* of DFT, and the development of efficient *algorithms* for practical calculations. Within the category 'foundations' we highlight the following:

1964   *Hohenberg–Kohn theorem and Kohn–Sham formulation*
      Hohenberg and Kohn (1964), Kohn and Sham (1965)

      Density functional theory is introduced. Shortly afterwards the concept of the density functional is employed in order to derive the Kohn–Sham equations. These equations form the backbone of DFT calculations.

1972   *Relativistic extension of density functional theory*
      von Barth and Hedin (1972), Rajagopal and Callaway (1973)

      The density functional theorem is extended to include relativistic effects. It becomes possible to study magnetic materials.

1980   *Local density approximation for the exchange and correlation energy*
      Ceperley and Alder (1980), Perdew and Zunger (1981)

      The exchange and correlation energy of the uniform electron gas is calculated using highly accurate 'quantum Monte Carlo' methods (i.e. essentially by solving the complete many-body Schrödinger equation). Subsequently these results are parametrized into an explicit expression for the 'exchange and correlation energy' as a function of density (see Chapter 3). This marks the beginning of DFT calculations for real materials, as opposed to idealized model systems.

1984 *Time-dependent density functional theory*
Runge and Gross

The DFT theorems are extended to the case of time-dependent potentials. This advance opens the way to the study of material properties in excited electronic states, especially optical properties.

1985 *First-principles molecular dynamics*
Car and Parrinello

Density functional theory is married to 'molecular dynamics' simulation techniques (i.e. computational methods whereby the motion of each atom is determined by solving Newton-like equations). From this point onwards it is possible to study both electrons and nuclei on the same footing. This marks the beginning of atomic-scale quantum-mechanical simulations.

1986 *Quasiparticle corrections for insulators*
Hybertsen and Louie

One well known difficulty met by DFT calculations is the underestimation of the band gap of insulators. This contribution demonstrates that accurate band gaps can be obtained by correcting DFT calculations using a method developed by Hedin (1965).

1987 *Density functional perturbation theory*
Baroni, Giannozzi and Testa

The concept of perturbation theory in quantum mechanics is applied to DFT. This is the beginning of systematic calculations of phonon dispersion relations and vibrational properties.

1988 *Towards quantum chemistry accuracy in the exchange and correlation energy*
Lee, Yang and Parr (1988), Becke (1993)

These two contributions demonstrate that DFT can aspire to be almost as accurate as quantum chemistry methods based on the many-body electron wavefunction. These works will shape the use of DFT in chemistry.

1991 *Hubbard-corrected density functional theory*
Anisimov, Zaanen and Andersen

Shortcomings of DFT calculations in the presence of atomic-like d or f electrons are corrected, and the systematic study of transition metal and rare-earth metal compounds starts.

1996 *The generalized gradient approximation to density functional theory*
Perdew *et al.* (1992), Perdew, Burke and Ernzerhof (1996)

The functional form of the exchange and correlation energy (Chapter 3) is improved by taking into account not only the density, but also its gradients. The accuracy of the total energy improves from ~1 eV in the local density approximation to ~0.1 eV in the generalized gradient approximation.

Within the category 'algorithms' we highlight the following milestones, which are related either to finding a way of separating core electrons from valence electrons, or to performing calculations for larger systems, i.e. hundreds to thousands of atoms:

1959   *Pseudopotentials*
       Phillips and Kleinman (1959), Cohen and Bergstresser (1966)

       The description of electrons in solids is restricted to the valence electrons. This
       is achieved by replacing the nuclear potentials by 'ionic pseudopotentials' (see
       Appendix E). At this stage the pseudopotentials are determined empirically,
       using a fit to measured optical spectra.

1979   *First-principles norm-conserving pseudopotentials*
       Hamann, Schlüter and Chiang (1979), Bachelet, Hamann and Schlüter (1982),
       Kleinman and Bylander (1982)

       Ionic pseudopotentials can now be obtained directly within DFT. This advance
       enables first-principles calculations without the core electrons, making them
       much faster than previous calculations.

1989   *Algorithmic improvements for studying large systems*
       Teter, Payne and Allan (1989), Kresse and Furthmüller (1996)

       Linear algebra techniques are benchmarked, adapted and improved in order to
       solve Kohn–Sham equations more efficiently.

1990   *Soft pseudopotentials and the projector-augmented wave method*
       Vanderbilt (1990), Troullier and Martins (1991), Blöchl (1994)

       Previous pseudopotentials suffered from the fact that they fluctuate strongly
       near the nucleus, and the description of such fluctuations is computationally
       expensive. With the introduction of 'soft' pseudopotentials this problem is
       resolved, and the study of first-row elements and transition metals becomes
       more accessible. The price to be paid is the loss of information on the electron
       wavefunctions in the vicinity of the nuclei. With the projector-augmented wave
       method of Blöchl it becomes possible to retrieve such information.

1991   *Linear-scaling DFT calculations*
       Yang (1991), Mauri, Galli and Car (1993), Ordejón *et al.* (1993)

       In the Kohn–Sham formulation of DFT (Chapter 3) the computational cost
       of the calculations scales approximately as the cube of the number of atoms.
       These contributions show that it is possible to employ alternative formulations
       whereby the computational cost scales linearly with the number of atoms. This
       marks the beginning of DFT calculations for very large systems.

Due to space limitations there are many omissions in this essential timeline, and
apologies go to those who feel that their contribution is not fairly represented. The
above timeline does not reflect the development and impact of DFT in the area of
quantum chemistry, which is thoroughly discussed in the book by Koch and Holthausen
(2001).

Besides the specifics, the take-home message of this section is that, while the use of
DFT in materials science has been gaining momentum during the past two decades,
relentless work on the foundations and practical uses of this method span almost half
a century.

1.4   Reasons behind the popularity of density functional theory

In Section 1.1 it was said that DFT is an *enabling technology* for materials modelling. We can now try to be more specific by pausing on a few key strengths of this theory.

Transferability
An important advantage of using a computational modelling technique based on DFT is that the methodology is universal. In other words one can use the same technique in order to describe widely different classes of materials. This is very convenient since it implies that the experience one gains in using DFT methods is completely transferable. The transferability of DFT affords us a substantial saving of effort when compared to methods where a new model is required for each new class of materials considered.

Simplicity
The Kohn–Sham equations (Chapter 3) are simple and intuitive, and appeal to one's tendency to think of electrons as independent particles. As we will see, these equations establish a most natural link between elementary quantum mechanics (e.g. the Schrödinger equation for the hydrogen atom) and materials science.

Reliability
Despite their simplicity, the Kohn–Sham equations yield most often very accurate results, to the point that in some cases we can calculate material properties which match experimental measurements within a few percent. The possibility of making direct and quantitative comparison with experiment is a distinctive strength of DFT. In particular, as discussed in Section 1.2, in some cases it is also possible to 'predict' material properties before even carrying out the measurements.

Software sharing
The popularity of DFT has received a significant boost with the introduction of standardized software, the introduction of online platforms for collaborative software development, and also the adoption of the open-source software model.
Today DFT is a truly global enterprise, in the sense that DFT software and methods are developed, used and tested by a global community, in a spirit similar to the Wikipedia project. The very large size of the community is also responsible for the fast prototyping and uptake of new advances in the field.

Reasonable starting point
The remarks above may give the wrong impression that DFT is free of shortcomings. In fact, it is well established that techniques based on DFT do not work for every material or every property. For instance, the van der Waals binding of biological molecules is not described correctly, the optical absorption spectra are red-shifted and transition metal oxides sometimes are incorrectly predicted to be metallic.
The reason why DFT stands the test of time is that, even though it fails in describing certain properties, it still represents a *reasonable starting point* for more sophisticated calculations. Indeed, it is often the case that, while plain DFT provides incorrect answers, some 'post-DFT' corrections can be devised in order to restore its predictive power.

1.5   Atomistic materials modelling and emergent properties

Given the predictive power of first-principles materials modelling, it is natural to ask whether in a not-so-distant future it will be possible to study phenomena and systems involving many millions of atoms at the quantum-mechanical level.

With only slight exaggeration, we could reformulate this question by asking whether it will be possible to 'reduce' materials science to the solution of one complicated Schrödinger equation. Is this sort of 'reductionist' vision realistic and feasible in practice?

Certainly sooner or later it will be possible to run quantum-mechanical simulations for very large systems and long time scales. However, it is not guaranteed that a brute-force approach will help us identify and understand the most important physical mechanisms.

In order to be more explicit on this, let us consider the case of biological macromolecules, e.g. proteins. With some abuse of language, proteins can be thought of as 'hierarchical materials', in the sense that different structural motifs and functions emerge at different length scales. This is illustrated in Figure 1.7, where we show one of the two rotors of ATP synthase, the enzyme that synthesizes adenosine triphosphate (ATP) in mitochondria. This protein consists of a chain of small amino-acid molecules,

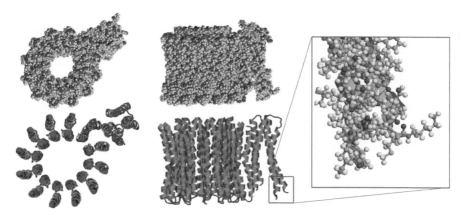

Fig. 1.7   Different structural models of the F_0 complex of the ATP synthase enzyme (Rastogi and Girvin, 1999). In the top row we have a top view (left) and a side view (centre), with all the atoms shown. The colour code is: O (red), C (grey), N (blue), S (yellow). The bottom row shows the same views, this time using a schematic representation of the secondary structure given by the alpha helices. The inset shows a comparison between the two views, where the ball-and-stick model is superimposed on the alpha helices. In order to understand the functions of this unit we need to acknowledge the coexistence of different organizational levels, i.e. a hierarchical structure. The F_0 motor is composed of 12 protein chains, clearly visible in the bottom left corner. This nanorotor is embedded in the membrane of cell mitochondria, and operates like a cylinder rotating around its axis during the synthesis of ATP (von Ballmoos *et al.*, 2009). The ATP synthase and its operation are examples of emergent structures and functions.

which constitute the primary structure (inset). The chain folds to form a secondary structure, the so-called alpha helices, represented by the curly ribbons. Together the helices define the tertiary structure, i.e. the rotor. The combination of two rotors and a stator (not shown in the figure) defines the quaternary structure of ATP synthase (von Ballmoos *et al.*, 2009).

We know that the system in the figure consists of 16,649 atoms, out of which 5,339 are C atoms, 1,339 O, 1,250 N, 102 S and 8,619 H atoms (Rastogi and Girvin, 1999). Let us imagine that we could perform a first-principles simulation of this very large system. Given the elemental composition, would we be able to 'predict' the structure in Figure 1.7 using DFT calculations? The answer is probably 'no'. In fact, such a large system will admit a vast multitude of atomistic configurations of similar energy. As a consequence, algorithms designed to find the most stable configuration will need to explore an extremely large number of possibilities, and so run for an extremely long time.

From this example we can guess that, in general, as the number of atoms in a system increases, the size of the configuration space grows exponentially. In this situation new structures, new functions and even new phenomena are likely to emerge, precisely as in the case of the hierarchical structures and functions of proteins. The notion that complex *emergent* structures and phenomena can arise as we increase the system size is best expressed by the words of Anderson (1972): 'more is different'.

These considerations should help us understand that a reductionist's approach to computational materials science may prove inadequate for studying complex systems. Going back to our initial question, materials modelling cannot be reduced to merely solving very difficult equations. Organizing observations, formulating hypotheses and testing them are and will always remain the keys to understanding materials.

2
Many-body Schrödinger equation

In this chapter we start from elementary concepts in quantum mechanics, and using a minimal set of prerequisites we develop the fundamental equations of the quantum theory of materials. This chapter should be regarded as a bridge between the quantum theory of one particle in an external potential, as given for example by the Schrödinger equation for the hydrogen atom, and the formalism required to study many interacting electrons and nuclei, also called *electronic structure theory*.

In view of developing an intuitive understanding of the key ideas underlying the quantum theory of materials, the discussion presented in this chapter is mostly based on heuristic arguments. Rigorous formal derivations of important equations are given in Appendices A and B.

2.1 The Coulomb interaction

The starting point for understanding the properties of materials at the atomic scale is to acknowledge that they are just complicated collections of electrons and nuclei. For this purpose it is helpful to keep in mind the following symbolic equation:

$$\text{materials} = \text{electrons} + \text{nuclei}.$$

Materials hold together owing to a fine balance between, on the one hand, the repulsive Coulomb interactions between pairs of electrons and pairs of nuclei, and, on the other hand, the attractive Coulomb interactions between electrons and nuclei. If we have two electrons at a distance d_{ee}, then from classical electrostatics it is known that the energy associated with their Coulomb repulsion is (Jackson, 1998):

$$E_{ee} = \frac{e^2}{4\pi\epsilon_0 d_{ee}}, \tag{2.1}$$

e being the electron charge and ϵ_0 the permittivity of vacuum. The repulsion energy between two nuclei with atomic number Z at a distance d_{nn} reads similarly:

$$E_{nn} = \frac{Z^2 e^2}{4\pi\epsilon_0 d_{nn}}. \tag{2.2}$$

The energy associated with the Coulomb attraction between an electron and a nucleus at a distance d_{en} is:

$$E_{en} = -\frac{Z e^2}{4\pi\epsilon_0 d_{en}}, \tag{2.3}$$

and obviously carries a negative sign since the interaction is attractive. The interactions described by eqns 2.1–2.3 form the backbone of the quantum theory of materials, as will become clear throughout this book.

2.2   Many-body Schrödinger equation

From introductory quantum mechanics textbooks (e.g. Merzbacher, 1998) we learn that, in order to understand the behaviour of quantum particles, we need to determine the corresponding wavefunction $\psi(\mathbf{r})$ for every point $\mathbf{r} = x\mathbf{u}_x + y\mathbf{u}_y + z\mathbf{u}_z$ in the region of interest by solving a Schrödinger equation (Schrödinger, 1926). Here $\mathbf{u}_x, \mathbf{u}_y$ and $\mathbf{u}_z$ denote the unit vectors along the Cartesian axes. If we restrict our discussion to stationary electronic states, then we need to consider the time-independent version of the Schrödinger equation (the time-dependent version of this equation will be discussed in Chapter 10).

The time-independent Schrödinger equation takes the following symbolic form:

$$(\text{ kinetic energy} + \text{potential energy})\, \psi = E\,\psi, \qquad (2.4)$$

where E is the energy eigenvalue for the stationary state described by the wavefunction ψ. The probability of finding the particle at the point $\mathbf{r}$ is $|\psi(\mathbf{r})|^2$. In the case of one electron in a potential energy landscape $V(\mathbf{r})$, eqn 2.4 can be written explicitly as:

$$\left[\frac{\mathbf{p}^2}{2m_e} + V(\mathbf{r}) \right] \psi(\mathbf{r}) = E\,\psi(\mathbf{r}), \qquad (2.5)$$

where m_e is the electron mass, and the quantum-mechanical momentum operator is given by:

$$\mathbf{p} = -i\hbar\nabla, \qquad \nabla = \mathbf{u}_x\frac{\partial}{\partial x} + \mathbf{u}_y\frac{\partial}{\partial y} + \mathbf{u}_z\frac{\partial}{\partial z}, \qquad (2.6)$$

with $\hbar$ the reduced Planck constant. Let us call ψ_0 the lowest-energy solution of eqn 2.5. At equilibrium the system is in its lowest-energy configuration, i.e. the electron occupies precisely the state ψ_0. The electron charge distribution at equilibrium is therefore given by $|\psi_0(\mathbf{r})|^2$. So far there is nothing new with respect to elementary quantum mechanics.

Suppose now that we add one electron to this system, and we want to determine the new distribution of the electronic charge. The Pauli principle tells us that we can accommodate the new electron in the same eigenstate ψ_0, provided the two electrons have opposite spins (this statement will become clear in Chapter 11, when the electron spin will be discussed). In this configuration the electronic charge distribution becomes $2|\psi_0|^2$, i.e. we have added $|\psi_0|^2$ to the original distribution. However, if we now consider eqn 2.1 we realize that the two electrons will tend to repel each other. It turns out that this repulsive interaction modifies both the shape of ψ_0 and the potential term, V, in eqn 2.5; therefore our first approach to the quantum mechanics of two electrons is not quite accurate.

In order to develop a realistic and useful theory of materials we need to properly describe systems with many electrons and many nuclei. To this end, we need to slightly modify our starting point, eqn 2.4.

When we discuss many electrons and many nuclei together we need to introduce a so-called *many-body* wavefunction, Ψ, which depends on the positions of each electron and each nucleus in the system. In the case of N electrons with coordinates $\mathbf{r}_1, \mathbf{r}_2, \cdots, \mathbf{r}_N$ and M nuclei with coordinates $\mathbf{R}_1, \mathbf{R}_2, \cdots, \mathbf{R}_M$ we have:

$$\Psi = \Psi(\mathbf{r}_1, \mathbf{r}_2, \cdots, \mathbf{r}_N; \mathbf{R}_1, \mathbf{R}_2, \cdots, \mathbf{R}_M). \tag{2.7}$$

In the same way as $|\psi(\mathbf{r})|^2$ represents the probability of finding one electron at position $\mathbf{r}$, $|\Psi(\mathbf{r}_1, \ldots, \mathbf{r}_N; \mathbf{R}_1, \ldots, \mathbf{R}_M)|^2$ represents the probability of *simultaneously* finding electron number 1 at point $\mathbf{r}_1$, electron number 2 at point $\mathbf{r}_2$, and so forth.

In many applications we are interested only in the total electronic charge density. This is intuitively defined as the probability of finding *any* electron (regardless of its 'label') at position $\mathbf{r}$, and is obtained as follows. The probability $P(\mathbf{r}_1 = \mathbf{r})$ of finding electron number 1 at $\mathbf{r}$, while the other electrons can be anywhere, is the 'logical OR' combination of all the configurations where $\mathbf{r}_1 = \mathbf{r}$ and $\mathbf{r}_2, \ldots, \mathbf{R}_M$, span the entire volume of the material:

$$P(\mathbf{r}_1 = \mathbf{r}) = \int |\Psi(\mathbf{r}, \mathbf{r}_2, \ldots, \mathbf{r}_N; \mathbf{R}_1, \ldots, \mathbf{R}_M)|^2 d\mathbf{r}_2 \ldots d\mathbf{r}_N d\mathbf{R}_1 \ldots d\mathbf{R}_M. \tag{2.8}$$

The probability of finding any electron at position $\mathbf{r}$, i.e. the electron density, is then:

$$n(\mathbf{r}) = P(\mathbf{r}_1 = \mathbf{r}) + P(\mathbf{r}_2 = \mathbf{r}) + \cdots + P(\mathbf{r}_N = \mathbf{r}). \tag{2.9}$$

Now we need to remember that in quantum mechanics electrons are *indistinguishable* particles; therefore each term on the right-hand side of eqn 2.9 is given by eqn 2.8. As a result we can rewrite eqn 2.9 in a more compact fashion:

$$n(\mathbf{r}) = N \int |\Psi(\mathbf{r}, \mathbf{r}_2, \ldots, \mathbf{r}_N; \mathbf{R}_1, \ldots, \mathbf{R}_M)|^2 d\mathbf{r}_2 \ldots d\mathbf{r}_N d\mathbf{R}_1 \ldots d\mathbf{R}_M. \tag{2.10}$$

If the many-body wavefuntion, Ψ, is normalized to unity within the material:

$$\int |\Psi(\mathbf{r}_1, \mathbf{r}_2, \ldots, \mathbf{r}_N; \mathbf{R}_1, \ldots, \mathbf{R}_M)|^2 d\mathbf{r}_1 \ldots d\mathbf{r}_N d\mathbf{R}_1 \ldots d\mathbf{R}_M = 1, \tag{2.11}$$

then by combining this condition with eqn 2.10 we find, as we intuitively expected, that the integral of the electronic charge density throughout the whole material yields the number of electrons, N:

$$\int n(\mathbf{r}) d\mathbf{r} = N. \tag{2.12}$$

Now that the many-body wavefunction has been introduced, we can go back to eqn 2.4 and simply replace ψ by Ψ when we have many particles:

$$(\text{kinetic energy} + \text{potential energy})\,\Psi = E_{\text{tot}}\Psi, \tag{2.13}$$

where the eigenvalue, E_{tot}, now represents the *total energy* of the system in the quantum state specified by the many-body wavefunction Ψ. The kinetic energy needed in eqn 2.13 can be written as in the one-particle case (eqn 2.5), except that now we need to take into account N electrons and M nuclei:

$$\text{kinetic energy} = -\sum_{i=1}^{N} \frac{\hbar^2}{2m_e} \nabla_i^2 - \sum_{I=1}^{M} \frac{\hbar^2}{2M_I} \nabla_I^2, \tag{2.14}$$

where M_1, M_2, ...are the masses of the nuclei. In this case the derivatives in the Laplace operators ∇^2 are taken with respect to the coordinates of each particle. For example:

$$\nabla_1^2 \Psi = \frac{\partial^2 \Psi}{\partial x_1^2} + \frac{\partial^2 \Psi}{\partial y_1^2} + \frac{\partial^2 \Psi}{\partial z_1^2}. \tag{2.15}$$

For the potential energy term we can directly use eqns 2.1–2.3, counting all possible pairs of charges in the system. First, we have the Coulomb repulsion between electron pairs:

$$(\text{potential energy})_{\text{ee}} = \frac{1}{2} \sum_{i \neq j} \frac{e^2}{4\pi\epsilon_0} \frac{1}{|\mathbf{r}_i - \mathbf{r}_j|}. \tag{2.16}$$

Here the indices i and j run from 1 to N. The terms $i = j$ are excluded because an electron does not repel itself, and we divide by 2 in order to count only one contribution per pair.

Second, we have the Coulomb repulsion between pairs of nuclei:

$$(\text{potential energy})_{\text{nn}} = \frac{1}{2} \sum_{I \neq J} \frac{e^2}{4\pi\epsilon_0} \frac{Z_I Z_J}{|\mathbf{R}_I - \mathbf{R}_J|}. \tag{2.17}$$

Here the indices I and J run from 1 to M and the Z_I represent the atomic numbers. Third, we have the Coulomb attraction between electrons and nuclei:

$$(\text{potential energy})_{\text{en}} = -\sum_{i,I} \frac{e^2}{4\pi\epsilon_0} \frac{Z_I}{|\mathbf{r}_i - \mathbf{R}_I|}, \tag{2.18}$$

with i from 1 to N and I from 1 to M. At this point we can combine eqns 2.13 to 2.18 and write the *many-body Schrödinger equation*:

$$\left[-\sum_i \frac{\hbar^2}{2m_e} \nabla_i^2 - \sum_I \frac{\hbar^2}{2M_I} \nabla_I^2 + \frac{1}{2} \sum_{i \neq j} \frac{e^2}{4\pi\epsilon_0} \frac{1}{|\mathbf{r}_i - \mathbf{r}_j|} \right.$$
$$\left. + \frac{1}{2} \sum_{I \neq J} \frac{e^2}{4\pi\epsilon_0} \frac{Z_I Z_J}{|\mathbf{R}_I - \mathbf{R}_J|} - \sum_{i,I} \frac{e^2}{4\pi\epsilon_0} \frac{Z_I}{|\mathbf{r}_i - \mathbf{R}_I|} \right] \Psi = E_{\text{tot}} \Psi. \tag{2.19}$$

From this point onwards the summation indices i, j and I, J will always run from 1 to N and from 1 to M, respectively (unless otherwise stated). Equation 2.19 is almost everything that we need to know in order to study the behaviour of materials at equilibrium. If we were to be more rigorous, then we would include the time dependence, the interaction with external electromagnetic fields, and some corrections arising from the theory of relativity. We will come back to these points later in the book, in Chapters 10 and 11, respectively.

If we were able to solve eqn 2.19 and find the eigenstate with the lowest energy, which is called the *ground state* of the system, then we would be able to calculate many equilibrium properties of materials, from elastic properties to enthalpies of formation, thermal properties, and phase diagrams. The trouble is that the solution of eqn 2.19 for all but the simplest systems (e.g. small molecules) is very challenging, and in most cases it is still practically impossible.

We can make a quick estimate of the size of the task using a back-of-the-envelope calculation. A possible strategy for solving the partial differential eqn 2.19 is to discretize the space into a uniform mesh of points, and transform the problem into a linear system by using finite differences for the derivatives (this is discussed in Appendix C). Let us imagine that we wanted to describe electrons and nuclei in the unit cell of a crystal, say a semiconductor like silicon. The volume of the unit cell of silicon in the diamond structure is $a^3/4$ with a=5.43 Å. A sensible discretization of the unit cell volume would have points spaced by $\Delta x \sim 0.1$ Å. Such a grid would consist of $N_p = (a^3/4)/(\Delta x)^3 \sim$40,000 points. Counting four valence electrons for each of the two silicon atoms in the unit cell, as well as the two nuclei, i.e. $N + M = 10$ particles, the complete specification of a quantum state $\Psi_{Si}(\mathbf{r}_1, \ldots, \mathbf{r}_8; \mathbf{R}_1, \mathbf{R}_2)$ would require $N_p^{N+M} \sim 10^{46}$ complex numbers. Performing matrix operations with arrays of this size is obviously impossible. In general, the complexity of the solution of eqn 2.19 increases *exponentially* with the size of the system. This problem is referred to as the 'exponential wall' in the solution of the many-body Schrödinger equation.

During the past century many researchers have made substantial efforts to circumvent this exponential wall. As a result, today there exists a spectacular hierarchy of approximations to eqn 2.19, which allow us to study materials at the atomistic level with varying degrees of sophistication and accuracy.

As we will discover in the following sections, almost all troubles in the study of materials from first principles starting from the many-body Schrödinger equation arise from the Coulomb repulsion between electrons, eqn 2.1.

2.3   Atomic units

In the following we will try to simplify eqn 2.19 in order to get a feeling of what a solution would look like. Before doing so it is convenient to do some housekeeping with the units of measurement.

In eqn 2.19 the only quantities that need to be determined from experiments are the reduced Planck constant, $\hbar$, the electron mass, m_e, the nuclear masses, M_I, the electron charge, e, and the permittivity of vacuum, ϵ_0. The nuclear masses of naturally occurring elements are known, and range between 1 (H) and 236.3 (^{238}U) times the proton mass, m_p. From the US National Institute of Standards and Technology database (http://physics.nist.gov/cuu) we find:

$$\hbar = 1.05457163 \cdot 10^{-34} \, \text{J} \cdot \text{s},$$
$$m_e = 9.10938291 \cdot 10^{-31} \, \text{kg},$$
$$m_p = 1.67262164 \cdot 10^{-27} \, \text{kg},$$
$$e = 1.60217649 \cdot 10^{-19} \, \text{C},$$
$$\epsilon_0 = 8.85418782 \cdot 10^{-12} \, \text{F/m}.$$

All these quantities are *fundamental physical constants* and do not depend on the particular material under consideration. Since eqn 2.19 does not contain any empirical parameters, as could be obtained for instance from measurements, estimates or data-fitting procedures, the study of material properties starting from eqn 2.19 is referred to as a 'first-principles approach'.

When we face an equation as complicated as eqn 2.19 it is convenient to develop a sense of the order of magnitude of the energies involved. The simplest system of electrons and nuclei that we can think of is the hydrogen atom. In the fundamental state of the H atom the electron orbital has an average radius $a_0 \simeq 0.529$ Å. Using this value and eqn 2.3 we can calculate the average Coulomb energy for such an electron–proton pair (in absolute value):

$$E_{\text{Ha}} = \frac{e^2}{4\pi\epsilon_0 a_0}, \tag{2.20}$$

where 'Ha' stands for 'Hartree'. The coulomb energy for a pair of two protons or a pair of electrons at the same distance is also E_{Ha}; therefore we can guess that the typical size of the potential energies in eqn 2.19 is of the order of E_{Ha}.

In order to estimate the magnitude of the kinetic energies appearing in eqn 2.19 we can use a semi-classical argument. In the model of the H atom by Bohr (1913) the electron trajectories are quantized in such a way that the angular momentum in the fundamental state is given by

$$m_e v \, a_0 = \hbar, \tag{2.21}$$

with v the electron velocity. In addition the balance between the centrifugal force and the nuclear attraction in this orbit requires:

$$m_e \frac{v^2}{a_0} = \frac{e^2}{4\pi\epsilon_0 a_0^2}. \tag{2.22}$$

By combining eqns 2.20–2.22 we find the interesting results:

$$\frac{e^2}{4\pi\epsilon_0 a_0} = \frac{\hbar^2}{m_e a_0^2},$$

and

$$\frac{1}{2} m_e v^2 = \frac{1}{2} E_{\text{Ha}}.$$

This last equation states that the kinetic energy is also of the order of E_{Ha}. These simple estimates suggest that a natural unit of energy for the quantities appearing in eqn 2.19 is precisely E_{Ha}. For this reason it is convenient to divide every term of eqn 2.19 by this quantity:

$$\left[-\sum_i \frac{1}{2} a_0^2 \nabla_i^2 - \sum_I \frac{1}{2(M_I/m_e)} a_0^2 \nabla_I^2 + \frac{1}{2} \sum_{i \neq j} \frac{a_0}{|\mathbf{r}_i - \mathbf{r}_j|} \right.$$

$$\left. + \frac{1}{2} \sum_{I \neq J} Z_I Z_J \frac{a_0}{|\mathbf{R}_I - \mathbf{R}_J|} - \sum_{i,I} Z_I \frac{a_0}{|\mathbf{r}_i - \mathbf{R}_I|} \right] \Psi = \frac{E_{\text{tot}}}{E_{\text{Ha}}} \Psi. \tag{2.23}$$

This form of the many-body Schrödinger equation shows quite clearly that we can achieve a great simplification of the notation by *measuring* energies in units of E_{Ha}, distances in units of a_0, and masses in units of m_e. In the reminder of this book we will therefore use these units, unless stated otherwise:

$$1\,\text{Ha} = 27.2114\,\text{eV} = 4.3597 \cdot 10^{-18}\,\text{J},$$
$$1\,\text{bohr} = 0.529177\,\text{Å} = 0.529177 \cdot 10^{-10}\,\text{m},$$
$$1\,\text{a.u. of mass} = 9.10938291 \cdot 10^{-31}\,\text{kg},$$

where 'Ha' stands for Hartree and 'a.u.' for atomic unit. Taken together these units form the so-called *Hartree atomic units*. Strictly speaking, in order to completely specify the Hartree units system we need to set the value of four constants. In this case the missing constant is the electron charge, which is set to $e = 1$.

In Hartree atomic units the many-body Schrödinger equation (eqn 2.19) acquires the following elegant form:

$$\left[-\sum_i \frac{\nabla_i^2}{2} - \sum_I \frac{\nabla_I^2}{2M_I} - \sum_{i,I} \frac{Z_I}{|\mathbf{r}_i - \mathbf{R}_I|} + \frac{1}{2}\sum_{i \neq j} \frac{1}{|\mathbf{r}_i - \mathbf{r}_j|} + \frac{1}{2}\sum_{I \neq J} \frac{Z_I Z_J}{|\mathbf{R}_I - \mathbf{R}_J|} \right] \Psi = E_{\text{tot}} \Psi.$$

$$(2.24)$$

This is the most commonly used form of the many-body Schrödinger equation in first-principles materials modelling. This equation shows very clearly that the only external parameters needed in this approach are the atomic numbers, Z_I, and the atomic masses, M_I.

2.4 Clamped nuclei approximation

After having specified in eqn 2.24 the atomic numbers and masses corresponding to a given material, we need to start thinking about how to determine its solutions. As it stands, eqn 2.24 is too general, since it describes almost everything, from gases to liquids to solids. This means that the solution will not only be immensely complicated, but also rather useless. At this stage it is appropriate to narrow down the range of possibilities and consider molecules and solids. While in the study of liquids, gases and plasmas the nuclei can travel long distances, in the case of solids and molecules adsorbed on solid surfaces the nuclei typically remain at or near certain positions. Therefore, as a starting point, we can assume that the *nuclei are held immobile (clamped) in known positions*. The nuclei can form an ordered crystalline lattice, or an amorphous structure, or a molecular structure. Regardless of the actual shape, what matters is that nuclei cannot move much, so that we can fully concentrate on the electrons. This situation appears restrictive but it is actually a typical scenario. For example, in the case of crystals, the nuclei are almost immobile and very accurate information on their positions is provided by X-ray crystallography.

A word of caution is now in order before proceeding. The uncertainty principle of quantum mechanics prevents the nuclei from being perfectly immobile in their equilibrium positions, because the uncertainty in momentum, Δp, and position, Δx, must satisfy $\Delta x \Delta p \geq \hbar$. We will see in Chapter 4 how to take this into account in our theory.

For now we will be content to think that the nuclei are so heavy that in practice they cannot move. Therefore we can set $M_I = \infty$ in eqn 2.24. This choice implies that we can neglect the kinetic energy of the nuclei in eqn 2.24, and that the Coulomb repulsion between nuclei is simply a constant. For convenience we bring this constant to the right-hand side of eqn 2.24 by defining:

$$E = E_{\text{tot}} - \frac{1}{2}\sum_{I \neq J} \frac{Z_I Z_J}{|\mathbf{R}_I - \mathbf{R}_J|}. \tag{2.25}$$

This definition allows us to rewrite eqn 2.24 as follows:

$$\left[-\sum_i \frac{\nabla_i^2}{2} - \sum_{i,I} \frac{Z_I}{|\mathbf{r}_i - \mathbf{R}_I|} + \frac{1}{2}\sum_{i \neq j} \frac{1}{|\mathbf{r}_i - \mathbf{r}_j|} \right] \Psi = E\,\Psi. \tag{2.26}$$

Now we can regard the nuclear coordinates, $\mathbf{R}_I$, as external parameters, and consider Ψ as a function of the electron coordinates, while ignoring its dependence on the nuclear coordinates: $\Psi = \Psi(\mathbf{r}_1, \ldots, \mathbf{r}_N)$. This procedure becomes more transparent if we define the *Coulomb potential of the nuclei experienced by the electrons*:

$$V_{\text{n}}(\mathbf{r}) = -\sum_I \frac{Z_I}{|\mathbf{r} - \mathbf{R}_I|}, \tag{2.27}$$

so that the nuclear coordinates disappear completely from eqn 2.26:

$$\left[-\sum_i \frac{\nabla_i^2}{2} + \sum_i V_{\text{n}}(\mathbf{r}_i) + \frac{1}{2}\sum_{i \neq j} \frac{1}{|\mathbf{r}_i - \mathbf{r}_j|} \right] \Psi = E\,\Psi. \tag{2.28}$$

This is the fundamental equation of *electronic structure theory*. At this point we should start realizing that this equation looks somewhat similar to the single-particle Schrödinger equation (eqn 2.5) discussed in introductory quantum mechanics (Merzbacher, 1998). The main difference is that we now have several electrons and also their mutual Coulomb repulsion.

In order to make our life easier it is convenient to do some housekeeping with the symbols. We define the *many-electron Hamiltonian*:

$$\hat{H}(\mathbf{r}_1, \ldots, \mathbf{r}_N) = -\sum_i \frac{1}{2}\nabla_i^2 + \sum_i V_{\text{n}}(\mathbf{r}_i) + \frac{1}{2}\sum_{i \neq j} \frac{1}{|\mathbf{r}_i - \mathbf{r}_j|}, \tag{2.29}$$

so that eqn 2.28 can be rewritten using the compact expression:

$$\hat{H}\,\Psi = E\,\Psi. \tag{2.30}$$

In addition, by looking at eqn 2.29 it is natural to define the *single-electron Hamiltonian*:

$$\hat{H}_0(\mathbf{r}) = -\frac{1}{2}\nabla^2 + V_{\text{n}}(\mathbf{r}), \tag{2.31}$$

so that we can break up the many-electron Hamiltonian as follows:

$$\hat{H}(\mathbf{r}_1, \ldots, \mathbf{r}_N) = \sum_i \hat{H}_0(\mathbf{r}_i) + \frac{1}{2} \sum_{i \neq j} \frac{1}{|\mathbf{r}_i - \mathbf{r}_j|}. \tag{2.32}$$

Exercise 2.1 The simplest case of a many-body Schrödinger equation involving electrons and nuclei is provided by the He atom, which consists of one nucleus with $Z=2$ and two electrons. ▶Show that the time-independent Schrödinger equation for the He atom in the clamped-nuclei approximation is given by:

$$\left[-\frac{1}{2}\nabla_1^2 - \frac{2}{|\mathbf{r}_1|} - \frac{1}{2}\nabla_2^2 - \frac{2}{|\mathbf{r}_2|} + \frac{1}{|\mathbf{r}_1 - \mathbf{r}_2|} \right] \Psi(\mathbf{r}_1, \mathbf{r}_2) = E\,\Psi(\mathbf{r}_2, \mathbf{r}_2). \tag{2.33}$$

2.5   Independent electrons approximation

In order to solve eqn 2.28 researchers have been developing approximation methods since the late 1920s (Thomas, 1927; Fermi, 1928; Hartree, 1928; Pauling, 1928; Slater, 1929; Fock, 1930*b*). Let us imagine that we did not have access to any books, journals, or online resources discussing such methods. What would we do in order to solve this equation?

Probably our first step would be to simplify the problem. We already noted the analogy between eqn 2.5 and eqn 2.28. Now we can try and push this analogy further by imagining that we were allowed to eliminate from eqn 2.28 the term describing the Coulomb repulsion between electrons. Since this term is the only possible form of interaction between the electrons, if it were absent then the electrons would not 'see' each other. This rather dramatic simplification of the problem is called the *independent electrons approximation*. Using eqns 2.30 and 2.32 the Schrödinger equation within the independent electrons approximation becomes:

$$\sum_i \hat{H}_0(\mathbf{r}_i)\,\Psi = E\,\Psi. \tag{2.34}$$

Since the electrons are now independent, the probability $|\Psi(\mathbf{r}_1, \ldots, \mathbf{r}_N)|^2$ of finding electron number 1 at $\mathbf{r}_1$ *and* electron number 2 at $\mathbf{r}_2$ *and* ... electron number N at $\mathbf{r}_N$ must be given by the *product* of the individual probabilities $|\phi_i(\mathbf{r}_i)|^2$ of finding the i-th electron at the position $\mathbf{r}_i$. At this stage we do not know the functions ϕ_i yet, but at least we can guess that it should be possible to write the solution of eqn 2.34 as a product:

$$\Psi(\mathbf{r}_1, \mathbf{r}_2, \cdots, \mathbf{r}_N) = \phi_1(\mathbf{r}_1) \cdots \phi_N(\mathbf{r}_N). \tag{2.35}$$

Suppose now that the wavefunctions, ϕ_i, were obtained as the solutions of the single-electron Schrödinger equations:

$$\hat{H}_0(\mathbf{r})\phi_i(\mathbf{r}) = \varepsilon_i \phi_i(\mathbf{r}), \tag{2.36}$$

with ε_1 the smallest eigenvalue and $\varepsilon_1 < \varepsilon_2 < \cdots < \varepsilon_N$. In this case we could replace the trial solution of eqn 2.35 inside eqn 2.34 and see what happens:

$$\left[\sum_i \hat{H}_0(\mathbf{r}_i)\right] \phi_1(\mathbf{r}_1)\cdots\phi_N(\mathbf{r}_N) = E\,\phi_1(\mathbf{r}_1)\cdots\phi_N(\mathbf{r}_N). \qquad (2.37)$$

Since in this equation the single-electron Hamiltonian, $\hat{H}_0(\mathbf{r}_1)$, acts only on the function $\phi_1(\mathbf{r}_1)$, $\hat{H}_0(\mathbf{r}_2)$ acts on $\phi_2(\mathbf{r}_2)$ and so on, we can rewrite:

$$\left[\hat{H}_0(\mathbf{r}_1)\phi_1(\mathbf{r}_1)\right]\phi_2(\mathbf{r}_2)\cdots\phi_N(\mathbf{r}_N) + \phi_1(\mathbf{r}_1)\left[\hat{H}_0(\mathbf{r}_2)\phi_2(\mathbf{r}_2)\right]\cdots\phi_N(\mathbf{r}_N) + \cdots =$$
$$E\,\phi_1(\mathbf{r}_1)\cdots\phi_N(\mathbf{r}_N).$$

Therefore, using eqn 2.36, we find:

$$E = \varepsilon_1 + \varepsilon_2 + \cdots + \varepsilon_N. \qquad (2.38)$$

This result means that, in the independent electrons approximation, the lowest-energy configuration of the system is obtained when we fill the lowest-energy eigenstates of the single-particle equation (eqn 2.36) with one electron in each state, starting from the lowest eigenvalue. This corresponds to our physical intuition, and matches what we usually learn in solid-state physics courses (Kittel, 1976).

Exercise 2.2 We want to investigate the electronic ground state of the helium atom. As a first approximation we start from eqn 2.33, we neglect the Coulomb interaction between the electrons, and we use the approximation of independent electrons. Let us recall that the fundamental state of a hydrogenic atom with the nucleus at $\mathbf{r} = 0$ and the atomic number Z is (Bransden and Joachain, 1983):

$$\phi_{1s}(\mathbf{r}) = \frac{Z^{3/2}}{\sqrt{\pi}}\exp\left(-Z|\mathbf{r}|\right) \quad \text{with} \quad E_{1s} = -\frac{Z^2}{2}. \qquad (2.39)$$

▶ Using eqn 2.33 and the product form

$$\Psi(\mathbf{r}_1, \mathbf{r}_2) = \phi_{1s}(\mathbf{r}_1)\phi_{1s}(\mathbf{r}_2) \qquad (2.40)$$

show that the total energy of the He atom is $E=-4$ Ha. ▶ Show that, within the independent electrons approximation, the wavefunction in eqn 2.40 corresponds to the lowest-energy state. ▶ Write explicitly the ground-state wavefunction in terms of $\mathbf{r}_1$ and $\mathbf{r}_2$. □ Strictly speaking the above wavefunction is allowed by the Pauli principle only if the two electrons have opposite spins. This aspect will be clarified in Section 11.4.

The independent electrons approximation as written in eqn 2.35 carries two important drawbacks. The first relates to the fact that the wavefunction, Ψ, should obey Pauli's exclusion principle (Merzbacher, 1998), which requires that the function *changes sign* whenever we exchange two electrons, e.g. if we swap $\mathbf{r}_1$ and $\mathbf{r}_2$. In general, eqn 2.35 does not obey this rule. The second problem is that the Coulomb term eliminated from eqn 2.24 is actually of the same magnitude as the other terms, and therefore it cannot be ignored. We will address these two issues in the following sections.

2.6   Exclusion principle

Pauli's exclusion principle states that, since electrons are 'fermions', the many-body wavefunction, Ψ, must change sign if we exchange the variables of any two electrons (Merzbacher, 1998). As we will see in Chapter 11, the variable exchange refers both to the position and to the 'spin' of the electrons. For simplicity we ignore the spin here, and defer its discussion until Section 11.4. This principle is equivalent to the statement that no two electrons can occupy the same electronic state.

Let us consider a simple example: if we have two electrons, the wavefunction $\phi_1(\mathbf{r}_1)\phi_2(\mathbf{r}_2)$ with $\phi_1 \neq \phi_2$ does not satisfy this requirement; therefore we cannot really use eqn 2.35. However, the wavefunction:

$$\Psi(\mathbf{r}_1, \mathbf{r}_2) = \frac{1}{\sqrt{2}} [\phi_1(\mathbf{r}_1)\phi_2(\mathbf{r}_2) - \phi_1(\mathbf{r}_2)\phi_2(\mathbf{r}_1)] \qquad (2.41)$$

seems to work. In fact, by direct substitution we find $\Psi(\mathbf{r}_2, \mathbf{r}_1) = -\Psi(\mathbf{r}_1, \mathbf{r}_2)$. In addition, each one of the terms on the right-hand side of eqn 2.41 gives a total energy $E = \varepsilon_1 + \varepsilon_2$ when replaced in eqn 2.34; therefore any linear combination of these terms is also a legitimate solution for the same energy.

A compact way of writing eqn 2.41 is by using a matrix determinant:

$$\Psi(\mathbf{r}_1, \mathbf{r}_2) = \frac{1}{\sqrt{2}} \begin{vmatrix} \phi_1(\mathbf{r}_1) & \phi_1(\mathbf{r}_2) \\ \phi_2(\mathbf{r}_1) & \phi_2(\mathbf{r}_2) \end{vmatrix}, \qquad (2.42)$$

which is referred to as a *Slater determinant* (Slater, 1929). When we have more than two electrons we can construct a Slater determinant as in eqn 2.42, with the electron label increasing along the rows, and the orbital label increasing along the columns. For $N > 2$ the prefactor becomes $N!^{-1/2}$ instead of $2^{-1/2}$, in order to have the function correctly normalized: $\int |\Psi|^2 d\mathbf{r}_1 \dots \mathbf{r}_N = 1$.

Exercise 2.3 ▶ Using the wavefunction Ψ in eqn 2.41 show that, if the two solutions of the single-particle Schrödinger equation (eqn 2.36) are such that $\int \phi_1^*(\mathbf{r})\phi_2(\mathbf{r})d\mathbf{r} = 0$, then the electron charge density as defined by eqn 2.10 becomes:

$$n(\mathbf{r}) = |\phi_1(\mathbf{r})|^2 + |\phi_2(\mathbf{r})|^2. \qquad (2.43)$$

The property described by eqn 2.43 also holds in the most general case of N electrons, and we can state that, in the independent electrons approximation, the electron charge density is simply obtained by adding up the probabilities of finding electrons in each occupied state i:

$$n(\mathbf{r}) = \sum_i |\phi_i(\mathbf{r})|^2. \qquad (2.44)$$

This result is intuitive because, when the electrons are independent of each other, the probability of finding electron i at point $\mathbf{r}$ is precisely $|\phi_i(\mathbf{r})|^2$, and the charge density corresponds to the logical 'OR' of the probabilities of finding electrons 1, 2, ..., N at the same point $\mathbf{r}$.

Exercise 2.4 ▶Using the same approximations introduced in Exercise 2.2, show that the ground-state electron density, $n(\mathbf{r})$, of the He atom is given by:

$$n(\mathbf{r}) = \frac{16}{\pi} \exp\left(-4|\mathbf{r}|\right). \tag{2.45}$$

▶Verify that the direct evaluation of the density using eqn 2.10 and the simpler expression given in eqn 2.44 yield the same result. ▶Check that the integral of the density correctly yields two electrons. ▶Plot the radial probability, $4\pi r^2 n(r)$, of finding any electron at a distance $r = |\mathbf{r}|$ from the nucleus, and determine the distance at which this probability reaches its maximum.

2.7 Mean-field approximation

In Section 2.5 we mentioned that the approximation of ignoring the Coulomb repulsion between electrons in the many-body Schrödinger equation is too drastic. At the same time the notion of independent particles and the expression of the charge density in eqn 2.44 appeal to our intuition and are convenient for practical calculations. The question then is whether one can maintain a single-particle description, and take the Coulomb repulsion into account in some form.

If we completely forget about quantum mechanics for a moment, and go back instead to classical electrostatics, we should remember that a distribution of electronic charge, $n(\mathbf{r})$, such as the one given by eqn 2.44 will generate an electrostatic potential $\varphi(\mathbf{r})$ through Poisson's equation (Jackson, 1998):

$$\nabla^2 \varphi(\mathbf{r}) = 4\pi\, n(\mathbf{r}).$$

The electrons immersed in this electrostatic potential have, in Hartree units, a potential energy $V_{\mathrm{H}}(\mathbf{r}) = -\varphi(\mathbf{r})$, which is called the 'Hartree potential'. By definition also the Hartree potential satisfies Poisson's equation:

$$\nabla^2 V_{\mathrm{H}}(\mathbf{r}) = -4\pi\, n(\mathbf{r}). \tag{2.46}$$

The formal solution of this equation is:

$$V_{\mathrm{H}}(\mathbf{r}) = \int d\mathbf{r}'\, \frac{n(\mathbf{r}')}{|\mathbf{r} - \mathbf{r}'|}, \tag{2.47}$$

which simply means that each element of volume $d\mathbf{r}'$ has a charge $dQ = -n(\mathbf{r}')d\mathbf{r}'$ which generates a Coulomb potential at point $\mathbf{r}$ given by $dQ/|\mathbf{r} - \mathbf{r}'|$.

Since every electron in our system experiences the Hartree potential, we can improve upon eqn 2.36 by taking this extra term into account:

$$\left[-\frac{\nabla^2}{2} + V_{\mathrm{n}}(\mathbf{r}) + V_{\mathrm{H}}(\mathbf{r})\right] \phi_i(\mathbf{r}) = \varepsilon_i\, \phi_i(\mathbf{r}), \tag{2.48}$$

$$n(\mathbf{r}) = \sum_i |\phi_i(\mathbf{r})|^2, \tag{2.49}$$

$$\nabla^2 V_{\mathrm{H}}(\mathbf{r}) = -4\pi n(\mathbf{r}). \tag{2.50}$$

The novelty with respect to eqn 2.36 is that in eqn 2.48 we added V_{H}. Since the potential V_{H} is the 'average' potential experienced by each electron, we call this approach the *mean-field approximation*. Equations 2.48–2.50 must be solved simultaneously, i.e. the solutions ϕ_i of eqn 2.48 must be such that, if we use them to calculate V_H through eqns 2.49 and 2.50, then the resulting potential inserted in eqn 2.48 gives back the same solutions ϕ_i. For this reason we call this approach a *self-consistent field* method. This method was introduced by Hartree (1928), hence the subscript 'H' in the potential.

Equations 2.48–2.50 represent a major simplification of our initial task of solving the complete many-body Schrödinger equation (eqn 2.28), since one differential equation in $3N$ dimensions has been replaced by N three-dimensional equations. If we go back to the example of silicon from Section 2.2, we see that in this case the characteristic size of the arrays needed for describing the wavefunctions is $N \times N_p \sim 10^5$, instead of the previous 10^{46}. This figure corresponds approximately to 1 Mb of computer storage and is very manageable. The price to pay for this simplification is that the solutions ϕ_i of eqn 2.48 are coupled through the density, $n(\mathbf{r})$, by eqns 2.49 and 2.50, and the equations need to be solved using iterative numerical methods. These methods will be discussed in Chapter 3.

The mean-field approximation introduced here *heuristically* would be a very good one if the electrons were classical particles. In fact, self-consistent Poisson solvers are commonly employed in the computational study of astrophysical plasmas and charge transport in semiconductor devices. However, this approximation is not accurate enough for a quantitative study of materials at the atomic scale. In order to be quantitative we still need to add two ingredients to eqns 2.48–2.50: the 'exchange potential' and the 'correlation potential'. We will discuss these terms in the following sections.

Exercise 2.5 We want to refine our calculation of the ground state of the He atom in Exercise 2.2 by including the Coulomb repulsion between the electrons within the mean-field approximation. We start by evaluating the Hartree potential of eqn 2.46. ▶By using the density in eqn 2.45 as a first approximation to the exact electron density of the He atom, show that the Hartree potential is given by:

$$V_H(r) = \frac{2}{r} \left[1 - (1 + 2r) \exp\left(-4r\right) \right]. \tag{2.51}$$

For this exercise it is useful to remember that the radial part of the Laplace operator in spherical coordinates is:

$$\nabla^2 = \frac{1}{r^2} \frac{\partial}{\partial r} \left[r^2 \frac{\partial}{\partial r} \right], \tag{2.52}$$

and that in the limit $r \to \infty$ the Hartree potential should reduce to the electrostatic potential of a point charge corresponding to two electrons.
▶Determine the total potential, $V_n + V_H$, felt by the electrons within the mean-field approximation. □The total potential calculated in this way decays exponentially fast as one moves away from the He nucleus. This trend is actually incorrect, as very general considerations indicate that in the case of He the total potential should decay as $-1/r$ at large distance (Umrigar and Gonze, 1994). The wrong trend stems from the fact that we are neglecting an important ingredient, the exchange interaction, which will be introduced in the next section.

2.8   Hartree–Fock equations

In the previous sections we have seen that, if the electrons do not interact via the Coulomb repulsion, then we can write the many-body wavefunction as a Slater determinant, and obtain the single-particle wavefunctions as the solutions of a simpler single-particle Schrödinger equation (eqn 2.48).

It is now possible to make a further improvement using the following reasoning: the electrons do interact indeed, but perhaps this interaction is not too strong; therefore one can still look for a solution in the form of a Slater determinant. Given this premise, how does one find the single-particle wavefunctions, $\phi_i(\mathbf{r})$, needed in eqn 2.42? The answer to this question can be obtained using a 'variational principle' as follows. Let us consider the quantum state Ψ with the lowest energy. The energy, E, of this state is obtained by multiplying both sides of eqn 2.30 by Ψ^* and integrating over all the variables:

$$E = \int d\mathbf{r}_1 \ldots d\mathbf{r}_N \, \Psi^* \, \hat{H} \, \Psi. \tag{2.53}$$

This result can easily be derived starting from eqns 2.11 and 2.30. From now on, in order to make the notation lighter, we will write integrals like the one appearing in eqn 2.53 using the *Dirac notation*:

$$E = \langle \Psi | \hat{H} | \Psi \rangle. \tag{2.54}$$

In Appendix A it is shown that, if we minimize the energy E with respect to variations of the functions $\phi_i(\mathbf{r})$ in the Slater determinant of eqn 2.42, and require that these functions be orthonormal:

$$\frac{\delta E}{\delta \phi_i^*} = 0, \tag{2.55}$$

$$\int d\mathbf{r} \phi_i^*(\mathbf{r}) \phi_j(\mathbf{r}) = \delta_{ij}, \tag{2.56}$$

(where δ_{ij} is the Kronecker delta and is equal to 1 if $i = j$, 0 if $i \neq j$) then we obtain the so-called *Hartree–Fock equations* (Fock, 1930b):

$$\left[-\frac{\nabla^2}{2} + V_n(\mathbf{r}) + V_H(\mathbf{r}) \right] \phi_i(\mathbf{r}) + \int d\mathbf{r}' \, V_X(\mathbf{r}, \mathbf{r}') \, \phi_i(\mathbf{r}') = \varepsilon_i \, \phi_i(\mathbf{r}), \tag{2.57}$$

$$n(\mathbf{r}) = \sum_i |\phi_i(\mathbf{r})|^2, \tag{2.58}$$

$$\nabla^2 V_H(\mathbf{r}) = -4\pi n(\mathbf{r}). \tag{2.59}$$

When we compare these equations with eqns 2.48–2.50, we realize that here we now have an additional potential, V_X. The explicit expression of this potential is:

$$V_X(\mathbf{r}, \mathbf{r}') = -\sum_j \frac{\phi_j^*(\mathbf{r}') \phi_j(\mathbf{r})}{|\mathbf{r} - \mathbf{r}'|}, \tag{2.60}$$

where the sum runs over the occupied single-particle states (having the same spin as the wavefunction ϕ_i in eqn 2.57, as explained in Appendix A).

The good thing about the Hartree–Fock equations is that we moved from 'classical' electrons in the mean-field approximation of Section 2.7 to 'quantum' electrons here. The bad thing is that this refinement introduces the *non-local* potential $V_X(\mathbf{r}, \mathbf{r}')$ in the single-particle equations. The potential V_X is non-local in the sense that its evaluation involves an integration over the additional variable $\mathbf{r}'$, as shown by eqn 2.57. This complicates enormously the practical solution of the Hartree–Fock equations.

Physically the potential V_X arises precisely from Pauli's exclusion principle, and prevents two electrons from occupying the same quantum state. The function V_X in eqn 2.60 is called the *Fock exchange potential*. The derivation of eqn 2.57 is rather lengthy and is given in Appendix A.

Exercise 2.6 In the case of the He atom it can be shown that the exchange potential of eqn 2.60 for the electron ground state is given exactly by $V_X = -V_H/2$ (Umrigar and Gonze, 1994). ▶Show that the total potential $V_{tot} = V_n + V_H + V_X$ for electrons in He including the exchange term is:

$$V_{tot}(r) = -\frac{1}{r}\left[1 + (1 + 2r)\exp(-4r)\right]. \tag{2.61}$$

▶Show that the total potential has the following asymptotic behaviour:

$$V_{tot}(r) \sim -\frac{2}{r} \quad \text{for} \quad r \to 0, \tag{2.62}$$

$$V_{tot}(r) \sim -\frac{1}{r} \quad \text{for} \quad r \to \infty. \tag{2.63}$$

☐Equation 2.62 indicates that very close to the nucleus the electrons feel a potential corresponding to a hydrogenic atom with $Z = 2$. Equation 2.63 indicates that very far from the nucleus the electrons feel the potential of a hydrogenic atom with $Z = 1$. These observations suggests that the ground-state single-particle eigenfunctions of the Hartree–Fock equation for the He atom will resemble a hydrogenic $1s$ orbital in between the two solutions for an atom with $Z = 1$ and another one with $Z = 2$. Therefore we can find an approximate solution for the ground state of He by determining the value of an 'effective' atomic number, Z, which minimizes the energy of the orbital $\phi_{1s}(\mathbf{r}; Z)$ in eqn 2.39 when subject to the potential of eqn 2.61. Let us define the energy expectation value in the state $\phi_{1s}(\mathbf{r}; Z)$ as follows:

$$\langle E_{1s}\rangle_Z = \int d\mathbf{r}\, \phi_{1s}^*(\mathbf{r}; Z)\left[-\frac{1}{2}\nabla^2 + V_{tot}\right]\phi_{1s}(\mathbf{r}; Z) \tag{2.64}$$

with $\phi_{1s}(\mathbf{r}; Z)$ given by eqn 2.39, the Laplace operator in spherical coordinates given by eqn 2.52, and the total potential given by eqn 2.61. Show that, as a function of Z, the expectation value $\langle E_{1s}\rangle_Z$ is given by:

$$\langle E_{1s}\rangle_Z = \frac{Z^2}{2} - Z^2\left[\frac{1}{Z} + \frac{Z(Z+4)}{(Z+2)^3}\right]. \tag{2.65}$$

▶Verify that the 'effective atomic number' $Z = 1.6$ minimizes the expectation value $\langle E_{1s}\rangle_Z$.
▶Calculate the charge density corresponding to having two electrons (with opposite spins) in the orbital $\phi_{1s}(\mathbf{r}; Z)$, and compare your density with the exact result calculated by Umrigar and Gonze (1994) and shown in Figure 2.1.
☐The simple procedure outlined here allows us to determine approximate ground-state electron wavefunctions of the He atom without much effort. In a more refined treatment we

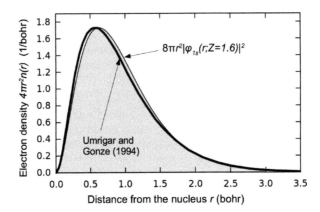

Fig. 2.1 Electron density in the ground state of He. For convenience we plot the radial probability $4\pi r^2 n(r)$ of finding any electron at the distance r from the nucleus. The thick line is the exact solution provided by Umrigar and Gonze (1994); the thin line corresponds to our simplified analysis yielding $n(r) = 2|\phi_{1s}(r; Z = 1.6)|^2$. The agreement between the exact calculation and our simplified treatment suggests that in the He atom the electrons feel on average a potential corresponding to a nucleus with a reduced charge $Z = 1.6$. We can think of this reduced charge as resulting from the shielding of the nuclear charge by one of the two electrons.

would need to consider that the effective Z should be determined in order to minimize the *total energy* as opposed to the electron *eigenvalues*. In fact, in the Hartree–Fock approximation, the total energy is the sum of the electron eigenvalues plus additional terms accounting for the electron–electron repulsion (see eqn A.5). On top of this we should also remember that the Hartree–Fock equations (eqn 2.57) are to be solved self-consistently. This could be done by re-calculating the electron density using the orbitals $\phi_{1s}(\mathbf{r}; Z = 1.6)$, then the Hartree potential as in Exercise 2.5, and finally the total potential in eqn 2.61. At this stage we should carry out the numerical minimization again, and determine a new effective atomic number, $Z^{(2)}$. This procedure should be repeated in order to generate the series $Z^{(3)}$, $Z^{(4)}$, ..., $Z^{(\infty)}$ until convergence. Although it is not too difficult to perform these iterations, we only state the final result: $Z^{(\infty)} = 1.84$ and $E^{(\infty)} = -2.82$ Ha. Our approximate Hartree–Fock energy is not too far from the exact numerical value, -2.9 Ha, obtained by Pekeris (1958).

Umrigar and Gonze (1994) give a breakdown of the exact total energy of He in its ground state. The single-particle energy neglecting the repulsion between electrons is -3.89 Ha, the Hartree energy is 2.05 Ha, and the exchange energy is -1.02 Ha. We can see that the Hartree and exchange contributions are very important here; in fact, if we neglect these effects we make an error in the total energy of He as large as 30 eV.

Apart from the numerical details, this exercise should help us understand that, firstly, the ideas of effective single-particle potentials and single-particle Schrödinger equations arise naturally when we try to solve the many-body Schrödinger equation. Secondly, electronic structure calculations are quite involved, even for the simplest possible system with two electrons and one nucleus. For any system with more than two electrons it is generally not possible to determine closed-form expressions, and we need to use numerical methods.

2.9 Kohn–Sham equations

We now make our final step towards the equations used in density functional theory. In this section we continue with our heuristic approach, while in Chapter 3 we provide a rigorous justification of density functional theory and of eqn 2.66 below.

To summarize our progress so far, in Section 2.5 we eliminated the Coulomb interaction between electrons and transformed the $3N$-dimensional many-body Schrödinger equation into N three-dimensional equations. In Section 2.7 we re-introduced the Coulomb repulsion between electrons using classical electrostatics, while assuming independent electrons. In Section 2.8 we added the exchange interaction in order to take into account the quantum nature of the electrons.

The only remaining element left out of the picture is the *correlation* between electrons. It is intuitive that, owing to the Coulomb repulsion, the probability of finding an electron somewhere will decrease if there is another electron nearby. Using the wavefunctions of the previous section this means $|\Psi(\mathbf{r}_1, \mathbf{r}_2)|^2 < |\phi_1(\mathbf{r}_1)\phi_2(\mathbf{r}_2)|^2$; therefore our trial product solution in eqn 2.35 may not be very accurate. Instead of abandoning this very convenient product form, we can proceed along the same line as in Section 2.8 and describe this repulsion by adding yet another component to the potential $V_n + V_H + V_X$ in the single-particle equations. We will call the additional component $V_c(\mathbf{r})$, where 'c' stands for correlation.

At this stage it is also convenient to replace the exchange potential, V_X, which is rather complicated, by a 'simplified' version which is designed to yield the same effect (as much as possible) but only depends on one space coordinate. We call $V_x(\mathbf{r})$ this simplified *local* exchange potential.

By assembling together all the components that we have discussed in this chapter, we arrive at the following single-particle equation:

$$\left[-\frac{\nabla^2}{2} + V_n(\mathbf{r}) + V_H(\mathbf{r}) + V_x(\mathbf{r}) + V_c(\mathbf{r}) \right] \phi_i(\mathbf{r}) = \varepsilon_i \, \phi_i(\mathbf{r}), \qquad (2.66)$$

where the potential of the nuclei is still given by eqn 2.27 and the Hartree potential is given by eqns 2.49 and 2.50.

Note that we did not specify $V_x(\mathbf{r})$ and $V_c(\mathbf{r})$ yet. Today we still do not know the exact form of these potentials; however, convenient and accurate approximations have been developed over the past few decades. We will discuss such approximations in Chapter 3. Equations like eqn 2.66 are called the *Kohn–Sham equations* and are central to first-principles materials modelling (Kohn and Sham, 1965).

At the end of this heuristic introduction it is important to remember that the potentials V_H, V_x, and V_c in eqn 2.66 all stem from the Coulomb repulsion between the electrons, i.e. the third term on the left-hand side of eqn 2.28. The emergence of several effective potentials is only a byproduct of our choice of working with single-particle wavefunctions, $\phi_i(\mathbf{r})$, instead of the original many-body wavefunction, $\Psi(\mathbf{r}_1, \ldots, \mathbf{r}_N)$.

3
Density functional theory

In Chapter 2 we have seen that the problem of determining the quantum states of a system with N electrons is extremely complex because it involves $3N$ Cartesian coordinates in the many-body wavefunction, $\Psi(\mathbf{r}_1, \mathbf{r}_2, \cdots, \mathbf{r}_N)$. By starting from the independent electrons approximation we have seen that it is possible, at least heuristically, to simplify the description of the many-electron system by using products of single-particle wavefunctions, $\phi_i(\mathbf{r})$, instead of the wavefunction Ψ.

In this chapter we need to address three questions: first, how to link rigorously the single-particle wavefunctions, ϕ_i, with the many-body wavefunction, Ψ; second, which equations the wavefunctions ϕ_i satisfy; and third, how to determine the total energy of the system, E in eqn 2.28.

As we will see in this chapter, DFT provides a very general framework for addressing these questions.

3.1 Total energy of the electronic ground state

In Chapter 2 we have seen that the total energy, E, of the many-electron system can be obtained through eqns 2.53 or equivalently 2.54, which we reproduce here for convenience:

$$E = \langle \Psi | \hat{H} | \Psi \rangle = \int d\mathbf{r}_1 \ldots d\mathbf{r}_N \, \Psi^*(\mathbf{r}_1, \ldots, \mathbf{r}_N) \, \hat{H} \, \Psi(\mathbf{r}_1, \ldots, \mathbf{r}_N). \qquad (3.1)$$

The Hamiltonian appearing in this expression is given by eqn 2.29:

$$\hat{H}(\mathbf{r}_1, \mathbf{r}_2, \cdots, \mathbf{r}_N) = -\sum_i \frac{1}{2} \nabla_i^2 + \sum_i V_{\mathrm{n}}(\mathbf{r}_i) + \frac{1}{2} \sum_{i \neq j} \frac{1}{|\mathbf{r}_i - \mathbf{r}_j|}. \qquad (3.2)$$

The structure of the Hamiltonian in this equation does not depend on the particular material under consideration; therefore any change in E must be associated with changes in the many-body wavefunction, Ψ. In technical jargon this simple observation is expressed by saying that E is a *functional* of Ψ; we indicate this property using square brackets:

$$E = \mathcal{F}[\Psi].$$

Exercise 3.1 What is a functional? Let us start with an example. Consider the function $g(x) = 3x^2$. In this case, for a given value of the input variable, x, we can calculate the value of the function $g(x)$ in the output. For example, with $x = 3$ we find $g = 27$ (number $\rightarrow$ number).

A 'functional' instead takes a function in input, and returns a number in output (function → number). We are certainly already familiar with one important functional, the definite integral. If we consider for instance the functional $\mathcal{F}[g] = \int_0^2 g(x)dx$, given the function $g(x)$, $\mathcal{F}[g]$ corresponds to its definite integral between $x = 0$ and 2. For the function $g(x)$ above we find $\mathcal{F}[g] = 8$. ▶Using Hartree atomic units show that the total-energy functional of a one-dimensional quantum harmonic oscillator of frequency ω is given by:

$$E[\psi] = \int_{-\infty}^{+\infty} \psi^*(x) \left(-\frac{1}{2}\nabla^2 + \frac{1}{2}\omega^2 x^2 \right) \psi(x)dx.$$

▶Evaluate this functional for $\psi(x) = (\omega/\pi)^{\frac{1}{4}}\exp(-\omega x^2/2)$.

The core concept of density functional theory is the observation that, if E is the lowest possible energy of the system, i.e. the energy of the *ground state*, then E is a functional of the *electron density* only:

$$E = F[n]. \tag{3.3}$$

This observation is quite remarkable because, while the energy of any quantum state is generally a functional of the entire wavefunction, $\Psi(\mathbf{r}_1, \mathbf{r}_2, \cdots, \mathbf{r}_N)$, which contains $3N$ variables, the ground-state energy depends only on $n(\mathbf{r})$, which is a function of three variables only. This observation is due to Hohenberg and Kohn (1964).

The key consequence of the above observation is that all that is needed for calculating the total energy E in the ground state is the electron density, n. The situation is more complicated for 'excited' states, i.e. all the quantum states except the one with the lowest energy. For excited states we need the complete many-body wavefunction, Ψ, in order to calculate the energy. In summary:

E is the energy of the ground state: $n(\mathbf{r}) \xrightarrow{F} E \quad E = F[n(\mathbf{r})]$

E is the energy of an excited state: $\Psi(\mathbf{r}_1,\ldots,\mathbf{r}_N) \xrightarrow{\mathcal{F}} E \quad E = \mathcal{F}[\Psi(\mathbf{r}_1,\ldots,\mathbf{r}_N)]$

3.1.1 Hohenberg–Kohn theorem

The statement that the total energy of a many-electron system is a functional of the electron density goes under the name of the Hoenberg–Kohn theorem (Hohenberg and Kohn, 1964). Since the proof of this theorem is instructive and simple enough, we outline it here. The proof is based on the following three premises, which will be justified later:

1) In the ground state the electron density determines uniquely the external potential of the nuclei, V_n, in eqn 3.2: $n \to V_n$.
2) In any quantum state the external potential, V_n, determines uniquely the many-electron wavefunction: $V_n \to \Psi$.
3) In any quantum state the total energy, E, is a functional of the many-body wavefunction through eqn 3.1: $\Psi \to E$.

By combining these premises we infer that, in the ground state, the density determines uniquely the total energy: $n \to V_n \to \Psi \to E$. This indicates that the total energy must be a functional of the density: $E = F[n]$.

Out of the three premises above, the third one corresponds to restating eqn 3.1 and does not require any proof. The second one simply means that if we change the positions (or atomic species) of the nuclei we will obtain a different many-body wavefunction. This is intuitive and does not require any proof. The first statement is not intuitive, but can be demonstrated by *reductio ad absurdum* (Hohenberg and Kohn, 1964). The idea is to start from the assumption that the same ground-state electron density can be obtained from two different external potentials, and show that this leads to a contradiction.

In order to outline the proof it is helpful to introduce the following symbolic notation for the kinetic energy and for the Coulomb energy in eqn 3.2:

$$\hat{T} = -\sum_i \frac{1}{2}\nabla_i^2, \quad \hat{W} = \frac{1}{2}\sum_{i\neq j}\frac{1}{|\mathbf{r}_i - \mathbf{r}_j|}. \tag{3.4}$$

Using this notation the total energy in eqn 3.1 becomes:

$$E = \langle\Psi|\sum_i V_n(\mathbf{r}_i)|\Psi\rangle + \langle\Psi|\hat{T} + \hat{W}|\Psi\rangle. \tag{3.5}$$

By using the relation between the electron density and the wavefunction in eqn 2.10 we discover that the first term can be rewritten in compact form using the electron density. We obtain:

$$E = \int d\mathbf{r}\, n(\mathbf{r})V_n(\mathbf{r}) + \langle\Psi|\hat{T} + \hat{W}|\Psi\rangle. \tag{3.6}$$

Exercise 3.2 ▶For a simple system composed of two electrons show that eqn 3.6 is indeed valid by using eqns 2.10 and 3.2. ▶Verify that this result also holds when we approximate the two-electron wavefunction $\Psi(\mathbf{r}_1, \mathbf{r}_2)$ using a Slater determinant as in eqn 2.42.

Now let us assume that Ψ is the ground-state wavefunction for the potential V_n, with energy E and density n. Suppose there exists another potential, $V_n' \neq V_n$, which generates the same density n, and let us call $\hat{H}'$, Ψ' and E' the Hamiltonian, the ground-state wavefunction and the ground-state energy corresponding to this new potential, respectively. Since Ψ is *not* the ground state of V_n' we can write:

$$\langle\Psi|\hat{H}'|\Psi\rangle > E'. \tag{3.7}$$

The expectation value on the left-hand side can be broken down into contributions from the kinetic energy, the Coulomb repulsion and the external potential, in analogy with eqn 3.6:

$$\langle\Psi|\hat{T} + \hat{W}|\Psi\rangle + \int d\mathbf{r}\, n(\mathbf{r})V_n'(\mathbf{r}) > E'. \tag{3.8}$$

If we now combine eqns 3.6 and 3.8 we find:

$$E - E' > \int d\mathbf{r}\, n(\mathbf{r})[V_n(\mathbf{r}) - V_n'(\mathbf{r})].$$

Since we did not make any assumptions about the external potentials, we can repeat the entire reasoning by simply starting from V_n instead of V_n' in eqn 3.7. In this case we would find:

$$E' - E > \int d\mathbf{r}\, n(\mathbf{r})[V_n'(\mathbf{r}) - V_n(\mathbf{r})].$$

As a last step we can add up the last two equations, obtaining $0 > 0$. This is obviously a contradiction. Therefore the premise that two different potentials, $V_n \neq V_n'$, lead to the same ground-state density, n, must be false. This demonstrates the first step of the Hohenberg–Kohn theorem and completes the proof.

It is important to keep in mind that this simple proof crucially relies on the fact that the energy of the ground state is the lowest possible energy of the system, and all other states are higher in energy. The same proof would not hold for a generic quantum state.

3.2   Kohn–Sham equations

The Hohenberg–Kohn theorem tells us that the total energy of many electrons in their ground state is a functional of the electron density. However, this theorem does not say anything about how to construct such functional. While the exact form of this functional is still unknown, since the original work by Hohenberg and Kohn a number of very useful approximations have been developed.

By comparing eqns 3.3 and 3.6 we can certainly rewrite this functional as follows:

$$F[n] = \int d\mathbf{r}\, n(\mathbf{r})V_n(\mathbf{r}) + \langle \Psi[n]|\hat{T} + \hat{W}|\Psi[n]\rangle. \tag{3.9}$$

Here we see that the first term in the functional is already explicitly dependent on the density, n; however, there are two extra terms (kinetic energy and Coulomb energy) for which the dependence on the density is only implicit. The idea of Kohn and Sham (1965) was to split these implicit terms into the kinetic and Coulomb energy of *independent electrons* as in eqn 2.48, plus an extra term which accounts for the difference:

$$E = F[n]$$

$$\overbrace{= \int d\mathbf{r}\, n(\mathbf{r})V_n(\mathbf{r}) - \underbrace{\sum_i \int d\mathbf{r}\, \phi_i^*(\mathbf{r})\frac{\nabla^2}{2}\phi_i(\mathbf{r})}_{\text{Kinetic energy}} + \underbrace{\frac{1}{2}\iint d\mathbf{r}\, d\mathbf{r}'\, \frac{n(\mathbf{r})n(\mathbf{r}')}{|\mathbf{r} - \mathbf{r}'|}}_{\text{Hartree energy}} + \underbrace{E_{xc}[n]}_{\text{XC energy}}.}^{\text{Total energy in the independent electrons approximation}} \tag{3.10}$$

where the first term is labelled *External potential*.

The extra term, E_{xc}, contains everything that is left out and is called the *exchange and correlation* energy. Equation 3.10 simply breaks down the unknown functional of the density, F, into the sum of known contributions taken from the independent electrons approximation, and an unknown contribution, the exchange and correlation energy. In practice the strategy is to collect everything that we do not know in one place, in the hope that this unknown part will not be too large. If we knew the exchange and correlation energy, $E_{xc}[n]$, then we could calculate the total energy of the system in its ground state, $E = F[n]$, using the electron density. The remaining question is therefore how to actually determine the electron density.

It turns out that the ground-state density, n_0, is precisely the function that minimizes the total energy, $E = F[n]$. This property is called the 'Hohenberg–Kohn variational principle' and can be expressed as follows (Hohenberg and Kohn, 1964):

$$\left. \frac{\delta F[n]}{\delta n} \right|_{n_0} = 0. \tag{3.11}$$

This property is analogous to the variational principle that led us to write the Hartree–Fock equations in Chapter 2. As in that case, stating that a functional derivative must be zero leads to an equation for the wavefunctions, $\phi_i(\mathbf{r})$, which can be used to construct the density, as in eqn 2.44. In fact, if we require these wavefunctions to be orthonormal (i.e. to satisfy eqn 2.56) then the Hohenberg–Kohn variational principle leads to:

$$\left[-\frac{1}{2}\nabla^2 + V_n(\mathbf{r}) + V_H(\mathbf{r}) + V_{xc}(\mathbf{r}) \right] \phi_i(\mathbf{r}) = \varepsilon_i \phi_i(\mathbf{r}), \tag{3.12}$$

where the external nuclear potential, V_n, the Hartree potential, V_H, and the kinetic energy, $-\nabla^2/2$, are identical to those in eqn 2.48. The extra term, V_{xc}, is given by:

$$V_{xc}(\mathbf{r}) = \left. \frac{\delta E_{xc}[n]}{\delta n} \right|_{n(\mathbf{r})}, \tag{3.13}$$

and is called the *exchange and correlation potential*. The steps required to obtain eqn 3.12 from eqn 3.11 are described in Appendix B. The set of equations given by eqn 3.12 are called *Kohn–Sham equations* and form the basis of the Kohn–Sham theory (Kohn and Sham, 1965). As we will see in the remainder of this book, this set of equations constitutes a very powerful tool for calculating many properties of materials starting from the first principles of quantum mechanics.

At this point it is useful to quote Kohn's Nobel Prize lecture (Kohn, 1999):

> 'The Kohn–Sham theory may be regarded as the formal exactification of Hartree theory. With the exact E_{xc} and V_{xc} all many-body effects are in principle included. Clearly this directs attention to the functional $E_{xc}[n]$. The practical usefulness of ground-state DFT depends entirely on whether approximations for the functional $E_{xc}[n]$ could be found, which are at the same time sufficiently simple and sufficiently accurate.'

In other words we know that there must be a functional $E_{xc}[n]$ which gives the *exact* ground-state energy and density using eqns 3.12 and 3.13; however, we do not know what this functional is. The problem therefore is to construct useful approximations to $E_{xc}[n]$.

3.3   The local density approximation

Since the introduction of the Kohn–Sham theory a great deal of effort has been devoted to constructing accurate exchange and correlation functionals, $E_{xc}[n]$, in order to solve the Kohn–Sham equations. As a result several approximate functionals are available today. Here we describe only the simplest functional, which goes under the name of the *local density approximation* to density functional theory (Ceperley and Alder, 1980; Perdew and Zunger, 1981). In order to introduce such a functional it is convenient to study the exchange and correlation energy of a very simple system, the *homogeneous electron gas*. This system is closely related to the 'free electron gas' that is encountered in introductory solid-state physics courses (Kittel, 1976), whereby a gas of electrons is constrained within a box and the potential of the nuclei is taken as constant. The additional complication of this model with respect to the free electron gas is that we also consider the Coulomb repulsion between the electrons. For the homogeneous electron gas it is possible to calculate the exchange energy exactly, and it is possible to determine the correlation energy using numerical techniques. The following sections are devoted to these two aspects.

3.3.1 Exchange and correlation energies of the electron gas

The free electron gas is the simplest model of electrons in a solid. In this model it is assumed that electrons do not interact with each other, that the potential due to the nuclei is simply a constant (which can be set to zero for convenience) and that the N electrons are contained in a large box of volume V. Under these assumptions the eigenstates and eigenvalues in Hartree units are given by (Kittel, 1976):

$$\phi_{\mathbf{k}}(\mathbf{r}) = \frac{1}{\sqrt{V}} \exp(i\mathbf{k} \cdot \mathbf{r}) \text{ and } \varepsilon_{\mathbf{k}} = \frac{|\mathbf{k}|^2}{2}. \tag{3.14}$$

These solutions represent stationary waves with wavevectors $\mathbf{k}$ (see Section 9.2). The eigenvalue of the highest occupied state is the Fermi energy, ϵ_{F}, and the corresponding wavevector is the Fermi wavevector, k_{F}, so that $\epsilon_{\mathrm{F}} = k_{\mathrm{F}}^2/2$. The beauty of the free electron gas model is that all its physical properties depend on one single parameter, the electron density, $n = N/V$. For example, the Fermi wavevector is related to the density by:

$$k_{\mathrm{F}} = (3\pi^2 n)^{\frac{1}{3}}. \tag{3.15}$$

As shown in Exercise 3.3, the exchange energy, E_{X}, of the electron gas can be obtained from the electron density using the following expression (in Hartree units):

$$E_{\mathrm{X}} = -\frac{3}{4} \left(\frac{3}{\pi}\right)^{\frac{1}{3}} n^{\frac{4}{3}} V. \tag{3.16}$$

This simple result is very important as it forms the basis of the local density approximation to DFT (Dirac, 1930; Slater, 1951).

Exercise 3.3 We want to determine the exchange energy of a homogeneous electron gas. In the case of an electron gas with zero net magnetic moment the exchange energy can be calculated as follows (Fetter and Walecka, 2003):

$$E_{\mathrm{X}} = -\frac{1}{2} 2 \sum_{i,j} \int_V d\mathbf{r} \int_V d\mathbf{r}' \frac{\phi_i^*(\mathbf{r})\phi_i(\mathbf{r}')\phi_j^*(\mathbf{r}')\phi_j(\mathbf{r})}{|\mathbf{r} - \mathbf{r}'|}, \tag{3.17}$$

where the indices i, j run over the occupied electronic states and the integrals are performed over the volume of the box, V. The factor of $1/2$ is to avoid counting twice the pairs (i, j)

and (j, i), and the factor of 2 is to take into account the fact that in the case of zero net magnetic moment each state is occupied by two electrons, one with spin up and one with spin down (see Chapter 11; eqn 3.17 corresponds to the generalization to many electrons of the last term in eqn A.5). Since in eqn 3.14 we have a continuous spectrum of eigenvalues, we can replace the sums over the occupied states by integrals over the wavevectors:

$$\sum_i \to \frac{V}{(2\pi)^3} \int_{|\mathbf{k}| \leq k_{\mathrm{F}}} d\mathbf{k}.$$

▶By using this replacement and the eigenfunctions of eqn 3.14 for the $\phi_i(\mathbf{r})$ inside eqn 3.17, show that:

$$E_{\mathrm{X}} = -\frac{1}{(2\pi)^6} \int_{|\mathbf{k}| \leq k_{\mathrm{F}}} d\mathbf{k} \int_{|\mathbf{k}'| \leq k_{\mathrm{F}}} d\mathbf{k}' \int_V d\mathbf{r} \int_V d\mathbf{r}' \frac{e^{-i(\mathbf{k}-\mathbf{k}')\cdot(\mathbf{r}-\mathbf{r}')}}{|\mathbf{r} - \mathbf{r}'|}. \tag{3.18}$$

☐This integral looks rather complicated; however, if we make convenient changes of variables its evaluation is not difficult. Firstly, we may note that in eqn 3.18 there appears only the relative position of two electrons, $\mathbf{r}-\mathbf{r}'$. Secondly, we can use dimensionless variables by scaling the wavevectors with k_{F} and the positions with k_{F}^{-1}. The resulting changes of variables are:

$$\mathbf{u} = k_{\mathrm{F}}(\mathbf{r} - \mathbf{r}'), \quad \mathbf{q} = \frac{\mathbf{k}}{k_{\mathrm{F}}}, \quad \mathbf{q}' = \frac{\mathbf{k}'}{k_{\mathrm{F}}}.$$

The corresponding changes of the volume elements in the integrals are:

$$d\mathbf{r}' = \frac{d\mathbf{u}}{k_{\mathrm{F}}^3}, \quad d\mathbf{k} = k_{\mathrm{F}}^3 \, d\mathbf{q}, \quad d\mathbf{k}' = k_{\mathrm{F}}^3 \, d\mathbf{q}'.$$

▶Using the new variables and considering that $\int_V d\mathbf{r} = V$, show that the expression for the exchange energy in eqn 3.18 simplifies as follows:

$$E_{\mathrm{X}} = -\frac{C}{(2\pi)^6} k_{\mathrm{F}}^4 V, \tag{3.19}$$

where C is a numerical dimensionless constant given by:

$$C = \int_{|\mathbf{q}| \leq 1} d\mathbf{q} \int_{|\mathbf{q}'| \leq 1} d\mathbf{q}' \int du \frac{e^{-i(\mathbf{q}-\mathbf{q}')\cdot\mathbf{u}}}{|\mathbf{u}|}.$$

☐In the evaluation of this constant the last integral on the right-hand side is performed in the limit of $V \to \infty$. The constant C can be evaluated explicitly by using a geometric construction, and the result is $16\pi^3$ (Fetter and Walecka, 2003). ▶By combining eqns 3.19 and 3.15, show that we obtain the exchange energy of the electron gas given in eqn 3.16. ☐Note that no explicit integration was required, and we arrived at the key result in eqn 3.19 by using simple dimensional scaling relations.

In contrast to the exchange energy, for the correlation energy of the electron gas we do not have a simple analytic expression such as eqn 3.16. Nonetheless, it has been possible to calculate the correlation energy for this simple model by solving directly the many-particle Schrödinger equation using stochastic numerical methods (Ceperley and Alder, 1980). The correlation energy of the electron gas can be extracted from the data of Ceperley and Alder by removing the known kinetic, Hartree and exchange contributions from the calculated total energies. The data calculated by Ceperley and Alder were subsequently parametrized by Perdew and Zunger (1981), and the resulting expression for the correlation energy is as follows (in the case of zero net magnetic moment):

$$E_{\rm C} = nV \cdot \begin{cases} 0.0311\ln r_s - 0.0480 + 0.002\, r_s \ln r_s - 0.0116\, r_s & \text{if } r_s < 1, \\[2mm] \dfrac{-0.1423}{1 + 1.0529\sqrt{r_s} + 0.3334\, r_s} & \text{if } r_s \geq 1. \end{cases} \qquad (3.20)$$

In this expression we introduced for ease of notation the Wigner–Seitz radius, r_s. This quantity is simply defined as the radius of the sphere occupied on average by each electron:

$$\frac{V}{N} = \frac{4\pi}{3} r_s^3 = \frac{1}{n}.$$

A detailed explanation of why the functional forms used in eqn 3.20 are appropriate for parametrizing the data of Ceperley and Alder (1980) is beyond the scope of this book. Here we mention only that these functional forms are obtained from the exact asymptotic expansions of the correlation energy for $r_s \to 0$ and $r_s \to \infty$, respectively. Figure 3.1 shows the exchange and correlation energies of the homogeneous electron gas for typical densities found in semiconductors. We note that the correlation energy is consistently an order of magnitude smaller than the exchange energy throughout the range of electron densities shown in Figure 3.1. If we compare $E_{\rm X} + E_{\rm C}$ with the total kinetic energy of a free electron gas, we discover that these contributions are of the same order of magnitude, and therefore cannot be neglected.

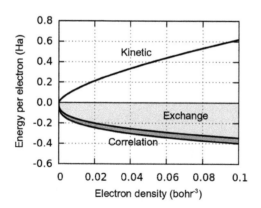

Fig. 3.1 Exchange and correlation energies per electron, $E_{\rm X}/N$ and $E_{\rm C}/N$, respectively in the homogeneous electron gas as a function of the electron density. The top of the range displayed along the horizontal axis corresponds approximately to the density in the middle of Si–Si bonds in crystalline silicon in the diamond structure (Cohen, 2000). The exchange energy (light grey) corresponds to eqn 3.16, and the correlation energy (dark grey) is given by eqn 3.20. The kinetic energy of the free electron gas in absence of Coulomb repulsion is also shown for comparison.

3.3.2 The electron gas as a local approximation to a real material

In Section 3.2 we have seen that the key ingredient for using density functional theory is the exchange and correlation energy functional $E_{xc}[n]$ (eqn 3.13), and we have indicated that this functional is still unknown. In Section 3.3.1 we calculated the exchange energy of the simplest model of interacting electrons, the homogeneous electron gas, and we discussed a numerical parametrization for the correlation energy (eqns 3.16 and 3.20). In this section we want to use the results obtained for the electron gas in order to obtain a practical approximation for studying electrons in real materials.

While the electron density in materials may not resemble at all the homogeneous electron gas, we can use this simple model in order to describe the exchange and correlation energy in those regions where the density is slowly varying. This concept is illustrated in Figure 3.2. By decreasing the width of the rectangular regions in Figure 3.2 to infinitesimal volume elements, it becomes natural to associate each volume element $d\mathbf{r}$ with a homogeneous electron gas having *local density* $n(\mathbf{r})$ at point $\mathbf{r}$. In analogy with the schematic representation in Figure 3.2, each volume element $d\mathbf{r}$ will contribute an exchange and correlation energy:

$$dE_{xc} = \frac{E_{xc}^{\text{HEG}}[n(\mathbf{r})]}{V} \, d\mathbf{r}, \qquad (3.21)$$

where 'HEG' stands for homogeneous electron gas, and $E_{xc}^{\text{HEG}}[n(\mathbf{r})]$ is obtained by adding up the exchange energy and the correlation energy of eqns 3.16 and 3.20 calculated for density $n(\mathbf{r})$ at point $\mathbf{r}$. The exchange and correlation energy of the entire system can then be obtained by adding up the individual contributions from each volume element:

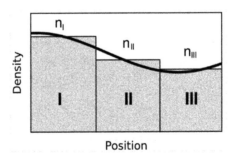

Fig. 3.2 Schematic representation of the electron density, $n(\mathbf{r})$, in a solid or a molecule along a given direction (thick line). In this case we can partition the system into three regions, I, II and III. For each region we approximate E_{xc} using eqns 3.16 and 3.20, by considering homogeneous electron gases of densities n_{I}, n_{II} and n_{III}, respectively (step-like function). In the absence of better approximations, the exchange and correlation energy of the entire system may be obtained by adding up the contributions from each of these regions. This idea is at the core of the local density approximation to density functional theory.

$$E_{xc} = \int_V dE_{xc} = \int_V \frac{E_{xc}^{\text{HEG}}[n(\mathbf{r})]}{V} d\mathbf{r}.$$

For example, using eqn 3.16 we can calculate the exchange energy of the entire system as:

$$E_x = -\frac{3}{4}\left(\frac{3}{\pi}\right)^{\frac{1}{3}} \int_V n^{4/3}(\mathbf{r}) d\mathbf{r}. \tag{3.22}$$

The corresponding expression for the correlation energy based on eqn 3.20 is slightly more involved; however the underlying concept remains the same. The approximation defined by eqn 3.21 is referred to as the *local density approximation* to density functional theory (abbreviated as LDA). The idea of using the local electron density in order to 'map' a real system into a homogeneous electron gas model has been around for a long time, and is actually older than density functional theory itself (Thomas, 1927; Fermi, 1928; Dirac, 1930; Slater, 1951).

3.3.3   The exchange and correlation potential

The availability of a practical approximation for the exchange and correlation energy, as given by the LDA in eqn 3.21, allows us to obtain the final ingredient needed for solving the Kohn–Sham equations, namely the exchange and correlation potential. In fact, given the LDA exchange and correlation energy, $E_{xc}[n]$ in eqn 3.21, we can obtain V_{xc} by using the functional derivative in eqn 3.13.

As shown in Exercise 3.4 below, the result of this functional derivative for the exchange interaction has the following very simple expression:

$$V_x(\mathbf{r}) = -\left(\frac{3}{\pi}\right)^{\frac{1}{3}} n^{\frac{1}{3}}(\mathbf{r}). \tag{3.23}$$

This expression is very appealing because, in order to obtain the exchange potential at point $\mathbf{r}$, we need to know only the density at the same point, $n(\mathbf{r})$. The expression for the correlation potential is slightly more complicated but also in that case only the local density is needed; i.e. the correlation potential at a given point is determined by the density at the same point.

Exercise 3.4 In order to derive the Kohn–Sham equations it is useful to introduce the concept of the *functional derivative* (Parr and Yang, 1989). Given the functional F of the function $g(\mathbf{r})$, an arbitrary function $h(\mathbf{r})$, and a real parameter ϵ, the functional derivative of F with respect to g, $\delta F/\delta g$, is the function which satisfies the following property:

$$\int d\mathbf{r}\, h(\mathbf{r}) \frac{\delta F}{\delta g}(\mathbf{r}) = \frac{d}{d\epsilon} F[g(\mathbf{r}) + \epsilon\, h(\mathbf{r})]\bigg|_{\epsilon=0}. \tag{3.24}$$

For example, consider the following functional:

$$I[g] = \int d\mathbf{r}\, g^2(\mathbf{r}).$$

By applying the definition in eqn 3.24 we find:

$$\int d\mathbf{r}\, h(\mathbf{r})\, \frac{\delta I}{\delta g}(\mathbf{r}) = \frac{d}{d\epsilon} \int d\mathbf{r}\, [g(\mathbf{r}) + \epsilon\, h(\mathbf{r})]^2 \bigg|_{\epsilon=0} = \int d\mathbf{r}\, 2\, h(\mathbf{r})\, g(\mathbf{r}).$$

Therefore, in this example we have:

$$\frac{\delta I}{\delta g} = 2\, g(\mathbf{r}).$$

▶Using the LDA exchange energy from eqn 3.22:

$$E_x = -\frac{3}{4}\left(\frac{3}{\pi}\right)^{\frac{1}{3}} \int_V n^{4/3}(\mathbf{r})d\mathbf{r},$$

determine the exchange potential, V_x, of eqn 3.23 using the functional derivative as in eqn 3.13:

$$V_x(\mathbf{r}) = \frac{\delta E_x[n]}{\delta n}\bigg|_{n(\mathbf{r})}.$$

Exercise 3.5 ▶Using the approximate electron density for the ground state of the He atom determined in Exercise 2.6, show that the LDA exchange potential of He is, in Hartree units:

$$V_x(r) = -1.36\, \exp\left(-1.07\, r\right). \tag{3.25}$$

▶Using the same electron density and remembering that in the case of He we have $V_X = -V_H/2$ (see Exercise 2.6), show that the exchange potential of He in the Hartree–Fock approximation is:

$$V_X(r) = -\frac{1}{r}\left[1 - (1 + 1.6\, r)\exp\left(-3.2\, r\right)\right]. \tag{3.26}$$

▶Plot the exchange potentials in the LDA and in the Hartree–Fock approximations. Identify and discuss the qualitative differences between the two curves.
☐Generally speaking, in the case of finite systems such as molecules and clusters, the LDA yields an exchange potential which is too short-ranged. This is a well-known deficiency of the LDA, and can manifest itself as the inability of the LDA to predict stable negatively charged atoms or molecules. This is particularly important in the atomistic study of biological processes (Burke, 2012).

The vast majority of modern first-principles calculations on materials uses the LDA or some slightly more elaborate exchange and correlation functionals which also take into account the slopes of the electron density (Langreth and Mehl, 1983; Becke, 1988; Perdew *et al.*, 1992; Perdew *et al.*, 1996). The development of useful exchange and correlation functionals constitutes a research field in its own right, but its discussion is beyond the scope of this introduction. The interested reader is referred to a review of the subject by Burke (2012).

3.4 Self-consistent calculations

In Sections 3.1–3.3 we introduced the basic concepts of density functional theory and the Kohn–Sham equations, and anticipated that this theoretical apparatus can be used to calculate the total energy, E, and the electron density, $n(\mathbf{r})$, of materials in their ground state. The questions that remain to be answered now are: how do we actually solve the Kohn–Sham equations; and once obtained the solutions of the Kohn–Sham equations, how do we calculate the total energy?

In order to answer the first question it is convenient to rewrite here the Kohn–Sham equations (eqn 3.12) and each term appearing in them:

$$\left[-\frac{1}{2}\nabla^2 + V_{\text{tot}}(\mathbf{r})\right]\phi_i(\mathbf{r}) = \varepsilon_i\phi_i(\mathbf{r}), \tag{3.27}$$

$$V_{\text{tot}}(\mathbf{r}) = V_{\text{n}}(\mathbf{r}) + V_{\text{H}}(\mathbf{r}) + V_{xc}(\mathbf{r}), \tag{3.28}$$

$$V_{\text{n}}(\mathbf{r}) = -\sum_I \frac{Z_I}{|\mathbf{r} - \mathbf{R}_I|}, \tag{3.29}$$

$$\nabla^2 V_{\text{H}}(\mathbf{r}) = -4\pi n(\mathbf{r}), \tag{3.30}$$

$$V_{xc}(\mathbf{r}) = \frac{\delta E_{xc}[n]}{\delta n}(\mathbf{r}), \tag{3.31}$$

$$n(\mathbf{r}) = \sum_i |\phi_i(\mathbf{r})|^2. \tag{3.32}$$

The nuclear and Hartree potentials, V_{n} and V_{H}, in eqn 3.28 come from eqns 2.27 and 2.46, respectively. The exchange and correlation potential of eqn 3.31 has been given explicitly in Section 3.3. We can see that eqn 3.27 is our standard single-particle Schrödinger equation, and can be solved as a standard eigenvalue problem. However, in order to determine the eigenfunctions, $\phi_i(\mathbf{r})$, and the eigenvalues, ε_i, we first need to know the total potential, $V_{\text{tot}} = V_{\text{n}} + V_{\text{H}}(\mathbf{r}) + V_{xc}(\mathbf{r})$. The complication here is that V_{H} and V_{xc} depend on the density, n, and the density depends on the unknown eigenfunctions, ϕ_i, through eqn 3.32. In other words, each solution ϕ_i depends implicitly on all other solutions, ϕ_j, describing the occupied electronic states. The fact that all the solutions ϕ_i are linked with each other in eqns 3.27–3.32 implies that they must be determined *self-consistently*. As in Section 2.7, 'self-consistency' means that, if we insert the solutions ϕ_i inside eqn 3.32 to calculate the density, determine the corresponding potential V_{tot} using eqn 3.28, and solve the Schrödinger equation (eqn 3.27) again, then we find as a solution the same function ϕ_i from which we started.

The practical procedure for solving the Kohn–Sham equations, eqns 3.27–3.32, is the following: we start by specifying the nuclear coordinates, in such a way that we can calculate the nuclear potential, V_{n}, from eqn 3.29. Typically this information is available from crystallography data (in Chapter 4 we will discuss how to find the atomic coordinates when these are not available). In principle we could try to solve eqn 3.27 using V_{n} as a first approximation to V_{tot}; however, this is too crude an approximation, and it is more convenient to 'guess' a possible electron density, $n(\mathbf{r})$, in order to determine a preliminary approximation to the Hartree and exchange and correlation potentials. A simple but very useful approximation is to construct the first guess for the electron density by adding up the densities corresponding to completely isolated atoms, but arranged in the atomic positions corresponding to the material under consideration. Using the density we obtain initial estimates of the Hartree and exchange and correlation potentials, $V_H + V_{xc}$, and from there the total potential, V_{tot}, needed in eqn 3.27. At this point we can proceed with the numerical solution of the Kohn–Sham equations. This can be done for example by discretizing the space into a mesh of points and representing the Laplace operator using finite difference formulas. Some common numerical procedures used to solve eqn 3.27 are discussed in Appendix C. By solving the Kohn–Sham equations we obtain the new wavefunctions, ϕ_i, which can in turn be used to construct a better estimate of the density, n, and the total potential, V_{tot}. This process is then repeated until the new density matches the old density within a desired tolerance, at which point we say that we have 'achieved self-consistency'. This procedure is illustrated in Figure 3.3.

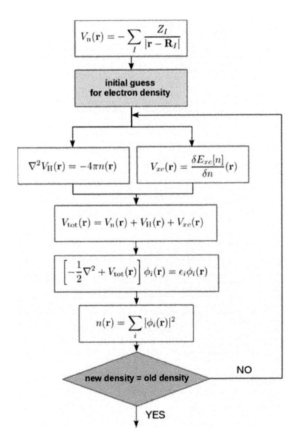

Fig. 3.3   Schematic flow-chart for finding self-consistent solutions of the Kohn–Sham equations (eqns 3.27–3.32). The equality sign in the conditional symbol means that the electron density at this iteration and the density at the previous iteration differ by less than a desired tolerance. While conceptually it makes sense to compare densities in order to check for self-consistency, in practical calculations it is often more convenient simply to compare the total energies evaluated using eqn 3.10 at two successive iterations.

Once we have obtained the electron density in the ground state, $n(\mathbf{r})$, it is possible to calculate the total energy E of the system using eqn 3.10. In the following chapters we will see how to use the ground-state total energy for calculating a variety of material properties.

3.5   Remit of density functional theory and limitations

Numerical algorithms for solving the Kohn–Sham equations are currently implemented in a large number of software packages, many of which are freely available online. These packages allow one to comfortably execute calculations on molecules and crystals with small unit cells even on personal computers. However, most of these packages are designed and optimized in order to be executed on large parallel computers, ranging from small research clusters (of the order of 100 processors) to national high-performance computing facilities (of the order of 100,000 processors). Today the combination of DFT with high-performance computing provides us with a very powerful tool for understanding and predicting material properties starting from the first principles of quantum mechanics.

Despite the success and widespread use of DFT, it is important to bear in mind that it addresses the electronic ground state of materials. Stated otherwise, DFT is not designed to describe *electronic excitations* and *non-equilibrium* phenomena. For example, the Kohn–Sham formulation of DFT is able to predict accurately the equilibrium structures of semiconductors, but it cannot be used for predicting with quantitative accuracy their corresponding optical properties.
A non-exhaustive list of material properties which can be calculated with good accuracy using DFT would include:

- equilibrium structures,
- vibrational properties and vibrational spectra,
- binding energies of molecules and cohesive energies of solids,
- ionization potential and electron affinity of molecules,
- band structures of metals and semiconductors.

On the other hand, material properties which cannot be calculated reliably using DFT include:

- electronic band gaps of semiconductors and insulators,
- magnetic properties of Mott–Hubbard insulators (mostly systems with atomic-like localized d or f electronic states)
- bonding and structure in sparse matter, e.g. proteins, where van der Waals forces are important.

From this list it should be clear that DFT has a number of limitations and cannot be expected to be a 'universal' tool for calculating material properties. Nevertheless, even in those cases where DFT performs poorly, it turns out that the Kohn–Sham eigenfunctions and eigenvalues represent a useful starting point for more refined (and more time-consuming) approaches. As an example, for predicting band gaps it is common to start from the Kohn–Sham equations and subsequently apply 'quasiparticle' corrections (Aryasetiawan and Gunnarsson, 1998), as will be discussed in Section 9.5. Similarly, for describing the magnetic properties of Mott–Hubbard insulators, it is convenient to use the Kohn–Sham eigenstates and eigenvalues as a starting point for carrying out more advanced calculations based on the 'dynamical mean-field theory' (Georges *et al.*, 1996).

4
Equilibrium structures of materials: fundamentals

In Chapters 2 and 3 we discussed the ground state of a system of electrons under the assumption that the nuclear coordinates are fixed. In particular, we have seen in Chapter 3 how to calculate the total energy, E, and the electron density, $n(\mathbf{r})$, for such a system using DFT. What we have learned in Chapter 3 is useful when the nuclear coordinates are available from experiment, but sometimes we are confronted with situations where experimental information on the nuclear positions is incomplete or unavailable. In these cases it is important to be able to predict the equilibrium structure of a material starting from the first principles of quantum mechanics.

In this chapter and in Chapter 5 we discuss how to use DFT to determine the equilibrium structures of molecules and solids. We limit the discussion to zero-temperature structures since this is often a good enough approximation for describing material properties at room temperature. The study of equilibrium structures at high temperature will be discussed at the end of Chapter 8.

The equilibrium structure corresponds to the situation where the total force acting on each nucleus vanishes, so that the positions of the nuclei do not change. Our task is therefore: (i) to understand how to determine the forces experienced by the nuclei using DFT, for a given set of nuclear coordinates; and (ii) to find the nuclear coordinates for which such forces are identically zero. In this chapter we discuss the basic theoretical tools which are needed for calculating forces. These techniques are of general validity and can be used with any theory of the electronic structure; in other words, they are not limited to DFT. In Chapter 5 we will discuss some illustrative examples of equilibrium structures of molecules, crystals and surfaces calculated using DFT.

4.1 The adiabatic approximation

In this section our goal is to understand how to compute the forces on the nuclei within DFT. For this purpose we need to re-introduce nuclei in our formalism and therefore we go back to eqn 2.24:

$$\left[-\sum_i \frac{\nabla_i^2}{2} - \sum_I \frac{\nabla_I^2}{2M_I} - \sum_{i,I} \frac{Z_I}{|\mathbf{r}_i - \mathbf{R}_I|} + \frac{1}{2}\sum_{i \neq j} \frac{1}{|\mathbf{r}_i - \mathbf{r}_j|} + \frac{1}{2}\sum_{I \neq J} \frac{Z_I Z_J}{|\mathbf{R}_I - \mathbf{R}_J|} \right] \Psi = E_{\text{tot}} \Psi.$$

(4.1)

Here, $\Psi(\mathbf{r}_1, \ldots, \mathbf{r}_N, \mathbf{R}_1, \ldots, \mathbf{R}_M)$ depends on the positions of the electrons and of the nuclei. Attempting a direct solution of eqn 4.1 is essentially an impossible task for most systems except a few extremely simple cases such as the H_2 molecule. Nevertheless, we can still capitalize on what we have learned in Chapter 3 and try to understand some qualitative features of its solutions. In fact we already know how to find the ground state of the system in the case of fixed nuclei. It is therefore convenient to see the problem in eqn 4.1 as one where the nuclei are 'almost' immobile while electrons are moving around.

Exercise 4.1 The approximation that in the ground state the nuclei remain close to their equilibrium positions while the electrons are more mobile can be understood by examining the standard textbook example of the harmonic oscillator. Consider the following one-dimensional Schrödinger equation for a particle in a parabolic potential well (in atomic units):

$$-\frac{1}{2M}\frac{d^2\psi(x)}{dx^2} + Cx^2\psi(x) = E\psi(x). \tag{4.2}$$

▶Show that the lowest-energy state is:

$$\psi(x) = \frac{1}{\sqrt{2\pi}\sigma}\exp\left(-\frac{x^2}{2\sigma^2}\right) \quad \text{with } \sigma = (2CM)^{-1/4}.$$

☐This result indicates that the ground-state wavefunction of a particle in a parabolic potential is a gaussian. The standard deviation of this gaussian scales with the particle mass as $\sigma \sim M^{-1/4}$. A parabolic potential can be considered as a first approximation for the potential felt by an electron in the middle of an interatomic bond, or as an approximation for the potential felt by a nucleus around its equilibrium position.

Exercise 4.2 Consider the parabolic potential in Exercise 4.1 with the force constant set to $C=1$ Ha/bohr^2. This value corresponds approximately to the potential felt by C nuclei in diamond around their equilibrium positions. ▶Calculate the spatial extent of the ground-state wavefunction for an electron in this potential, then repeat the calculation for a C nucleus and for a Pb nucleus (remember that in atomic units the electron mass is $M = 1$). The wavefunctions corresponding to an electron and to the Pb nucleus are shown in Figure 4.1.

From Exercises 4.1 and 4.2 it should be clear that the approximation of nuclei 'almost immobile' is a sensible one because, as a consequence of the heavier mass, the ratio of kinetic to potential energy for nuclei is much smaller than in the case of electrons. This observation suggests that we may try to separate the complete Schrödinger equation (eqn 4.1) into two equations, one for the electrons and one for the nuclei. For this purpose we write the total wavefunction, Ψ, as a product of an electron-only wavefunction, $\Psi_{\mathbf{R}}$, and a nuclear-only wavefunction, χ:

$$\Psi(\mathbf{r}_1,\ldots,\mathbf{r}_N,\mathbf{R}_1,\ldots,\mathbf{R}_M) = \Psi_{\mathbf{R}}(\mathbf{r}_1,\ldots,\mathbf{r}_N)\,\chi(\mathbf{R}_1,\ldots,\mathbf{R}_M). \tag{4.3}$$

In this expression the subscript $\mathbf{R}$ is used as a reminder that the wavefunction $\Psi_{\mathbf{R}}$ depends parametrically on the set of nuclear coordinates $\mathbf{R}_1,\ldots,\mathbf{R}_M$. With reference to Figure 4.1 we can see how this 'guess' for the total wavefunction is sensible: if we first determine the electron wavefunction for a set of fixed nuclear positions, and then we allow the nuclei to move according to their quantum-mechanical wavefunction, they will remain very close to their original coordinates and therefore the electron-only

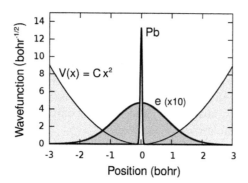

Fig. 4.1 Ground-state wavefunctions of an electron (labelled 'e') and of a Pb nucleus ('Pb') in the parabolic potential given by eqn 4.2 (thin line). Due to the much heavier mass of the Pb nucleus, the Pb wavefunction is localized in a very narrow region of space compared with the electron.

wavefunction will not be affected too much. It is important to keep in mind that these considerations apply only for materials in their equilibrium configuration.

At this stage it is convenient to rewrite eqn 2.28 by indicating explicitly the set of nuclear coordinates, $\mathbf{R}$:

$$\left[-\sum_i \frac{\nabla_i^2}{2} + \sum_i V_n(\mathbf{r}_i; \mathbf{R}) + \frac{1}{2} \sum_{i \neq j} \frac{1}{|\mathbf{r}_i - \mathbf{r}_j|} \right] \Psi_\mathbf{R} = E_\mathbf{R} \, \Psi_\mathbf{R}, \qquad (4.4)$$

where the subscript $\mathbf{R}$ in the total electronic energy indicates that it is a function of all the nuclear coordinates, $E_\mathbf{R} = E(\mathbf{R}_1, \ldots, \mathbf{R}_M)$. By using eqns 4.3 and 4.4 inside eqn 4.1 we find immediately:

$$E_\mathbf{R} \Psi_\mathbf{R} \, \chi + \left[-\sum_I \frac{\nabla_I^2}{2M_I} + \frac{1}{2} \sum_{I \neq J} \frac{Z_I Z_J}{|\mathbf{R}_I - \mathbf{R}_J|} \right] \Psi_\mathbf{R} \, \chi = E_{\text{tot}} \Psi_\mathbf{R} \, \chi. \qquad (4.5)$$

If we now multiply both sides by $\Psi_\mathbf{R}^*$, integrate over the electronic variables $\int d\mathbf{r}_1 \ldots d\mathbf{r}_N$, and use the normalization condition in eqn 2.11, we obtain:

$$\left[-\sum_I \frac{\nabla_I^2}{2M_I} + \frac{1}{2} \sum_{I \neq J} \frac{Z_I Z_J}{|\mathbf{R}_I - \mathbf{R}_J|} + E(\mathbf{R}_1, \ldots, \mathbf{R}_M) \right] \chi = E_{\text{tot}} \, \chi. \qquad (4.6)$$

This result is quite impressive, since we have obtained a many-body Schrödinger equation for the nuclei alone. The effect of the electrons is contained inside the total electronic energy, $E(\mathbf{R}_1, \ldots, \mathbf{R}_M)$, which acts as an 'effective potential' for the nuclei. The replacement of the complete equation for electrons and nuclei (eqn 4.1) by two separate equations (eqns 4.4 and 4.6) corresponds to a decoupling of the dynamics of the electrons from that of the nuclei. This decoupling goes under the name of the *Born–Oppenheimer approximation* (Born and Oppenheimer, 1927). From eqn 4.6 we can see that the total effective potential experienced by the nuclei is the sum of their Coulomb repulsion and of the total energy of the electrons at fixed nuclear positions. This observation will be used in Section 4.2 to define a practical procedure for calculating the forces acting on nuclei.

Before proceeding it is useful to try and understand what is being neglected in the Born–Oppenheimer approximation. The idea of writing the total wavefunction for electrons and nuclei in the product form, eqn 4.3, is always legitimate. However, in general we have no guarantee that the electronic component, $\Psi_{\mathbf{R}}$, is an eigenstate of the electron-only Schrödinger equation (eqn 4.4).

In the most general case, even if $\Psi_{\mathbf{R}}$ is not an eigenstate of eqn 4.4, it can certainly be written as a linear combination of eigenstates. If we call these eigenstates $\Psi_{\mathbf{R}}^{(n)}$ with $n = 0, 1, \ldots$ we have:

$$\Psi_{\mathbf{R}} = \sum_n c_{\mathbf{R}}^{(n)} \Psi_{\mathbf{R}}^{(n)}, \tag{4.7}$$

where the coefficients $c_{\mathbf{R}}^{(n)}$ are to be determined. In a rigorous treatment we should use precisely this linear combination inside eqn 4.1.

By using only the lowest-energy term of the linear combination in eqn 4.7 we are implicitly making the following assumption: when the nuclear coordinates change, the electrons evolve from the electronic ground state associated with the initial nuclear coordinates to the electronic ground state associated with the final coordinates. Intuitively we can picture this process by imagining that the nuclei move so slowly that electrons always have enough time to re-adjust and remain in their lowest-energy state. In these conditions the electrons do not exchange energy with the nuclei; therefore their evolution is adiabatic. The notion of adiabatic electron–nuclear dynamics is analogous to the concept of adiabatic processes in thermodynamics (Fermi, 1956). For this reason the Born–Oppenheimer approximation is also referred to as the *adiabatic approximation.*

We may ask ourselves what the size of the error is that we make by neglecting all but the lowest-energy eigenstate in eqn 4.7. It can be shown on very general grounds that this error scales as $(M/m_e)^{-1/4}$, where M denotes a typical mass of the nuclei, in line with the observations that were made in Exercise 4.2 (Grimvall, 1981). For example, in the case of Pb the neglected terms would give a correction to the lowest-energy state of about 4%. We stress once again that the adiabatic approximation is valid only for materials at or near equilibrium. In this approximation it is not possible for nuclei to promote electronic transitions above the ground state. An important phenomenon which is not described by the adiabatic approximation is the *electron–phonon interaction*, which will be mentioned briefly in Section 10.3.2 (Grimvall, 1981).

4.2 Atomic forces

We can summarize the previous section by stating that, within the Born-Oppenheimer approximation, the stationary states of the nuclei can be described using the nuclear Hamiltonian appearing in eqn 4.6:

$$\hat{H}_{\mathrm{n}} = -\sum_I \frac{\nabla_I^2}{2M_I} + U(\mathbf{R}_1, \ldots, \mathbf{R}_M), \tag{4.8}$$

where the total potential energy of the nuclei is defined as:

$$U(\mathbf{R}_1, \ldots, \mathbf{R}_M) = \frac{1}{2} \sum_{I \neq J} \frac{Z_I Z_J}{|\mathbf{R}_I - \mathbf{R}_J|} + E(\mathbf{R}_1, \ldots, \mathbf{R}_M), \qquad (4.9)$$

and $E(\mathbf{R}_1, \ldots, \mathbf{R}_M)$ is the total energy of the electrons when the nuclei are held immobile in positions $\mathbf{R}_1, \ldots, \mathbf{R}_M$. The study of the Hamiltonian in eqn 4.8 can be rather complicated, but we can afford a considerable simplification if we take a careful look at Figure 4.1. The striking element of this figure is that, given the same potential energy profile, the nuclear wavefunction is much more compact than the electron wavefunction. We could almost say that, while the electron in Figure 4.1 behaves like a wave, the nucleus behaves like a point particle. We can formalize this observation by making the assumption that quantum mechanics is not necessary for describing the dynamics of nuclei, and we can revert to classical mechanics as a first approximation. It turns out that, in the study of equilibrium structures of materials, this approximation is a very good one for all but the lightest elements in the periodic table (e.g. hydrogen). The move from a quantum description of the nuclei to a classical description can be performed by replacing the quantum-mechanical momentum in eqn 4.8 by its corresponding classical momentum:

$$- i\nabla_I = -i\frac{\partial}{\partial \mathbf{R}_I} \quad \longrightarrow \quad \mathbf{P}_I. \qquad (4.10)$$

This replacement yields the following 'classical Hamiltonian' for the nuclei:

$$\hat{H}_{\mathrm{n}}^{\mathrm{clas}} = \sum_I \frac{\mathbf{P}_I^2}{2M_I} + U(\mathbf{R}_1, \ldots, \mathbf{R}_M). \qquad (4.11)$$

When the classical Hamiltonian for a set of particles is known, it is possible to use *Hamilton's equations of motion* (Goldstein, 1950; Gregory, 2006) in order to derive Newton's equations. The procedure for deriving Newton's equations starting from a Hamiltonian is illustrated in Exercise 4.3. If we perform the steps of Exercise 4.3 for the Hamiltonian in eqn 4.11 we find without difficulty:

$$M_I \frac{d^2 \mathbf{R}_I}{dt^2} = -\frac{\partial U}{\partial \mathbf{R}_I}. \qquad (4.12)$$

Since the left-hand side of this equation is the product of the mass and the classical acceleration of the I-th nucleus, we can identify the right-hand side with the force acting on that nucleus:

$$\mathbf{F}_I = -\frac{\partial U}{\partial \mathbf{R}_I}. \qquad (4.13)$$

This is a remarkable result which establishes the link between the total energy of electrons and nuclei and the equilibrium structures of materials. In fact this equation tells us that the equilibrium structures, i.e. those configurations where the nuclei have no acceleration and therefore are immobile in an inertial reference frame, correspond to the stationary points of the function $U(\mathbf{R}_1, \ldots, \mathbf{R}_M)$ in eqn 4.9. Equation 4.13 constitutes the main conceptual tool for determining materials structures from first principles. It is worth pointing out that eqn 4.13 is not limited to the framework of density functional theory, and carries general validity.

The function $U(\mathbf{R}_1, \ldots, \mathbf{R}_M)$ is often referred to as the *potential energy surface*, even though, strictly speaking, it is a hyper-surface in a space of dimensions $3M+1$. For example, the potential energy surface of a nucleus in a parabolic potential well in two Cartesian dimensions is a paraboloid, i.e. a surface in a three-dimensional space.

The equilibrium structures determined through the minimization of the potential energy surface, U, correspond to the case of zero temperature. The study of equilibrium structures at $T = 0$ K is useful since in many cases the structural properties of materials at ambient temperature ($25°C$) are essentially the same as those at $T = 0$ K. The study of equilibrium structures at $T > 0$ is slightly more involved and will be discussed in Section 8.5.

Exercise 4.3 Hamilton's equations of motion provide a very general and elegant framework for studying classical mechanics, and are also very useful for establishing the formal connection between quantum mechanics and classical mechanics (Goldstein, 1950). The classical Hamiltonian of one particle of mass m moving along one dimension in the potential energy $U(x)$ with position x and momentum p is given by:

$$\hat{H} = \frac{p^2}{2m} + U(x). \tag{4.14}$$

Hamilton's equations express the time evolution of the position and momentum of this particle in terms of derivatives of the Hamiltonian:

$$\frac{dx}{dt} = \frac{\partial \hat{H}}{\partial p}, \tag{4.15}$$

$$\frac{dp}{dt} = -\frac{\partial \hat{H}}{\partial x}. \tag{4.16}$$

▶Using eqns 4.15 and 4.16 show that the force acting on the particle at position x is given by:

$$F(x) = -\frac{\partial U}{\partial x}. \tag{4.17}$$

▶Now write the classical Hamiltonian associated with the Schrödinger equation in Exercise 4.1 (eqn 4.2). ▶For this classical Hamiltonian, write Hamilton's equations of motion, and verify that the particle is subject to the elastic restoring force $F = -2Cx$.

Exercise 4.4 The simplest possible application of eqn 4.13 for calculating equilibrium structures is for the hydrogen molecular ion H_2^+, which consists of one electron and two protons. In this exercise we want to estimate the bond length, d, of the H_2^+ molecule at equilibrium. ▶By setting the reference frame such that the two protons are located at $\pm d/2$ along the x axis, use eqn 2.28 to show that the electronic Hamiltonian for fixed nuclear positions is given by:

$$\hat{H} = -\frac{1}{2}\nabla^2 - \frac{1}{|\mathbf{r} - d\mathbf{u}_x/2|} - \frac{1}{|\mathbf{r} + d\mathbf{u}_x/2|}. \tag{4.18}$$

In order to find the electronic ground state for this Hamiltonian, as a first approximation the electron wavefunction can be obtained by using the following linear combination of atomic orbitals (Heitler and London, 1927; Lennard-Jones, 1929):

$$\psi(\mathbf{r}) = \frac{1}{\sqrt{2(1+S)}} \left[\phi_{1s}(\mathbf{r} - d\mathbf{u}_x/2) + \phi_{1s}(\mathbf{r} + d\mathbf{u}_x/2) \right], \tag{4.19}$$

where ϕ_{1s} is the fundamental state of the H atom with energy $E_{1s} = -0.5$ Ha, and the parameter S guarantees the normalization of the wavefunction in eqn 4.19:

$$S = \int d\mathbf{r}\, \phi_{1s}(\mathbf{r} - d\mathbf{u}_x/2) \phi_{1s}(\mathbf{r} + d\mathbf{u}_x/2). \tag{4.20}$$

▶Using eqn 2.54 show that the electron energy of the state in eqn 4.19 is given by:

$$E = E_{1s} + \frac{h + \gamma}{1 + S}, \tag{4.21}$$

with the definitions:

$$h = -\int d\mathbf{r}\, \frac{\phi_{1s}(\mathbf{r} - d\mathbf{u}_x/2)}{|\mathbf{r} - d\mathbf{u}_x/2|} \phi_{1s}(\mathbf{r} + d\mathbf{u}_x/2), \tag{4.22}$$

$$\gamma = -\int d\mathbf{r}\, \frac{|\phi_{1s}(\mathbf{r} - d\mathbf{u}_x/2)|^2}{|\mathbf{r} + d\mathbf{u}_x/2|}. \tag{4.23}$$

In the solid-state physics literature these parameters are referred to as the overlap integral (S), the bond integral (h), and the crystal-field term (γ) (Pettifor, 1995). ▶Show that in the limit of large H–H separation, $d \gg 1$, the evaluation of these integrals simplifies considerably and the results are:

$$S = 8\,e^{-d}, \quad h = -4\,e^{-d}, \quad \gamma = -\frac{1}{d}. \tag{4.24}$$

While these results are valid only for large H–H separation, they provide a qualitatively correct picture even for small bond lengths. The exact expressions for the integrals in eqns 4.20, 4.22 and 4.23 can be found in Heitler and London (1927) and Pauling (1928). ▶By using the results of eqn 4.24 calculate the total potential energy, U, of the H_2^+ molecule as a function of the H–H distance (eqn 4.9), determine the equilibrium bond length, and compare your result with the exact solution shown in Figure 4.2.

Exercise 4.5 The study of soft materials and biomaterials, such as plastics and proteins, requires the determination of equilibrium structures in systems which do not form strong bonds (such as ionic, metallic, or covalent bonds) and hold together due to weak *van der Waals* forces. In this case the energy required for breaking one bond is of the order of a few tens of meV. In this exercise we want to understand the origin of van der Waals forces in the simplest possible case, the H_2 molecule. At variance with Exercise 4.4, where the H_2^+ molecular ion has only one electron, here we have two electrons; therefore we will need to find some approximations for the two-electron wavefunction $\Psi(\mathbf{r}_1, \mathbf{r}_2)$ as for the He atom in Exercise 2.2.
▶Show that the electronic Hamiltonian of the molecule when the protons are fixed at the positions $\pm d\mathbf{u}_x/2$ is given by:

$$\hat{H} = \hat{H}_1 + \hat{H}_2 + \Delta\hat{H}, \tag{4.25}$$

with

$$\hat{H}_1 = -\frac{1}{2}\nabla_1^2 - \frac{1}{|\mathbf{r}_1 + d\mathbf{u}_x/2|}, \quad \hat{H}_2 = -\frac{1}{2}\nabla_2^2 - \frac{1}{|\mathbf{r}_2 - d\mathbf{u}_x/2|}, \tag{4.26}$$

and

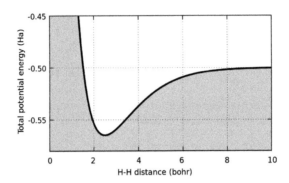

Fig. 4.2   Total potential energy, $U(d)$, of the hydrogen molecular ion H_2^+ as a function of the H–H separation, d, obtained using the adiabatic approximation, and using the linear combination of atomic orbitals in eqn 4.19. The integrals in eqns 4.20, 4.22 and 4.23 have been evaluated exactly following Pauling (1928). At a very large interatomic separation the energy tends to E_{1s}, as we intuitively expect by considering an isolated H atom and a proton very far away (there is no electrostatic interaction between the proton and the H atom because the atom is neutral). Since eqn 4.19 is only a first approximation to the actual electron wavefunction, the calculated equilibrium bond length, $d = 2.5$ bohr, overestimates the exact result, $d = 1.997$ bohr (Schaad and Hicks, 1970).

$$\Delta \hat{H} = \frac{1}{|\mathbf{r}_1 - \mathbf{r}_2|} - \frac{1}{|\mathbf{r}_1 - d\mathbf{u}_x/2|} - \frac{1}{|\mathbf{r}_2 + d\mathbf{u}_x/2|}. \tag{4.27}$$

We can recognize in $\hat{H}_1$ and $\hat{H}_2$ the Hamiltonians of two isolated hydrogen atoms. When d is very large the term $\Delta \hat{H}$ in eqn 4.27 goes to zero and therefore we can write a simple guess for the two-electron wavefunction at large H–H separation as follows:

$$\Psi_0(\mathbf{r}_1, \mathbf{r}_2) = \phi_{1s}(\mathbf{r}_1 + d\mathbf{u}_x/2)\phi_{1s}(\mathbf{r}_2 - d\mathbf{u}_x/2). \tag{4.28}$$

This guess does not obey the Pauli principle as described in Section 2.6, but this is not very important since the exchange interaction decreases exponentially fast with increasing d (Fermi, 1966). From our guess in eqn 4.28 we realize that Ψ_0 is large only when $\mathbf{r}_1$ and $\mathbf{r}_2$ are near their respective protons. This suggests that it will be useful to define electron coordinates relative to the nuclei:

$$\mathbf{s}_1 = \mathbf{r}_1 + d\mathbf{u}_x/2, \qquad \mathbf{s}_2 = \mathbf{r}_2 - d\mathbf{u}_x/2. \tag{4.29}$$

▶Using a Taylor expansion up to second order in $|\mathbf{s}_1|/d$ and $|\mathbf{s}_2|/d$, show that in the limit of large d the term $\Delta \hat{H}$ in eqn 4.27 becomes

$$\Delta \hat{H} = -\frac{1}{d} - \frac{3\,(\mathbf{s}_1 \cdot \mathbf{u}_x)(\mathbf{s}_2 \cdot \mathbf{u}_x) - \mathbf{s}_1 \cdot \mathbf{s}_2}{d^3}. \tag{4.30}$$

☐At this stage we need to find the ground-state wavefunction for the Hamiltonian of eqn 4.25 using the approximate potential in eqn 4.30. ▶Show that the trial function, Ψ_0, gives the total energy $E_0 = \langle \Psi_0|\hat{H}|\Psi_0\rangle = 2E_{1s} - 1/d$, and that the corresponding total potential energy is $U = 2E_{1s}$. For this assignment you should use Dirac's notation, should not perform integrals explicitly, and should note that the second term in eqn 4.30 is an odd function of $\mathbf{s}_1$ and $\mathbf{s}_2$.
☐Since U is a constant, when the electrons are in the quantum state Ψ_0 the two H atoms do not experience any force at large distance. It remains to see whether another wavefunction

Ψ can provide a total energy $E = \langle\Psi|\hat{H}|\Psi\rangle$ smaller than E_0, in which case the latter would represent a better estimate of the electronic ground state. Since the potential in eqn 4.30 corresponds to a dipole–dipole interaction (Jackson, 1998), it makes sense to consider a trial wavefunction which breaks the spherical symmetry of the $1s$ state of H. The simplest wavefunction in this class is $\Psi = \alpha\Psi_0 + \beta\Psi_1$ with

$$\Psi_1(\mathbf{r}_1, \mathbf{r}_2) = \phi_{2p_x}(\mathbf{r}_1 + d\mathbf{u}_x/2)\phi_{2p_x}(\mathbf{r}_2 - d\mathbf{u}_x/2), \tag{4.31}$$

and with ϕ_{2p_x} being the $2p_x$ wavefunction of H with eigenvalue E_{2p}. Since Ψ must be normalized we also have $\alpha^2 + \beta^2 = 1$.

▶Show that, in the limit of large d, and using $\Delta\hat{H}$ from eqn 4.30, the total energy $E = \langle\Psi|\hat{H}|\Psi\rangle$ associated with this wavefunction is:

$$E(\beta) = E_0 + 2(E_{2p} - E_{1s})\beta(\beta - c), \tag{4.32}$$

$$\text{with} \qquad c = (d/d_0)^{-3} \qquad \text{and} \qquad d_0 = \left[\frac{2|\langle\phi_{1s}|x|\phi_{2p_x}\rangle|^2}{E_{2p} - E_{1s}}\right]^{1/3}. \tag{4.33}$$

For this calculation use Dirac's notation, do not perform any integrals explicitly, and use the approximation $\beta \ll 1$ (the validity of this approximation for $d \gg 1$ can be checked a *posteriori*).

☐The total energy reaches a minimum at $\beta = c/2$ given by:

$$E = E_0 - \frac{E_{2p} - E_{1s}}{2(d/d_0)^6} < E_0. \tag{4.34}$$

Therefore the wavefunction $\Psi = \Psi_0 + (c/2)\Psi_1$ is a better estimate for the electronic ground state. Writing the wavefunction explicitly we find:

$$\Psi(\mathbf{r}_1, \mathbf{r}_2) = \Psi_0(\mathbf{r}_1, \mathbf{r}_2) + \frac{1}{2(d/d_0)^3}\Psi_1(\mathbf{r}_1, \mathbf{r}_2). \tag{4.35}$$

▶Using eqn 4.13 and the total energy in eqn 4.34, show that, at large H–H separation, the H atoms experience an attractive force which is proportional to $1/d^7$.

☐This is the so-called 'van der Waals' force. The calculation conducted here is rather simplified because we allowed only for the occupation of the $2p_x$ orbitals of the H atoms, while in practice any p-type orbitals would enter the description, in particular the $2p_y$ and $2p_z$. A complete calculation was performed by Eisenschitz and London (1930), and the exact result given by Pauling and Beach (1935) is $E = E_0 - 6.5/d^6$. The van der Waals force plays an important role in the physisorption of molecules on surfaces, in the cohesion of layered solids like graphite, and in protein folding.

Figure 4.3 shows the probability, $|\Psi(\mathbf{r}_1, \mathbf{r}_2)|^2$, of finding one electron around one H atom for a given position of the second electron. This figure shows clearly the concept of induced electron dipole, and captures the intuitive notion of van der Waals interaction. Since the wavefunction Ψ in eqn 4.35 cannot be expressed as a product of two independent wavefunctions, we say that the ground state is 'correlated'. As a consequence of this correlation at a distance, computational methods based on local approximations to DFT, such as the LDA of Section 3.3, do not predict reliably the binding in the presence of van der Waals forces. There is currently a large research activity aimed at extending the scope of DFT calculations to the case of van der Waals interactions; see, among others, Andersson *et al.* (1996) and Dion *et al.* (2004).

4.3   Calculating atomic forces using classical electrostatics

As we have seen in the previous section, the determination of equilibrium structures of materials requires the minimization of the total potential energy, U, with respect

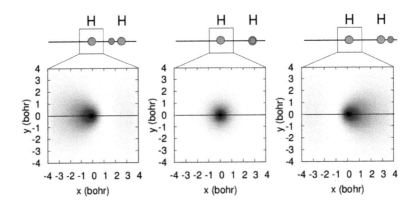

Fig. 4.3 Probability distribution $|\Psi(\mathbf{r}_1, \mathbf{r}_2)|^2$ of one electron in H_2, with Ψ given by eqn 4.35. In each plot $\mathbf{r}_2$ is fixed as shown in the schematic diagrams on top (the grey disc represents electron 2, the orange discs are the protons), and z_1 is set to 0. The H–H separation schematically shown in the diagrams is not to scale, and the prefactor of Ψ_1 in eqn 4.35 has been artificially exaggerated for clarity. From these plots we understand that the second electron pushes or pulls the first one as it fluctuates around its proton. These charge redistributions generate tiny electric dipoles which are at the origin of van der Waals forces. Note that without the extra term Ψ_1 in eqn 4.35 the three plots would all look identical to the one in the middle.

to all the $3M$ nuclear degrees of freedom. In a naïve approach we could calculate each derivative $\partial U / \partial \mathbf{R}_I$ using finite-difference formulas; for instance, we could slightly displace one atom along a Cartesian direction, calculate U, and take the difference with the energy in the undisplaced configuration. The calculation of all the forces in eqn 4.13 would therefore require $3M+1$ numerical evaluations of U. While the evaluation of the nucleus–nucleus repulsion term entering U in eqn 4.9 is straightforward, the calculation of the electron energy, $E(\mathbf{R}_1, \ldots, \mathbf{R}_M)$, for each nuclear configuration can prove very time-consuming.

As an example we can consider that one evaluation of the total potential energy, U, for a strand of DNA containing 2,606 atoms requires a calculation lasting 56 min using 64 processors (Skylaris *et al.*, 2005). Therefore the determination of the forces in this case would require $(3 \times 2606 + 1) \times 56$ min, corresponding to about 10 months of supercomputing time. This simple estimate is also based on the assumption that the calculation is performed using very efficient 'linear-scaling' DFT algorithms (Skylaris *et al.*, 2005), while standard algorithms would require considerably more time.

The challenge of evaluating atomic forces can be overcome by exploiting a very simple and general theorem which allows us to replace $3M+1$ total energy calculations by one single calculation.

This can be seen most easily in a one-dimensional example. If we consider the Hamiltonian, $\hat{H}$, of Exercise 4.4 and the eigenstate ψ with energy E depending parametrically on the internuclear separation d, we find:

$$\frac{\partial E}{\partial d} = \frac{\partial}{\partial d} \int d\mathbf{r}\, \psi^* \hat{H} \psi = \int d\mathbf{r} \left[\frac{\partial \psi^*}{\partial d} \hat{H} \psi + \psi^* \frac{\partial \hat{H}}{\partial d} \psi + \psi^* \hat{H} \frac{\partial \psi}{\partial d} \right].$$

Since by construction we have $\hat{H}\psi = E\psi$, the first and third terms on the right-hand side can be grouped and the equation simplifies as follows:

$$\frac{\partial E}{\partial d} = \int d\mathbf{r}\, \psi^* \frac{\partial \hat{H}}{\partial d} \psi + E \frac{\partial}{\partial d} \int d\mathbf{r}\, \psi^* \psi.$$

Since the eigenstate ψ is normalized, the second term is the derivative of a constant and cancels. Therefore we are left with:

$$\frac{\partial E}{\partial d} = \int d\mathbf{r}\, \psi^* \frac{\partial \hat{H}}{\partial d} \psi. \tag{4.36}$$

This result means that when we need to evaluate the derivative of an eigenvalue of the Hamiltonian with respect to a parameter, we can simply calculate the expectation value of the derivative of the Hamiltonian. The only assumption behind eqn 4.36 is that ψ is an eigenstate of the Hamiltonian; therefore the result is very general and does not depend of the specific system considered. This result is called the *Hellmann–Feynman theorem* (Hellmann, 1933; Feynman, 1939).

The Hellmann–Feynman theorem affords us a considerable simplification in the task of calculating atomic forces. In fact, by combining eqns 4.36 with eqns 3.6 and 2.27 we find:

$$\mathbf{F}_I = Z_I \left[\int d\mathbf{r}\, n(\mathbf{r}) \frac{\mathbf{r} - \mathbf{R}_I}{|\mathbf{r} - \mathbf{R}_I|^3} - \sum_{J \neq I} Z_J \frac{\mathbf{R}_J - \mathbf{R}_I}{|\mathbf{R}_J - \mathbf{R}_I|^3} \right]. \tag{4.37}$$

This equation tells us that the calculation of the forces for all the M atoms in our system can be performed by determining the electron density, $n(\mathbf{r})$, for one set of nuclear coordinates, as opposed to $3M+1$ calculations required by eqn 4.13.

Besides its practical importance in modern calculations of equilibrium structures of materials, eqn 4.37 is conceptually very interesting since it tells us that the *quantum-mechanical* force acting on each nucleus arises from the total electric field generated by the other nuclei and by the electron charge distribution. In other words, molecules and solids hold together owing to the *classical electrostatic equilibrium* between the nuclear charges and the electronic charge density, regardless of complicated many-body quantum effects.

In order to stress this aspect it is helpful to rewrite eqn 4.37 using eqn 2.47 for the Hartree potential and making explicit the *electrostatic field*, $\mathbf{E}_I$, at the position of the I-th nucleus:

$$\mathbf{F}_I = Z_I \mathbf{E}_I, \tag{4.38}$$

$$\mathbf{E}_I = \underbrace{\nabla V_{\mathrm{H}}(\mathbf{R}_I)}_{\text{Field of the electrons}} \underbrace{- \sum_{J \neq I} Z_J \frac{\mathbf{R}_J - \mathbf{R}_I}{|\mathbf{R}_J - \mathbf{R}_I|^3}}_{\text{Field of the nuclei}}. \tag{4.39}$$

This understanding of the structural equilibrium in materials was already clear during the early days of quantum mechanics. For example, the following observation is from Feynman (1939):

> 'The force on any nucleus (considered fixed) in any system of nuclei and electrons is just the classical electrostatic attraction exerted on the nucleus in question by the other nuclei and by the electron charge density distribution for all electrons [...]. Or finally, the force on a nucleus is the charge on that nucleus times the electric field there due to all the electrons, plus the fields from the other nuclei. This field is calculated classically from the charge distribution of each electron and from the nuclei.'

The Hellmann–Feynman theorem is a special case of a much more general theorem, which states that the derivative of order n of the electron eigenfunctions yields the derivative of order $2n+1$ of the eigenenergy E (Gonze, 1995). This '$2n+1$' theorem is very useful for computing non-linear properties of materials, such as electro-optic effects, Raman cross-sections and thermal expansion.

Exercise 4.6 We want to calculate the equilibrium distance of the H_2^+ molecule using the Hellmann–Feynman theorem as given by eqn 4.37. ▶By using the same approximations introduced in Exercise 4.4, namely that the electronic ground state is given by the linear combination of atomic orbitals in eqn 4.19, and that the H–H separation is very large, show that the force on each proton (along the direction of the bond and towards the other proton) is given by:

$$F(d) = \frac{16e^{-d} - 1}{2d^2}. \tag{4.40}$$

▶Compare the equilibrium bond length obtained using eqn 4.40 with the solution given by Pauling (1928) and with the exact result of Schaad and Hicks (1970).

4.4 How to find the equilibrium configuration using calculated forces

Since the Hellmann–Feynman theorem states that all we need to know in order to compute atomic forces is the ground-state electron density, $n(\mathbf{r})$, and since this quantity can be obtained using DFT as we have seen in Chapter 3, we now have a complete theoretical framework for calculating atomic forces starting from the first principles of quantum mechanics.

One final aspect that we need to address in this chapter is how can we use the calculated forces in order to find the equilibrium structures of materials. There exists a variety of very powerful computational methods for finding equilibrium structures. All these methods are designed to search for the minimum of the potential energy surface, $U(\mathbf{R}_1, \ldots, \mathbf{R}_M)$, using knowledge of the values of U and its slopes (which, according to eqn 4.13, correspond to minus the forces). This problem is analogous to finding the lowest point in a hilly landscape when the visibility is reduced by fog, with the difference that U is a landscape in $3M$ dimensions.

Popular methods for finding the minimum of the potential energy surface are, for instance, the 'steepest descent' and the 'conjugate gradients' methods (Nocedal and

Wright, 1999). Discussing such methods would require a detour in the world of numerical optimization. In order to keep things simple we prefer instead to illustrate a very simple minimization approach which appeals to our physical intuition and builds upon our previous knowledge of classical mechanics. Let us consider the simplest possible optimization problem, corresponding to one classical particle moving along a potential energy surface, $U(x)$, as in Exercise 4.3. Newton's equation of motion for this particle is:

$$M\frac{d^2x}{dt^2} = F(x), \quad \text{with} \quad F(x) = -\frac{\partial U}{\partial x}. \tag{4.41}$$

Let us take for simplicity $U(x) = Cx^2$, as in Exercise 4.1. In this case the equilibrium position of our particle is obviously $x = 0$, but let us imagine for the time being that we need to find the minimum of the potential energy, U, using our knowledge of the force, $F(x)$. If we release the particle with no velocity from an initial position $x_0 \neq 0$, the integration of eqn 4.41 would lead to periodic oscillations of our particle around the minimum. A simple way to 'drive' the particle towards the minimum consists of dissipating energy. This can be achieved for example by introducing an additional friction force in the equation of motion, as follows:

$$M\frac{d^2x}{dt^2} = -\frac{\partial U}{\partial x} - \frac{2M}{\tau}\frac{dx}{dt}. \tag{4.42}$$

The last term of eqn 4.42 can be interpreted physically by imagining that the particle is effectively moving inside a viscous fluid. The integration of eqn 4.42 yields:

$$x(t) = x_0 \left[\cos\omega t + \frac{\sin\omega t}{\omega\tau}\right] \exp\left(-\frac{t}{\tau}\right) \quad \text{with} \quad \omega^2 = \frac{2C}{M} - \frac{1}{\tau^2}. \tag{4.43}$$

Owing to the exponential decay in eqn 4.43, regardless of the starting point x_0, after a few oscillations the particle will eventually stop at the minimum of the potential energy surface. This very simple observation is at the basis of structural optimization methods based on 'damped molecular dynamics'. Obviously for finding the minimum of a simple parabolic potential the use of this method is unnecessary; however, in systems of practical interest, for which the calculation of $U(\mathbf{R}_1, \ldots, \mathbf{R}_M)$ at each nuclear configuration is very time-consuming, damped molecular dynamics can be a very effective strategy for determining equilibrium structures. In practice, Newton's equation is solved numerically by discretizing the time variable as $t_i = i\Delta t$ with $i = 0, 1, \ldots$, and by approximating the time derivatives using finite-difference formulas:

$$M\frac{x_{i+1} - 2x_i + x_{i-1}}{\Delta t^2} = F(x_i) - \frac{2M}{\tau}\frac{x_{i+1} - x_{i-1}}{2\Delta t}, \tag{4.44}$$

with $x_i = x(t_i)$. This simple strategy for discretizing Newton's equation of motion is usually referred to as *Verlet's algorithm* (Verlet, 1967). The terms in eqn 4.44 can be rearranged in order to express position x_{i+1} at time t_{i+1} in terms of the positions at previous times and the force at x_i:

$$x_{i+1} = \frac{2}{1+\eta}x_i - \frac{1-\eta}{1+\eta}x_{i-1} + \frac{\kappa}{1+\eta}F(x_i), \tag{4.45}$$

with the friction parameter $\eta = \Delta t/\tau$ and the time-step parameter $\kappa = \Delta t^2/M$. Equation 4.45 indicates that, after setting two initial conditions for the nuclear

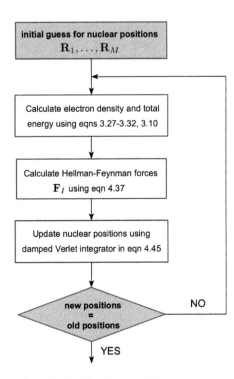

Fig. 4.4 Schematic flow-chart for finding the equilibrium structure of a material at $T=0$ K using DFT, the Hellmann–Feynman forces and the damped Verlet integrator. The equality sign in the conditional symbol means that the nuclear coordinates at this iteration and those at the previous iteration differ by less than a desired tolerance. Similarly to Figure 3.3, while conceptually we can check for convergence by monitoring the nuclear coordinates, it can be more convenient to monitor how the total energy of the system approaches its minimum value.

coordinates at t_0 and t_1 we can generate a series of configurations $x_0, x_1, x_2, \ldots$ that will converge to the minimum of the potential energy surface. All that is needed in this procedure is an evaluation of the forces at each set of coordinates, $F(x_i)$. By tuning the friction parameter and the time-step parameter it is possible to reduce the number of iterations needed for finding the minimum. These aspects are explored in Exercise 4.7. Despite its apparent simplicity, damped molecular dynamics is a very powerful tool and its use is very common in structural optimizations using DFT (Car and Parrinello, 1985). Improvements upon this concept can even compete with numerical optimization methods such as conjugate gradients (Bitzek *et al.*, 2006).

Exercise 4.7 We want to determine the equilibrium bond length of the H_2^+ molecule using the damped Verlet algorithm in eqn 4.45 and the approximate expression for the Hellmann–Feynman forces derived in Exercise 4.6. ▶After setting $x_0 = x_1 = d_0/2$, with d_0 an initial guess for the equilibrium bond length, calculate explicitly the series of eqn 4.45 for $i = 1, 2, \ldots$ in each of the following cases (d_0 and κ are in Hartree units): (1) $d_0 = 4$, $\kappa = 1$, $\eta = 0.25$; (2) $d_0 = 4$, $\kappa = 1$, $\eta = 0.05$; (3) $d_0 = 2$, $\kappa = 1$, $\eta = 0.25$; (4) $d_0 = 2$, $\kappa = 1$, $\eta = 0.05$.

▶Verify that the equilibrium bond length can be determined with a relative accuracy $|d_i - d_{i-1}|/d_i < 10^{-3}$ using (1) 25 iterations, (2) 58 iterations, (3) 21 iterations, and (4) 49 iterations, respectively. ▶Compare your result with the exact solution, $d_{min} = \log 16$, from Exercise 4.6. ▶Verify that the time-step parameter κ corresponds to $\Delta t \simeq 1$ fs.

In practical structural optimizations using DFT and involving many atoms (typically from 10 to 500) we will have an equation like eqn 4.45 for each atom and for each Cartesian direction. Each integration step will correspond to a set of nuclear coordinates, and for each new set of coordinates the forces will have to be computed using the Hellmann–Feynman theorem. This requires in turn the calculation of the DFT total energy, $E(\mathbf{R}_1, \ldots, \mathbf{R}_M)$, and electron density, $n(\mathbf{r})$, at fixed nuclear coordinates, as described in Chapter 3. Figure 4.4 illustrates schematically this iterative process.

5

Equilibrium structures of materials: calculations vs. experiment

---

In Chapter 4 we introduced a method for determining the equilibrium structures of materials at zero temperature. The key ingredient of this method is the electronic total energy, $E(\mathbf{R}_1, \ldots, \mathbf{R}_M)$, at fixed nuclear coordinates, and we have seen in Chapter 3 that this quantity can be calculated using DFT.

In this chapter we take a break from the theoretical formalism, and we illustrate some prototypical examples of structure determination using DFT. We will go through examples of increasing structural complexity, starting from a simple diatomic molecule and ending with a semiconductor surface.

The goal of this chapter is to provide an idea of the level of accuracy that can be expected from DFT calculations of equilibrium structures, and to show that in many cases it is possible to perform a *quantitative* comparison between theory and experiments.

We will also provide a brief description of experiments probing the structure of materials such as X-ray crystallography (Section 5.3) and scanning tunnelling microscopy (Section 5.5). This digression is useful in order to show that, not only we can reverse-engineer structures from experiment and then compare them with DFT calculations, but we can also use DFT to directly predict experimental data, for instance tunnelling maps, and compare calculations with the original measurements.

5.1 Structure of molecules

The simplest poly-atomic system that we can consider is a diatomic molecule. Let us consider the nitrogen molecule, N_2. Atomic nitrogen is a key element of amino acids and nucleic acids, hence it is very important for life on Earth. In molecular form nitrogen is an inert gas at ambient conditions and forms approximately 80% of the Earth's atmosphere.

Using the techniques described in Chapters 3 and 4 we can calculate the total potential energy, $U(d)$, of the N_2 molecule as a function of the N–N bond length, d. Since at very large interatomic separation the total energy must correspond to the sum of the energies of two isolated N atoms (each with energy E_N), it is convenient to introduce the *binding energy* of the molecule, $E_b(d)$:

$$E_b(d) = U(d) - 2E_N, \tag{5.1}$$

which quantifies how stable the N_2 molecule is relative to two isolated atoms at any given separation. By definition E_b should go to zero at infinite separation. If we denote the equilibrium bond length by d_{min}, $E_b(d_{min})$ corresponds to the amount of energy required to break the molecule into its two constituent atoms, i.e. the dissociation energy of the molecule.

The N atom has seven electrons in the configuration $1s^2 2s^2 p^3$, and therefore the N_2 molecule has two nuclei and 14 electrons. It turns out that, in the study of structural properties, the lowest-lying ($1s$) electrons do not contribute appreciably to the bonding; therefore we can effectively 'freeze' these electrons in the wavefunctions that they occupy in the isolated N atoms, and focus on the higher-energy ($2s^2 p^3$) electrons. This reasoning naturally leads us to identify a set of *core* electrons and a set of *valence* electrons for each atom. The effect of the frozen core electrons on the valence electrons can be taken into account by means of *pseudopotentials*. For simplicity in this chapter we will not discuss the theory of pseudopotentials, although an elementary introduction is given in Appendix E.

Figure 5.1 shows the binding energy, E_b, of the N_2 molecule as a function of N–N separation, calculated using the local density approximation to density functional theory, as described in Sections 3.3 and 3.4. It is worth noting the qualitative similarity between the curve for N_2 in Figure 5.1 and that for H_2^+ in Figure 4.2. This similarity suggests that the mechanisms underlying the binding of H_2^+ and N_2 are essentially the same, the only difference being that in N_2 the problem is considerably more complicated due to the presence of 14 electrons instead on only one electron for H_2^+. From Figure 5.1 we see that the calculated equilibrium bond length of N_2 is 1.102 Å, and the corresponding dissociation energy is 11.46 eV.

It is interesting to observe that in Figure 5.1 the E_b vs. d curve does not go to zero for large d; in other words, the energy of N_2 at large internuclear separation appears to be larger than the sum of the energies of two isolated N atoms. This finding is contrary to what we would expect intuitively, and can be explained as follows. The N_2 molecule around its equilibrium bond length is diamagnetic, i.e. it does not possess a net magnetic moment. The N atom is instead paramagnetic, as we can understand using Hund's first rule, which requires the three $2p$ orbitals to be occupied each with one electron of the same spin (Kittel, 1976). The binding energy, E_b, in Figure 5.1 has been obtained from eqn 5.1 by using $U(d)$ for the diamagnetic configuration of N_2 and E_N for the paramagnetic N atom; therefore, it is clear that in this case we must have $E_b(d = \infty) > 2E_N$. This consideration also indicates that the results of the calculation presented in Figure 5.1 are valid only near the equilibrium bond length, where the diamagnetic state is the one with the lowest energy. In general, one has to check whether, by allowing the system to have a net magnetization, it is possible to identify a state of lower energy. For the time being we will not worry about these aspects, and we defer the discussion of spin in DFT calculations to Chapter 11.

At this point we need to compare our calculations with experiment. Before quoting experimental values from the literature it is appropriate to ask ourselves how it is possible to measure the length of a molecular bond and the dissociation energy of the molecule. In the simplest case of a diatomic homonuclear molecule (such as N_2) information on the bond length comes from the analysis of rotational frequencies. In

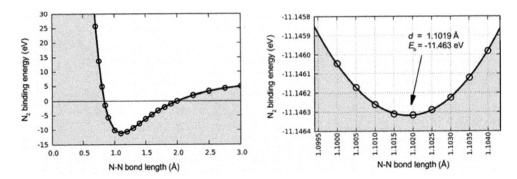

Fig. 5.1   Calculated (DFT/LDA) binding energy of the N_2 molecule as a function of N–N separation (circles). The panel on the right provides a zoom of the curve on the left around the minimum. A parabolic fit around the minimum (solid line in the right panel) allows us to determine an equilibrium bond length of 1.1019 Å and a dissociation energy of 11.463 eV. It is worth pointing out a few technical details: due to the frozen-core approximation, the shape of the curve in the left panel is not very accurate around and below 0.5 Å; for numerical convenience eqns 3.27–3.32 were solved in a cubic computational cell of size 10 Å using periodic boundary conditions. As a consequence of the periodic boundary conditions, this calculation truly corresponds to a simple cubic lattice of N_2 molecules; however, nearest-neighbour molecules are far enough that their interaction is negligible. More details on the use of periodic boundary conditions are given in Appendix C.

fact, by placing the centre of our reference frame in the middle of the N–N bond, the motion of the molecule can be decomposed into a rigid rotation and a bond compression or elongation. The total kinetic energy of the nuclei associated with the rotation will be $L^2/2I$, with L the angular momentum of the nuclei and $I = Md^2/2$, the moment of inertia (Goldstein, 1950; Gregory, 2006). This expression for the kinetic energy can be derived from eqn 4.11 by neglecting the potential energy. Using arguments similar to Bohr's model of the atom (Bohr, 1913), we can assume that the angular momentum is quantized in units proportional to $\hbar$, resulting in a quantization of the kinetic energy of rotation. The energy required to make a transition between two 'rotational' energy levels will be therefore $\sim \hbar^2/(Md^2)$. This is typically of the order of 0.1–1 meV (Bransden and Joachain, 1983). The precise value of this energy can be measured using rotational Raman scattering experiments, and from the measured transition energy it is possible to determine the bond length, d. More on Raman scattering will be said in Chapter 8. Regarding the dissociation energy, this can be measured by determining the frequency of the electromagnetic field necessary to break a bond, and is done by irradiating a gas with ultraviolet light.

The measured bond length and dissociation energy of the N_2 dimer are 1.098 Å and 9.76 eV, respectively (Lee and Handy, 1993). By comparing calculated and measured bond lengths we see that DFT/LDA can predict the experimental value within 0.4%, which is quite impressive. On the other hand, the calculated DFT/LDA dissociation energy is 17% larger than in experiment. While the dissociation energy is not very accurate, we should keep in mind that the calculation was performed by starting from the first principles of quantum mechanics and without using *any* empirical parameters. Generally speaking, extensive comparison between calculations and measurements indicate that DFT/LDA slightly underestimates, typically by about 1%, measured bond lengths in molecules and solids. At the same time the binding energy is typically too large by about 1 eV per bond. The use of more elaborate formulations of the exchange and correlation functional E_{xc} in place of the LDA functional of eqns 3.20 and 3.22 can improve considerably the calculated binding energies. For instance, functionals based on so-called 'generalized-gradient approximations' (GGA) yield a dissociation energy for N_2 of 10.44 eV, only 7% larger than in experiment (Furche, 2001).

5.2   Structure of crystals

The next step on the Jacob's ladder of complexity is the structure of a crystalline solid. Since typical crystal samples used in experiments contain billions of replicas of the basic *crystal unit cell* (think of a sample which is 1 mm wide and has a unit cell size of 10 Å), it is advantageous to model the system as an infinite repetition of this periodic unit. This strategy allows us to solve the Kohn–Sham equations (eqn 3.12) inside one periodic unit cell instead of dealing with a large crystal sample. The solution within a given unit cell is then matched to the equivalent solutions on all the other cells by imposing periodic boundary conditions. While this approach may sound rather involved, in practice it is very easy to implement, by expanding the Kohn–Sham wavefunctions and the electron density using Fourier series. The interested reader will find more information on the use of periodic boundary conditions in DFT calculations for crystals in Appendix C. As a very simple example of crystal structure determination let us consider silicon. Si is one of the most abundant elements on Earth, as 90% of the crust is composed of silicate minerals. Arguably the most important application of silicon in contemporary society is as a substrate and electron channel in integrated circuits, which form the basis of essentially every electronic device in use. At ambient conditions Si crystallizes into the diamond structure, consisting of two interpenetrating face-centred cubic (fcc) primitive lattices. The primitive lattice vectors are $\mathbf{a}_1 = \frac{1}{2}a(\mathbf{u}_x + \mathbf{u}_y)$, $\mathbf{a}_2 = \frac{1}{2}a(\mathbf{u}_y + \mathbf{u}_z)$ and $\mathbf{a}_3 = \frac{1}{2}a(\mathbf{u}_z + \mathbf{u}_x)$, with a the lattice parameter. The primitive unit cell contains two atoms (called the 'basis') which occupy positions $\mathbf{R}_1 = 0$ and $\mathbf{R}_2 = \frac{1}{4}a(\mathbf{u}_x + \mathbf{u}_y + \mathbf{u}_z)$. All the other atoms are generated by translations of this basis using linear combinations of the primitive lattice vectors (Kittel, 1976). Figure 5.2 shows a ball-and-stick representation of silicon, highlighting the primitive lattice vectors and the basis. Each Si atom has four nearest neighbours arranged in a tetrahedral configuration. Since the primitive lattice vectors and the atomic coordinates of the basis depend solely on the lattice parameter, a, the total potential energy in eqn 4.9 will be a function of this parameter only: $U = U(a)$. Similar to our discussion of the N_2 molecule in Section 5.1, in the case of crystalline solids it is convenient to introduce the *cohesive energy* of the solid, defined as:

$$E_{\mathrm{c}}(a) = \frac{U(a)}{M} - E_{\mathrm{Si}}, \tag{5.2}$$

where M is the total number of atoms in the crystal, U/M is the total potential energy per Si atom, and E_{Si} is the total energy of an isolated Si atom. Equation 5.2 quantifies

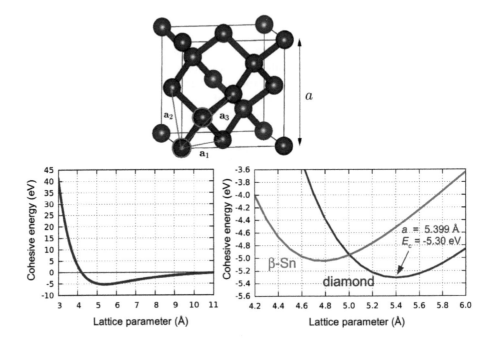

Fig. 5.2 Silicon unit cell and cohesive energy. Top panel: ball-and-stick representation of the Si crystal in the diamond structure. The primitive lattice vectors, the lattice parameter, a, and the two atoms forming the basis (blue circles) are indicated. Bottom left panel: cohesive energy as a function of the lattice parameter, a, calculated using DFT/LDA. Note the similarity with the binding energy curve of the N_2 molecule in Figure 5.1. Bottom right panel: zoom of the cohesive energy curve in the left panel around the equilibrium lattice parameter. The minimum of the curve yields the equilibrium lattice parameter, $a = 5.399$ Å, and the cohesive energy, $E_c = 5.30$ eV. Also shown is the cohesive energy calculated for the β-Sn phase of silicon (red). This phase can be obtained from the structure shown in the top panel by shrinking the vertical axis by a factor $c/a = 0.5516$. From the comparison between the curves for the β-Sn and for the diamond structures we see that the diamond structure is more stable at zero pressure and temperature, since it has the largest cohesive energy.

how stable the Si crystal is relative to a collection of isolated atoms. As in the case of the molecular binding energy, E_c should go to zero when the lattice parameter, a, becomes very large. As discussed in Section 4.2, the stationary points of U, and hence of E_c, yield the equilibrium structure of the material at zero temperature. Since we are assuming that our crystal is isolated, i.e. no external forces are acting upon it, this equilibrium configuration corresponds to the case of zero applied pressure or strain. The Si atom has 14 electrons in the configuration $(1s^2 2s^2 p^6)3s^2 p^2$; therefore, each unit cell will have two Si nuclei and 28 electrons. In practice, by freezing the core, as already done for the N atom in Section 5.1, we are left with the valence $3s^2 p^2$ electrons, and so the DFT calculations will only involve two nuclei and eight electrons (Appendix E).

Figure 5.2 shows the calculated cohesive energy of silicon in the diamond structure as a function of the lattice parameter. It is interesting to note the similarity between the dependence of the cohesive energy on the lattice parameter and the molecular binding energy curve of N_2 in Figure 5.1. This similarity is a particular example of a more general binding law in materials known as the 'universal binding energy curve' (Rose *et al.*, 1981; Graves and Parr, 1985). The calculated equilibrium lattice parameter is 5.399 Å, only 0.6% smaller than the experimental value of 5.43 Å. This is again very impressive, and together with the discussion made for N_2 in Section 5.1 it suggests that DFT/LDA does a very good job in predicting structural properties from first principles. At this point we may want to ask how lattice parameters are determined in experiment, but for the sake of clarity we will postpone the answer to this question until Section 5.3.

The cohesive energy of a solid can be measured by determining the heat of sublimation of a solid in equilibrium with its own vapour in a vacuum chamber, and this is accomplished by monitoring the variation of the vapour pressure with temperature. The calculated cohesive energy of silicon is 5.30 eV, and overestimates the experimental value of 4.62 eV by 15% (Farid and Godby, 1991). As in the case of the N_2 molecule we find that the cohesive energy is too large by about 1 eV in DFT/LDA. However, as already noted for N_2, the use of a more refined exchange and correlation functional can improve substantially the agreement between DFT calculations and experiment. For instance, GGA calculations yield a cohesive energy of 4.59 eV, which is only 1% smaller than in experiment (Moll *et al.*, 1995).

In our discussion so far we implicitly assumed that we may perform DFT calculations of solids without worrying about the underlying crystal structure, which is possibly already known from experiment. While this is indeed the case in many practical applications, in general we may not know what is the most stable structure of a crystalline solid. In this case the optimization procedure shown in Figure 5.2 must be repeated for every possible crystal structure. For example, the right panel of Figure 5.2 shows the cohesive energy curve for the β-Sn phase of silicon. The β-Sn phase of silicon is the stable phase at high pressure (above 10 GPa), and its structure can be derived from the ball-and-stick model of Figure 5.2 by shrinking the vertical axis by approximately a factor of two. The calculated curves in Figure 5.2 clearly show that the diamond structure is more stable than the β-Sn structure. If we were to determine the structure of silicon without any input from experiment we should calculate curves similar to those of Figure 5.2 for all the possible phases of silicon. Detailed DFT calculations of this type have been conducted and, as expected, indicate that the diamond structure is the most stable phase of silicon at ambient conditions (Yin and Cohen, 1982*b*).

What we should retain from this discussion of bulk solids is that: (i) crystal structures can be predicted quite successfully using DFT; (ii) if the crystal structure is not known, the calculation procedure can become rather involved. Yet the calculations described so far are for very small systems and can be executed on a standard laptop computer within 10–20 s for each data-point.

Table 5.1 Calculated DFT/LDA cohesive energies of silicon in the diamond structure and in the β-Sn structure as a function of the lattice parameter, a. For simplicity the axial ratio of the β-Sn structure was kept fixed at the experimental value $c/a=0.5516$.

	diamond	β-Sn		diamond	β-Sn		diamond	β-Sn
a	E_c	E_c	a	E_c	E_c	a	E_c	E_c
Å	eV	eV	Å	eV	eV	Å	eV	eV
4.0	2.962	-2.861	4.7	-3.972	-5.034	5.4	-5.300	-4.517
4.1	1.378	-3.504	4.8	-4.384	-5.044	5.5	-5.293	-4.380
4.2	0.037	-4.005	4.9	-4.703	-5.015	5.6	-5.246	-4.237
4.3	-1.091	-4.386	5.0	-4.943	-4.954	5.7	-5.175	-4.091
4.4	-2.033	-4.665	5.1	-5.115	-4.869	5.8	-5.084	-3.942
4.5	-2.814	-4.856	5.2	-5.220	-4.765	5.9	-4.977	-3.793
4.6	-3.454	-4.975	5.3	-5.290	-4.647	6.0	-4.858	-3.643

Exercise 5.1 Table 5.1 reports the calculated cohesive energies of silicon as a function of the lattice parameter for the diamond structure and the β-Sn phase, respectively. According to eqn 5.2 the cohesive energy is the total potential energy per atom in the crystal, relative to the energy of the isolated atom. The calculations were performed using the DFT/LDA as described in relation to Figure 5.2. ▶Determine the lattice parameters and the cohesive energies of silicon in the two phases at equilibrium, and verify that at ambient conditions the diamond structure is more stable than the β-Sn structure by 0.25 eV/atom. ▶Compare your results with the experimental data of Hu *et al.* (1986).

Exercise 5.2 The data reported in Table 5.1 not only provide information about structural stability at zero pressure, but can also be used for predicting pressure-induced phase transitions. At non-zero pressure and low temperature (i.e. ambient temperature in the example considered here) the relevant thermodynamic potential for studying structural equilibria is the *enthalpy* of the system (Wallace, 1998). The enthalpy is defined as $h = U+pV$, with p the pressure and V the crystal volume. The pressure term, pV, describes the mechanical coupling of the crystal with the environment, and at any given pressure p the most stable structure is the one with the lowest enthalpy. ▶Using Table 5.1 plot the enthalpy of silicon (in eV) as a function of volume per atom (in Å^3), both for the diamond structure and for the β-Sn structure, for the pressures $p_1=0.01$ GPa, $p_2=8$ GPa, and $p_3=11$ GPa. The conversion between units of pressure is 1 eV/Å^3 = 160.2 GPa. ▶Identify the equilibrium structure at each one of the three pressures p_1, p_2, and p_3. ▶Verify that the *critical pressure* at which the two phases are equally stable is 9.6 GPa, and compare this calculated DFT/LDA value with the experimental datum of Hu *et al.* (1986). This exercise is an example of the use of DFT/LDA calculations to study structural phase transitions and phase diagrams. More examples will be discussed in Chapters 6 and 8.

5.3   Comparison of DFT structures with X-ray crystallography

In the previous section we have presented a comparison between the measured lattice parameters of a crystal and those calculated using DFT. The question that we left unanswered was how to determine lattice parameters from experiment. This question is part of a more general question, that is how to determine the structure of a crystal. The key experimental probe for this purpose is X-ray diffraction (XRD). XRD is so crucial to understanding crystals that usually one of the first chapters of solid-state physics textbooks is devoted to this technique. In this section we discuss only the aspects which are relevant for comparing DFT calculations and experiment, and the reader is referred to the books by Kittel (1976) or Ashcroft and Mermin (1976) for a detailed introduction, and to Born and Wolf (1999) for a thorough presentation.

The basic idea of XRD is to shine electromagnetic radiation onto a sample and analyse the radiation scattered by this sample and collected by a detector. In order to resolve spatial features with a characteristic size of the order of the interatomic spacing in a crystal, for instance the lattice parameter a, the Laue condition for electron diffraction states that we need electromagnetic waves with wavelength smaller than $2a$ (Kittel, 1976). Therefore resolving features of size between 0.1 and 100 Å requires the use of X-rays, i.e. radiation in the energy range 100 eV to 100 keV. The state-of-the-art X-ray sources in use nowadays are synchrotrons, as they provide high luminosity and tunable wavelengths.

In an XRD measurement a monochromatic electromagnetic wave propagating along the wavevector $\mathbf{k}$ hits the sample, and the field scattered along the wavevector $\mathbf{k}'$ behind the sample is recorded. The outcome of the measurement is the intensity map of the scattered field, $F(\mathbf{q})$, with $\mathbf{q} = \mathbf{k}' - \mathbf{k}$. This quantity is very useful as it is simply related to the electronic charge density, $n(\mathbf{r})$, in the sample.

The quantum-mechanical theory of XRD is slightly beyond an undergraduate presentation, but the interested reader can grasp the basic concepts by consulting Johnson (1974) and Johnson (1975). There are three key ideas in this theory.

(i) The first one is that X-ray fields with energy $\hbar\omega > 100$ eV oscillate with a period $2\pi/\omega < 0.05$ fs. During this time interval the nuclei of the crystal remain essentially at rest, since they typically oscillate with a period between 10 fs and 1 ps (see Exercises 7.2, 7.8 and 7.11). During the same time interval the electrons undergo collective oscillations with an amplitude determined by the frequency-dependent dielectric function. According to Drude's law (Ashcroft and Mermin, 1976) this dielectric function is given, in atomic units, by

$$\epsilon(\omega) = 1 - \frac{4\pi n}{\omega^2}. \tag{5.3}$$

We can take this result for granted for the time being, and we reserve a proper discussion of dielectric functions in solids to Chapter 10. The scattered field, resulting from the electron oscillations, is proportional to last term of eqn 5.3.

(ii) The second key idea in the quantum theory of XRD is that the correct dielectric 'constant' to be used for evaluating the scattered field does actually depend on position, $\mathbf{r}$, in the sample; therefore, in eqn 5.3 we need to use the position-dependent electron density, $n(\mathbf{r})$.

Taking together the previous two observations, in the most basic approximation the intensity, $F(\mathbf{q})$, of the electromagnetic field propagating in the direction $\mathbf{k}' = \mathbf{k} + \mathbf{q}$ is proportional to the *structure factor*, $S(\mathbf{q})$:

$$F(\mathbf{q}) \propto S(\mathbf{q}) = \int d\mathbf{r}\, n(\mathbf{r})\, e^{-i\mathbf{q}\cdot\mathbf{r}}, \tag{5.4}$$

where the integral extends to the whole crystal and the exponential term describes the phase difference between the incoming and the outgoing X-ray waves. We stress that here we have given only plausibility arguments for the validity of eqn 5.4. The rigorous derivation can be worked out by combining the presentations of Born and Wolf (1999) and Johnson (1975), although it is rather involved. While most textbooks start the discussion of XRD directly from eqn 5.4, we find that this approach may lead to confusion about how the scattered field is being generated and why it is proportional to the electron density. It is hoped that the discussion around eqn 5.3 will help clarify these aspects.

(iii) The third key idea of XRD is that, since the crystal is perfectly periodic, it is possible to expand the electron charge density, $n(\mathbf{r})$, using a Fourier series. The Fourier transform of the density in a three-dimensional crystal can conveniently be written as follows:

$$n(\mathbf{r}) = \sum_{\mathbf{G}} e^{i\mathbf{G}\cdot\mathbf{r}} n_{\mathbf{G}}, \tag{5.5}$$

where the reciprocal lattice vectors, $\mathbf{G}$, are generated from the primitive vectors of the reciprocal lattice, $\mathbf{b}_1, \mathbf{b}_2, \mathbf{b}_3$:

$$\mathbf{G} = m_1\mathbf{b}_1 + m_2\mathbf{b}_2 + m_3\mathbf{b}_3, \text{ with } m_1, m_2, m_3 \text{ integers.} \tag{5.6}$$

Here it is assumed that the notion of reciprocal lattice is known from solid-state physics textbooks such as Kittel (1976) or Ashcroft and Mermin (1976); however, for convenience a brief reminder is given in Appendix D. By inserting eqn (5.5) into eqn (5.4) we obtain:

$$S(\mathbf{q}) = \sum_{\mathbf{G}} n_{\mathbf{G}} \int d\mathbf{r}\, e^{i(\mathbf{G}-\mathbf{q})\cdot\mathbf{r}}. \tag{5.7}$$

It will be shown in Exercises 5.3 and 5.4 that integrals such as $\int d\mathbf{r} \exp(i\mathbf{q}\cdot\mathbf{r})$ can be rewritten as follows:

$$\int d\mathbf{r}\, e^{i\mathbf{q}\cdot\mathbf{r}} = (2\pi)^3\delta(\mathbf{q}), \tag{5.8}$$

where $\delta(\mathbf{q})$ indicates the *Dirac delta function*. The Dirac delta can be thought of as a function sharply peaked around $\mathbf{q} = 0$, vanishing away from this point, and such that it integrates to 1 over the whole space. An introduction to the Dirac delta is provided in Exercise 5.3. By combining eqns 5.7 and 5.8 we find:

$$S(\mathbf{q}) = (2\pi)^3\sum_{\mathbf{G}} n_{\mathbf{G}}\, \delta(\mathbf{q} - \mathbf{G}). \tag{5.9}$$

This important result means that in an XRD experiment the diffraction pattern consists of a series of peaks corresponding to the directions $\mathbf{k}' = \mathbf{k} + \mathbf{G}$, with intensity proportional to $n_{\mathbf{G}}$. In other words XRD provides information on two aspects: (i) the reciprocal lattice vectors, $\mathbf{G}$, of the crystal, which allow one to determine the direct lattice and in particular the lattice parameters, i.e. to reconstruct the crystal structure; and (ii) the Fourier components, $n_{\mathbf{G}}$, of the electron charge density, which allow one to reconstruct the position-dependent electron density distribution in the crystal, $n(\mathbf{r})$, through eqn 5.5.

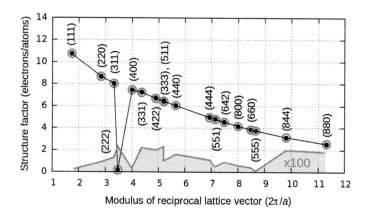

Fig. 5.3 Comparison of measured (black discs) and calculated (grey circles) static structure factors of silicon in the diamond structure. The values in the figure are taken from Lu *et al.* (1993) and are expressed in terms of electrons per atom (Table IV of Lu *et al.*, columns two and four). In this case the DFT/LDA calculations include explicitly the Si $1s^2 2s^2 p^6$ core electrons, as this is important for a meaningful comparison with experiment. The shaded area indicates the deviation between the measured and calculated values, enhanced by a factor of 100 for clarity. The relative deviation is consistently smaller than 2%.

Since the core quantity in density functional theory is the electron density, $n(\mathbf{r})$, we can expect that DFT will do well in determining structure factors through the Fourier components, $n_{\mathbf{G}}$. The comparison in Figure 5.3 between the structure factor of silicon in the diamond structure obtained from XRD and from DFT/LDA calculations shows that this is indeed the case. In fact we can see that the calculated structure factors show very little deviation from experiment: the agreement is within 2%.

The important message to retain at the end of this section is that, by using first-principles calculations based on DFT, we can perform a direct comparison between theory and raw experimental data. This is especially helpful for assigning spectral features in certain delicate cases where semi-empirical processing of XRD data may introduce spurious information.

The discussion of XRD presented in this section neglects important 'dynamic' effects arising from the vibrations of the nuclei. A detailed analysis of these effects and of how to compare DFT calculations with XRD measurements is given by Lu *et al.* (1993).

Exercise 5.3 The Dirac delta is a close analogue of the Kronecker delta, δ_{mn} (which is 1 for $m = n$ and 0 otherwise). The main difference is that the Dirac delta applies to continuous variables, while the Kronecker delta works for integers. In one dimension the Dirac delta, $\delta(x)$, is a function sharply peaked around $x = 0$. This function can be defined as a Lorentzian function, $L_\gamma(x)$, in the limit of vanishing width, γ:

$$\delta(x) = \lim_{\gamma \to 0} L_\gamma(x), \text{ with } L_\gamma(x) = \frac{1}{\pi} \frac{\gamma}{x^2 + \gamma^2}. \qquad (5.10)$$

▶Verify that for each value of γ the function L_γ is normalized to 1 along the real axis. ▶Plot the function L_γ for (1) $\gamma = 10$, (2) $\gamma = 1$, (3) $\gamma = 10^{-2}$, and observe how the function becomes a very sharp peak around $x = 0$ as the width, γ, decreases, while the area under the peak remains unchanged. □Given a function $f(x)$ of the real variable x, the Dirac delta can be used to single out the value of this function at x_0 as follows:

$$\int_{-\infty}^{+\infty} dx\, f(x)\delta(x - x_0) = f(x_0). \tag{5.11}$$

▶Derive eqn 5.11 starting from the definition in eqn 5.10 and using the residue theorem, after noting that the integrand can be written as:

$$f(x)\delta(x - x_0) = \frac{1}{2\pi i} \lim_{\gamma \to 0} \left[\frac{f(x)}{x - x_0 - i\gamma} - \frac{f(x)}{x - x_0 + i\gamma} \right].$$

A quick reminder on the residue theorem can be found in Spiegel *et al.* (2009). If a course on complex analysis has not been undertaken yet, eqn 5.11 can also be derived using the following equivalent definition of the Dirac delta function:

$$\delta(x) = \lim_{\gamma \to 0} \sqcap_\gamma(x), \text{ with } \sqcap_\gamma(x) = \begin{cases} 1/\gamma & \text{if } |x| < \gamma/2, \\ 0 & \text{otherwise.} \end{cases} \tag{5.12}$$

Exercise 5.4 ▶Using the definition of Dirac delta in eqn 5.10, derive the following result:

$$\int_{-\infty}^{+\infty} dx\, e^{iqx} = 2\pi\delta(q). \tag{5.13}$$

The following identity is helpful:

$$\frac{1}{1 + x^2} = \frac{1}{2} \int_{-\infty}^{+\infty} du \exp(iux - |u|).$$

▶Generalize eqn 5.13 to the case of three dimensions in order to justify eqn 5.8.

5.4 Structure of surfaces

Taking one further step in the direction of structural complexity, the next hurdle after crystalline solids is the structure of their surfaces. Cutting a crystalline solid along one of its crystallographic planes exposes one of its surfaces. By maintaining the sample in ultra-high vacuum (UHV), i.e. at a pressure around 10^{-7} Pa, it is possible to avoid surface contamination by impurities, oxygen, or water for a few hours, and hence perform experiments in order to investigate the structure of the exposed surface. Since the atoms at the surface are no longer in the middle of a perfect crystal, the forces on the nuclei associated with the atoms underneath the surface are not balanced by equal and opposite forces on the vacuum side. As a result, atoms at and near the surface are no longer in equilibrium positions. In order to restore the equilibrium the atoms at the surface must undergo some rearrangement, a phenomenon referred to as *surface reconstruction*.

DFT is particularly suited for the study of surface reconstruction as the underlying theory makes no assumptions whatsoever about the equilibrium configuration, and the solution of the full quantum-mechanical problem, eqn 2.24, within the approximations discussed in Chapters 2–4 naturally leads to the identification of the most stable structure.

Unit cell   Periodic replicas

Surface
atoms

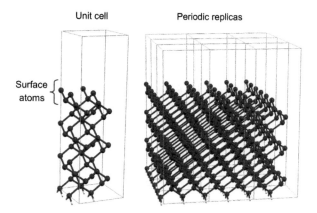

Fig. 5.4 Setting up a DFT calculation for studying the reconstruction of the Si(001) surface. The left panel shows the computational cell containing the periodically repeated structure which corresponds to a piece of silicon crystal. The right panel shows an expanded view of the periodic structure. The coloured spheres represent Si atoms, as in Figure 5.2, and the small white spheres represent H atoms.

We illustrate the study of surface reconstruction using DFT by considering the Si(001) surface as a prototypical system. The Si(001) surface is very important in electronics since the wafers used for manufacturing integrated circuits are obtained by cutting monocrystalline silicon ingots along the (001) crystallographic plane. The silicon surface is subsequently oxidized and in the final field-effect transistors the transport of electrons takes place in a very thin layer (thickness 1–3 nm) right below the surface (Hu, 2009). Owing to its primary role in the electronics industry, the investigation of the atomistic structure of the Si(001) surface has been very topical in surface science. Figure 5.4 shows the setup for a DFT calculation of the Si(001) surface. The initial geometry for performing the structural optimization is obtained by carving a block of atoms out of a bulk silicon crystal. This block contains 52 Si atoms arranged in 13 layers. The bottom layer is bonded to 8 H atoms in order to 'passivate' dangling bonds. The Si atoms belonging to the four bottom layers are kept fixed in the positions they would occupy in a perfect crystal. This expedient allows us to mimic a situation where the crystalline layer underneath the surface is very thick. The surface of interest is the top one, and we expect that this surface will undergo some structural rearrangement. The right panel shows an expanded view of the periodic structure: while the actual calculations are performed on the system in the left panel, the use of periodic boundary conditions allows us to describe a system which is effectively infinite in the plane of the surface. For reasons of computational convenience the system is also periodic along the vertical direction, but the unit cell is kept large enough to accommodate a vacuum layer of several Å and thus avoid spurious interactions between periodic replicas of the slabs.

In order to determine the equilibrium structure of the Si(001) surface we can perform a DFT optimization of the atomic coordinates using methods similar to the one described in Section 4.4. In this case the calculations are considerably more time-consuming than those described in Section 5.2 for the case of bulk silicon. As a crude rule of thumb we can estimate the computer time required for a standard DFT calculation by considering that the number of floating-point operations (flops) scales as the third power of the number of atoms, M, in the computational cell: $\text{flops}(M) \propto M^3$. By comparing the size of the system in Figures 5.4 and 5.2 we can estimate that the surface calculation will take approximately $(60/2)^3 = 27,000$ times longer than the calculation for the bulk silicon crystal. Using the timing provided in Section 5.2 we can therefore estimate that a surface calculation would require something of the order of 75 hours on a laptop computer. For this reason it is more convenient to employ a 'parallel computing cluster', i.e. a computer where the floating-point operations are distributed among many computing cores. For example by using a 32-core computer it would be possible to perform the same calculation within a couple of hours.

The outcomes of the structural optimization are shown in Figure 5.5. When we start from the Si(001)-(1×1) surface where all the atoms are in the same positions as in the

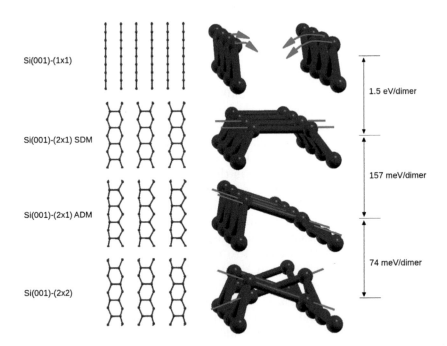

Fig. 5.5 Si(001) surface reconstructions calculated using DFT and the computational setup described in Figure 5.4. The panels show only the first and second surface layers counting from the vacuum side. The top views in the left column show the dimer rows in the (1×1), the (2×1)-SDM, the (2×1)-ADM, and the (2×2) reconstructions, respectively. The panels in the right column show an oblique view for each structure. The red lines are guides to the eye indicating the tilting of dimers. On the right we also show the relative gain in total energy when the surface structure evolves from the top to the bottom row.

silicon crystal, the configuration is very energetic and the system is unstable against distortions. In order to reduce the energy of the system, atoms belonging to parallel rows bend towards each other in such a way as to form 'dimers'. This structural rearrangement is indicated as Si(001)-(2×1) SDM reconstruction in Figure 5.5 (SDM stands for symmetric dimer model) and is accompanied by a large energy gain of 1.5 eV/dimer. The formation of dimers can be rationalized using elementary chemical bonding arguments: each Si atom carries four valence electrons. The outermost surface atoms accommodate two valence electrons each in bonding orbitals with the layer of atoms underneath, while the remaining two electrons occupy one surface 'dangling bond' each. This leads to four half-filled orbitals for every pair of surface Si atoms. In these conditions it is energetically more favourable for the structure to rearrange in such a way to have only three orbitals occupied, two dangling ones and a third one in between the two atoms of a dimer. The dangling bonds will be singly occupied while the dimer bond will be doubly occupied. The double occupation of a dimer bond instead of two dangling bonds is energetically favourable. The mathematical formalism supporting these hand-waving considerations can be found, among others, in the book by Pettifor (1995). A clear discussion of these aspects is also provided by Kaxiras (2003).

In the Si(001)-(2×1) SDM reconstruction of Figure 5.5, each dimer carries two identical half-occupied dangling bonds, and it is possible to further reduce the total energy of the system by tilting the dimer in such a way as to make these orbitals inequivalent. In the tilted configuration, denoted by Si(001)-(2×1) ADM (asymmetric dimer model) in Figure 5.5, since the two dangling bonds are inequivalent they must have different energies, hence the two electrons will both occupy the orbital with the lowest energy. The calculation shows that in this configuration approximately four electrons localize around the higher Si atom of the dimer, while only three electrons approximately localize around the lower Si atom. This is an example of a 'broken symmetry' leading to a stabilization of the structure.

Our broken-symmetry argument can be extended to pairs of adjacent dimers belonging to the same row. In fact by now we can guess that it should be possible to further lower the energy of the system by alternating the tilt angle along the dimer rows of Figure 5.5. The ensuing stabilization leads to the Si(001)-(2×2) reconstruction. Interestingly the mechanism for the alternation of dimers *along* the dimer chains is very similar to the Peierls' distortion (Peierls, 1955) taking place in one-dimensional systems such as polyacetylene.

We can even try to identify surface reconstructions more stable than the Si(001)-(2×2) in Figure 5.5. In fact if we were to consider adjacent *dimer rows* which are mirror images of each other instead of being related by a translation as in Si(001)-(2×2), we would lower the total energy by an additional 3 meV/dimer. This additional structure is denoted by Si(001)-(4×2) and is shown in Figure 5.8.

This progression shows that as structural distortions involve increasingly longer distances (neighbouring atoms, neighbouring dimers, neighbouring dimer rows) the energy gain becomes smaller and smaller, decreasing from 1.5 eV/dimer to 3 meV/dimer. It is also interesting to observe that surface reconstruction does not alter significantly the length of Si–Si bonds. In fact the bonds between the dimer atoms and those between the dimer and the Si atoms underneath range from 2.32 Å to 2.37 Å; that is, they differ from the calculated bond length in bulk Si (2.34 Å) by 1% at most. The equilibrium surface reconstruction Si(001)-(4×2) predicted by DFT calculations has been confirmed by experiments (Perdigão *et al.*, 2004). However, it should be noted that, given the small energy difference between the (2×1)-ADM, the (2×2) and the (4×2) reconstructions, the combined effects of temperature, strain and doping may lead to the coexistence of surface domains with different reconstructions. The detailed nature of the Si(001) surface reconstruction has been debated intensely in the scientific literature (Yokoyama and Takayanagi, 2000; Perdigão *et al.*, 2004; Manzano *et al.*, 2011).

So far we have considered a typical use of DFT for determining equilibrium structures of surfaces entirely from first principles, and we have mentioned that DFT predictions are in good agreement with experiment. The question that arises naturally at this point is how do we probe surface reconstruction in experiment.

It turns out that it is possible to investigate surface structures with atomic resolution using scanning tunnelling microscopy (STM). In the following section we introduce the basics of STM and show how we can combine DFT calculations and STM experiments in order to determine surface structures.

5.5   Comparison of DFT surface reconstructions with STM

The scanning tunnelling microscope is a very powerful experimental tool for probing the structure of surfaces at the atomic scale. It would not be an exaggeration to state that the development of STM by Binnig *et al.* (1982) marked the beginnings of what we now refer to as nanoscience and nanotechnology. One of the distinctive aspects of STM is that its principle is general enough that it can be used to investigate a variety of systems, ranging from atomically sharp surfaces to molecular adsorbates and nanoscale assemblies. The tunnelling microscope can also be used for directly manipulating atoms and molecules at the nanoscale.

As shown schematically in Figure 5.6, in an STM experiment an atomically sharp metallic tip is placed in close proximity to a conducting surface (a few Å away), and by applying a bias voltage the electrons belonging to the tip can tunnel into the surface (or vice-versa, depending on the polarity of the bias). The resulting tunnelling current is very sensitive to the tip–sample distance, hence a current measurement provides direct information on the surface topography. Typical tips are made of tungsten, platinum or platinum/iridium, and the tip position is controlled by a piezoelectric actuator. Typical tunnelling currents are between 10 pA and 10 nA. The main technological challenge in the realization of STM is to stabilize the tip–sample distance to within less than 0.1 Å. This requires the system to be isolated from external vibrations such as sound or vibrations of the building, and this is achieved through the use of vibration isolation dampers, such as pneumatic suspensions and eddy current dampers. A detailed account of the technology behind STM can be found in the book by Wiesendanger (1994). In this section we simply discuss those aspects which are essential for understanding how to calculate STM maps using DFT and how to perform meaningful comparisons with experiment.

The tunnelling current from the tip to the surface can be calculated from first principles using a theory developed by Bardeen (1961) for two metal layers separated by a vacuum gap. This theory relates the current to the local Fermi levels in the layers, the respective density of electronic states, and the spatial overlap between their electronic wavefunctions. Here we will first write the expression for the current within Bardeen's theory and then discuss each term separately. We will provide plausibility arguments without attempting a derivation of the theory. In Hartree atomic units the STM tunnelling current can be expressed as:

$$I = 2\pi \sum_i \sum_j f(E_{\text{T},i} - E_{\text{T},\text{F}})[1 - f(E_{\text{S},j} - E_{\text{S},\text{F}})] |M_{ij}|^2 \delta(E_{\text{T},i} - E_{\text{S},j}). \quad (5.14)$$

In this expression $E_{\text{T},i}$ and $E_{\text{S},j}$ are the single-particle electronic eigenvalues of the tip (T) and the sample (S), respectively. The quantities $E_{\text{T},\text{F}}$ and $E_{\text{S},\text{F}}$ are the local Fermi levels in the tip and sample respectively, which are offset by the voltage, V:

$$E_{\text{S},\text{F}} = E_{\text{T},\text{F}} + V. \quad (5.15)$$

Here we are using the convention that the voltage source has the positive end on the tip side, as in Figure 5.6; therefore, under a negative bias electrons tunnel from tip to sample. The $f(E)$ functions in eqn 5.14 are the Fermi–Dirac thermal occupations:

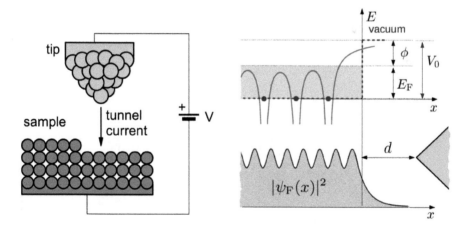

Fig. 5.6 Idealized schematic of an STM experiment (left), and tunnelling of the sample wavefunctions (ψ) outside of the surface within the simplest possible model, i.e. the free electron gas (right). The panel on the right is rotated by 90° clockwise with respect to the schematic on the left. The blue dots and red lines indicate the nuclei of the sample and the associated potential. The corresponding potential energy profile experienced by the electrons in the sample can be approximated by a step function of height V_0 (dashed blue line). The electrons in the sample occupy single-particle eigenstates up to the Fermi energy, E_F, and are separated from the vacuum level by the 'work function', ϕ (see Section 9.3), in such a way that $V_0 = E_\text{F} + \phi$. d is the tip–sample distance and x is the coordinate along the direction perpendicular to the surface.

$$f(E) = \frac{1}{\exp(E/k_BT) + 1},$$ (5.16)

with k_B the Boltzmann constant and T the temperature. The Fermi–Dirac function gives the probability that, on average, an electronic state with energy E be occupied at temperature T (Ashcroft and Mermin, 1976). The quantities M_{ij} in eqn 5.14 are the matrix elements of the so-called 'current operator' and are defined as follow:

$$M_{ij} = \frac{1}{2} \int d\mathbf{S} \cdot (\psi_{T,i}^* \nabla \psi_{S,j} - \psi_{S,j} \nabla \psi_{T,i}^*),$$ (5.17)

with $\psi_{T,i}$ and $\psi_{S,j}$ the single-particle wavefunctions of electrons in the tip and the sample, respectively, and the integral running over a surface which separates the sample from the tip. The explicit expression for the matrix elements in eqn 5.17 is not essential for our present discussion, and it is given only for completeness. The last term in eqn 5.14 is the Dirac delta function introduced in Exercise 5.3.

We can understand eqn 5.14 heuristically as follows. The Dirac delta, $\delta(E_{T,i} - E_{S,j})$, is non-zero only when $E_{T,i} \simeq E_{S,j}$, hence it guarantees that the energy of the electron is conserved during the tunnelling process. The number of electrons which tunnel across the vacuum gap is proportional to the number of occupied electronic states in the tip, $\sum_i f(E_{T,i} - E_{T,F})$. The tunnelling process is allowed only if the states on the other side are unoccupied, hence the current is also proportional to the number of unoccupied states in the sample, $\sum_j [1 - f(E_{S,j} - E_{S,F})]$. On top of this, the square of the matrix elements, $|M_{ij}|^2$, selects only those tunnelling events which involve a

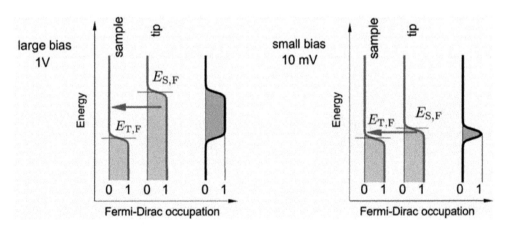

Fig. 5.7 This scheme illustrates the conditions for electron tunnelling from an STM tip to a metallic sample. A necessary condition for tunnelling is that the electron leaves an occupied state of the tip and ends up in an empty state of the sample with the same energy. The grey lines indicate the Fermi–Dirac thermal occupations, and the black line is the product $f(E - E_{T,F})[1 - f(E - E_{S,F})]$ appearing in eqn 5.14. Tunnelling is allowed only when this product is non-zero. The comparison between large (left) and small (right) bias shows that, in the latter case, only electrons with energy at or near the Fermi level participate in the current.

constructive interference of the wavefunctions in the tip and the sample. The bias V in eqn 5.15 acts to offset the local Fermi levels, thereby modifying the 'window of opportunity' for the electrons to tunnel. This is shown schematically in Figure 5.7.

Equation 5.14 can be simplified considerably by making a few simple and fairly general assumptions about the shape of the electronic wavefunction in the tip, $\psi_{T,i}$. Tersoff and Hamann (1985) found that, by assuming a spherical tip, the matrix element, M_{ij}, becomes proportional to the magnitude of the wavefunction of the sample at the centre $\mathbf{r}_T$ of the sphere:

$$M_{ij} = \text{const} \cdot \psi_{S,j}(\mathbf{r}_T). \tag{5.18}$$

This result can also be obtained heuristically by imagining that the wavefunctions in the tip are sharply peaked around its centre, $\mathbf{r}_T$. This extreme simplification allows us to carry out the integration in eqn 5.17 with no effort and recover immediately the Tersoff–Hamann result in eqn 5.18. Equation 5.18 is very appealing because it affords us a considerable simplification of the whole problem. In fact, when we focus our discussion on the case of low temperature and small bias voltage, the expression for the tunnelling current drastically simplifies as follows:

$$I = \text{const} \cdot \sum_j |\psi_{S,j}(\mathbf{r}_T)|^2 \delta(E_{S,j} - E_{S,F}). \tag{5.19}$$

Equation 5.19 states that the tunnelling current in an STM experiment at small bias is proportional to the total magnitude, $\sum_j |\psi_{S,j}|^2$, of the square modulus of the electron wavefunctions from the sample at the centre of the tip, evaluated for those eigenstates with energy at the common Fermi level, $E_{S,j} \approx E_{S,F}$. The derivation of eqn 5.19 from eqn 5.14 is carried out in Exercise 5.5.

At this point we are in a good position to ask what makes STM so useful for characterizing surfaces at the atomic scale. The answer is that the tunnelling current is extremely sensitive to the tip–surface distance. This sensitivity can simply be understood using eqn 5.19. Since the electronic states of the sample decay exponentially outside of its surface (see Exercise 5.6), eqn 5.19 implies that the current decreases exponentially when the tip is pulled away from the surface. The *exponential* dependence of the current on the tip–sample distance is the founding principle of STM.

Exercise 5.5 Here we want to derive eqn 5.19 starting from Bardeen's expression for the tunnelling current (eqn 5.14). First we define the *density of electronic states* of the tip as follows:

$$\rho_T(E) = \sum_i \delta(E - E_{T,i}). \tag{5.20}$$

The density of states quantifies how many electronic states exist in the tip per unit energy (Kittel, 1976). ▶Using eqns 5.15 and 5.18, and exploiting the property of the Dirac delta expressed by eqn 5.11, show that eqn 5.14 can be rewritten as:

$$I = \text{const} \cdot \int dE \, \rho_T(E) \, f(E - E_{S,F} + V)[1 - f(E - E_{S,F})] \times \sum_j |\psi_{S,j}(\mathbf{r}_T)|^2 \delta(E_{S,j} - E).$$

In the limit of low temperature the Fermi–Dirac distribution in eqn 5.16 resembles a step function. ▶Show that in this limit the previous equation reduces to:

$$I = \text{const} \cdot \int_{E_{S,F}}^{E_{S,F}+|V|} dE \; \rho_T(E) \times \sum_j |\psi_{S,j}(\mathbf{r}_T)|^2 \delta(E_{S,j} - E).$$

This expression can be simplified further if we make the approximation that the density of states of the tip does not change much with the energy:

$$I = \text{const} \sum_{j:E_{S,j}>E_{S,F}}^{E_{S,F}+|V|} |\psi_{S,j}(\mathbf{r}_T)|^2, \tag{5.21}$$

where the sum is restricted to those states with energy $E_{S,j}$ between $E_{S,F}$ and $E_{S,F} + |V|$. ▶Use eqn 5.21 to show that at very small voltage (i.e. in the limit of $V \to 0$) we recover the simplified expression for the tunnelling current given in eqn 5.19.

Exercise 5.6 In this exercise we want to relate the STM current to the work function of a metallic sample. We consider a simplified model of a metal consisting of a free electron gas, as shown in Figure 5.6. The electrons can be considered to be confined by a step-wise potential along the direction perpendicular to the sample surface (x coordinate), with the surface at $x = 0$. Therefore we make the approximation that the effective potential experienced by the electrons is vanishing for $x < 0$ and is equal to $V_0 = E_F + \phi$ for $x > 0$. Here E_F is the metal Fermi level and ϕ the corresponding work function as indicated in Figure 5.6 (see Section 9.3). ▶Show that the single-particle Schrödinger equation for an electron with energy equal to the Fermi level in the region $x > 0$ is given by:

$$-\frac{1}{2}\frac{d^2\psi}{dx^2} + (E_F + \phi)\psi(x) = E_F\psi(x).$$

▶Show that the electronic wavefunction, $\psi(x)$, for this electron outside the sample is given by:

$$\psi(x) = \text{const} \cdot \exp\left[-\sqrt{2\phi}\,x\right].$$

Since, by construction, this state has its energy at the Fermi level, we can calculate the tunnelling current using eqn 5.19 and including only this state in the sum. ▶Show that the resulting tunnelling current is:

$$I = \text{const} \cdot \exp\left[-2\sqrt{2\phi}\,d\right], \tag{5.22}$$

where d is the distance between tip and sample. This last expression shows very clearly the exponential sensitivity of the STM tunnelling current to the tip–sample distance.

Exercise 5.7 Suppose that an STM experiment is being performed in order to determine the topography of a gold surface. Initially the measured current is 1 nA. ▶Show that by moving the tip 3 Å away from the surface the current drops below 1 pA. This simple example provides a fairly good idea of how sensitive the STM current is to atomic-scale displacements of the tip.

While Exercises 5.6 and 5.7 are useful for developing a simple and intuitive understanding of STM experiments, the electron gas model is not adequate for the study of real surfaces. Indeed, calculations of STM tunnelling maps require a detailed knowledge of the shape and energetics of electron wavefunctions outside the sample. The good news is that the formalism introduced in this section, and in particular

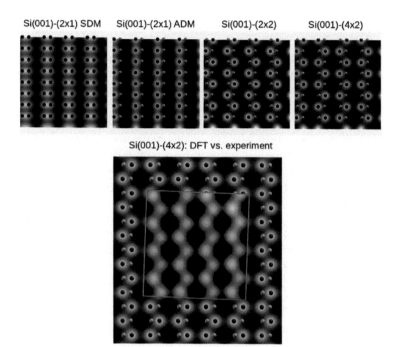

Si(001)-(2x1) SDM   Si(001)-(2x1) ADM   Si(001)-(2x2)   Si(001)-(4x2)

Si(001)-(4x2): DFT vs. experiment

Fig. 5.8 Calculated STM maps of the Si(001) surface reconstructions obtained in Sec. 5.4 and Figure 5.5. The maps were calculated using DFT/LDA and eqn 5.21. The centre of the tip, $\mathbf{r}_T$, is in a plane parallel to the surface, at a distance of 5 Å from the topmost Si layer. The voltage is $V=-1$ V using the convention of Figure 5.6. Bright spots represent regions of larger current. The Si atoms of the topmost layer are shown, with dark blue spheres indicating the top atom in buckled dimers, and light blue the bottom atoms. Top panels: the four different reconstructions in Figure 5.5 carry clear and distinct signatures in the calculated maps. In particular the Si(001)-(2×1) reconstructions show linear patterns while Si(001)-(2×2) and Si(001)-(4×2) exhibit zigzag patterns. Bottom panel: the map calculated for the predicted equilibrium geometry Si(001)-(4×2) is in very good agreement with the STM image measured by Yokoyama and Takayanagi (2000), and shown as the inset in red. A recent account of progress in this area can be found in Manzano *et al.* (2011). Experimental STM image courtesy of T. Yokoyama, adapted by permission from the American Physical Society: *Physical Review Letters* (Yokoyama and Takayanagi, 2000), copyright 2000.

eqns 5.14 and 5.21, is very general and can be combined with DFT in order to calculate realistic STM maps.

The most common approach for the calculation of STM maps consists of using the Kohn–Sham eigenstates, ϕ_j, and eigenvalues, ε_j, of Section 3.4 as a 'best guess' for the electron wavefunctions, $\psi_{S,j}(\mathbf{r})$, and energies, $E_{S,j}$, needed in eqn 5.21. In other words, we can set

$$\psi_{S,j}(\mathbf{r}) = \phi_j(\mathbf{r}) \quad \text{and} \quad E_{S,j} = \varepsilon_j, \tag{5.23}$$

with ϕ_j and ε_j from eqn 3.27. This replacement is rather delicate since in principle DFT/LDA is a theory of the electronic density in the ground state, and is not designed to describe processes such as electron tunnelling. In fact, in DFT/LDA the Kohn–Sham eigenstates are only auxiliary functions which do not carry a physical meaning on their own. Despite these caveats, it turns out that Kohn–Sham wavefunctions are often good enough for calculating tunnelling maps, and hence are extensively used for this purpose. The use of Kohn–Sham states in the calculation of material properties will be discussed in greater detail in Chapters 9 and 10.

Figure 5.8 shows typical DFT/LDA calculations of STM maps for the surface reconstructions shown in Figure 5.5. The colour maps are obtained from eqn 5.21 by calculating the term on the right-hand side for r_T on a plane parallel to the surface. The agreement between the calculated STM map and the topography measured by Yokoyama and Takayanagi (2000) indicates that the equilibrium geometry of the Si(001) surface determined in Section 5.4 is correct.

The comparison in Figure 5.8 allow us to draw two conclusions: (i) DFT/LDA is able to reproduce the equilibrium geometry at zero temperature of the Si(001) surface; (ii) it is possible to *validate* an atomistic surface model by first performing a 'virtual' STM experiment on the atomistic model, and then comparing the outcome with an actual experiment. These two conclusions are rather general and make DFT calculations a powerful tool for investigating the structure of complex surfaces.

DFT calculations can also be used for *reverse-engineering* experimental STM maps. For example, we could calculate STM maps for several possible surface models, and then select the most realistic model based on the similarity between calculated and measured topographies. Illustrative examples of this approach and more details on how to interpret and calculate STM images can be found in the review by Hofer (2003).

6
Elastic properties of materials

In Chapter 5 we have seen examples of calculations of equilibrium structures of molecules, solids and surfaces using DFT. In the case of solids we started from the assumption that the material under consideration is perfectly isolated from the environment, i.e. there are no external forces acting on it (see discussion following eqn 5.2). In this chapter we want to understand how to study the deformations of solids resulting from the application of external mechanical forces. These material properties are conveniently investigated by measuring 'stress–strain curves', which describe the relation between the *stress*, i.e. the set of mechanical forces applied to a sample, and the *strain*, i.e. the ensuing change in shape and size. Note that the notion of mechanical stress has already made its debut in this book. In fact in Exercise 5.2 we have seen that the stable structure of silicon depends on the applied pressure.

6.1 Elastic deformations

In terms of mechanical properties, solids can be classified broadly into two categories, *brittle* or *ductile*, as shown in Figure 6.1. The portion of the stress-strain curve where the stress, σ, is proportional to the strain, ϵ, is referred to as the *elastic regime*. In the elastic regime we define the Young modulus, Y, as the proportionality coefficient

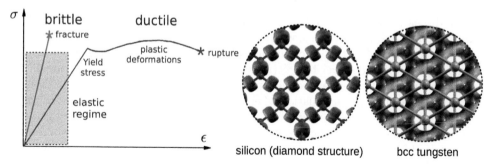

Fig. 6.1 The left panel shows schematic stress–strain curves for brittle and ductile solids. The elastic regime corresponds to the portion of the diagram where the strain is proportional to the stress. The ball-and-stick models on the right show the valence electron density distribution in silicon (brittle) and tungsten (ductile). Si atoms form covalent bonds and the charge is localized between nearest-neighbour atoms. W atoms form metallic bonds and the charge is spread all over the crystal. In both cases the density isovalue corresponds to 0.3 electrons/Å^3.

$Y = \sigma/\varepsilon$. Brittle solids exhibit an elastic response until they undergo fracture at a critical tensile stress. Typical examples of brittle solids are ceramics like alumina and semiconductors like silicon. At the atomic scale brittle solids typically exhibit highly directional 'covalent bonds'. This is a way to say that the electronic charge density concentrates along the lines connecting nearest-neighbour atoms (Figure 6.1).

The stress–strain curve of ductile solids shows an elastic behaviour until the 'yield stress' is reached. Beyond this critical stress, irreversible plastic deformations take place, until rupture is achieved. Platinum, copper and tungsten are common examples of ductile solids. The ductile behaviour is typically associated with the presence of 'metallic bonds', whereby the electronic charge density is spread all over the crystal unit cell (Figure 6.1).

Brittle fracture and plastic deformations are very complex phenomena which are related to the presence of dislocations and grain boundaries. The understanding of these phenomena from an atomistic point of view is still an active research area (see Section 1.2.4) and is beyond the scope of the present book. In this chapter we limit our discussion to the elastic regime of Figure 6.1. Here we want to focus on how to employ first-principles quantum-mechanical methods to study the deformation of materials; therefore we give only an elementary account of elasticity in solids. A comprehensive presentation of the theory of elasticity can be found in the classic book by Timoshenko (1951). A clear and detailed introduction to the behaviour of materials beyond the elastic regime is given by Dieter (1986).

6.2   Intuitive notions of stress and strain using computer experiments

The starting point for studying stress–strain relations in solids within the elastic regime is the definition of stress in terms of atomistic properties. Here, in order to develop an intuitive understanding of the concept of stress, we resort to an extremely simplified 'computer experiment' as shown in Figure 6.2. In the figure we have the same slab of silicon discussed in Figure 5.4. The slab on the left is in equilibrium and there are no external forces acting on it. We now stretch the slab along the direction perpendicular to its surfaces, by holding the nuclei of the bottom layer fixed, and increasing the height of the nuclei in the top layer by 5% (middle panel of Figure 6.2). By allowing all the other nuclei to find their new equilibrium geometry using DFT and the methods of Section 4.4, we end up in a situation where the DFT forces on all nuclei are vanishing, except for those nuclei belonging to the outermost layers. In fact in this configuration the slab is strained and external forces are necessary to maintain the new equilibrium. The *internal forces* on the surface nuclei, which arise from the other nuclei and electrons in the slab, are compensated exactly by the *external forces*, indicated as green arrows in Figure 6.2.

Exercise 6.1 A DFT/LDA calculation of the silicon nano-film in the middle panel of Figure 6.2 indicates that the external forces required to maintain a strain of 5% along the direction perpendicular to the surfaces are 0.682 eV/Å on each surface atom, directed as shown in the figure. The computational cell and the calculation parameters are the same as in Figure 5.4, the transverse size of the tetragonal cell is 7.635 Å, and there are four Si atoms in each layer. ▶ Using the familiar notion of stress as total force per unit area, show that the

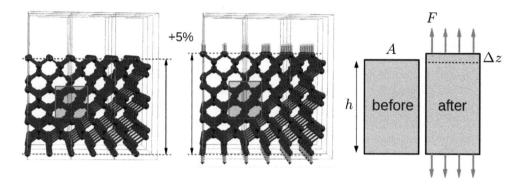

Fig. 6.2 Stretching a silicon nano-film in a computer experiment. The slab on the left is the same as in Figure 5.4, except the hydrogen atoms passivating the surfaces are not shown, for clarity. The slab in the central panel is obtained by holding the bottom layer fixed, increasing the height of the top layer by 5%, and holding the nuclei of the top layer in the new positions. Subsequently the positions of all the other nuclei are optimized until the system reaches its minimum-energy configuration. The only nuclei for which the internal forces are non-vanishing are those belonging to the surface layers. In order for the nuclei to remain immobile these forces are balanced by external forces shown by green arrows. In the actual DFT calculation the action of the external forces is implicitly contained in the *constraints on the nuclear positions*. By releasing these constraints the system would revert to its minimum-energy configuration on the left. It is important to observe that inside the unit cells represented as shaded rectangles the forces on every nucleus are vanishing, regardless of the applied stress. The panel on the right provides a schematic representation of the system, with A the transverse area of the computational cell, h the thickness of the slab, F the sum of the forces acting on the surface nuclei, and Δz the displacement of the top layer.

applied stress corresponds to 7.497 GPa. The Young modulus, Y, of the film can be obtained as $\sigma = Y\epsilon$. ▶Calculate the Young modulus and compare with the experimental value of 166 GPa (McSkimin and Andreatch, 1964) and with previous DFT/LDA calculations of bulk silicon yielding 159 GPa (Nielsen and Martin, 1983). ▶Formulate an hypothesis to explain the difference between your calculations and those of Nielsen and Martin (1983).

While computer experiments like that outlined in Figure 6.2 can help us seize the atomistic meaning of stress using the familiar notion of force per unit area, for reliable DFT calculations on solids we need a more general and rigorous procedure. In fact this simple experiment on an idealized 'nano-film' incorporates a great deal of information about the exposed surfaces, while in practical applications we are most often interested in describing the elasticity of bulk solids. More fundamentally we can observe that each unit cell in the slabs of Figure 6.2 undergoes a deformation, and hence is subject to a stress, even though the forces on each nucleus inside the unit cell are vanishing. This suggests that it should be possible to calculate the stress by working directly with the crystallographic unit cell instead of resorting to the expedient of using a nano-film.

In what follows we use a simple heuristic argument to derive rather general relations between stress, strain and the total energy of the solid. The first step of our argument is to use the Born–Oppenheimer approximation and the classical approximation for the nuclei as described in Sections 4.1 and 4.2. Within this framework the internal forces on the nuclei are completely determined by the potential energy surface, $U(\mathbf{R}_1, \ldots, \mathbf{R}_M)$, using eqn 4.13. The second step is to remind ourselves the *work–energy theorem* of classical mechanics (Halliday *et al.*, 1997). According to this theorem the work of external forces on a system equals the increase in its total potential and kinetic energies. Since we are considering a static deformation whereby the nuclei evolve from one equilibrium configuration to another, the total kinetic energy does not change, and the external work only modifies the total potential energy. With reference to Figure 6.2 we can therefore calculate the change, ΔU, of the total potential energy as the work of the external forces:

$$\Delta U = \int_{h}^{h+\Delta z} F(z)dz, \tag{6.1}$$

where F is the sum of the external forces acting on the atoms of the topmost layer, which are displaced from height h to $h + \Delta z$. The bottom layer is fixed, hence the work done by the external forces there is zero. Since by definition the applied stress corresponds to the external force per unit area, $\sigma = F/A$, and the displacement of the top layer can be expressed in terms of the strain, $\epsilon = \Delta z/h$, we can rewrite eqn 6.1 as follows:

$$\Delta U = \Omega \int_{0}^{\epsilon} \sigma \, d\epsilon, \tag{6.2}$$

with $\Omega = Ah$ the volume of one periodic repeat unit of the slab. The last step is to observe that we are in the elastic regime and therefore by definition the stress is proportional to the strain (see Figure 6.1). By introducing the elastic constant, C, such that

$$\sigma = C \, \epsilon, \tag{6.3}$$

we obtain:

$$\frac{\Delta U}{\Omega} = \frac{1}{2} C \, \epsilon^2. \tag{6.4}$$

Therefore a simple derivative yields the stress in terms of the total energy:

$$\sigma = \frac{1}{\Omega} \frac{\partial U}{\partial \epsilon}. \tag{6.5}$$

Despite the simplicity of our heuristic derivation, eqn 6.4 and eqn 6.5 are rather general results and constitute the basis for practical DFT calculations of the elastic properties of solids.

The interesting aspect of eqn 6.5 is that the surfaces of the nano-film and the external forces are no longer in the picture, and the stress is defined purely in terms of the potential energy surface.

While in our derivation we have implicitly assumed that strain and stress lie along the same direction, i.e. perpendicular to the exposed surfaces in the example in Figure 6.2, in the most general situation they do not need to be collinear. The most general definition using 'tensorial' notation will be presented in eqns 6.9–6.11 of the next section.

For the reader interested in the mathematical aspects, a rigorous approach to the concept of stress within an atomistic point of view can be formulated by using the *virial theorem* (Tsai, 1979). The approximation of classical nuclei that we used in our heuristic derivation can also be removed by starting from an entirely quantum-mechanical formulation of the problem (Fock, 1930a; Nielsen and Martin, 1985). These rigorous approaches provide a solid ground for eqn 6.5 and for its more general version, eqn 6.10 below, and will be left as further reading.

6.3 General formalism for the elastic properties of solids

In the most general case the linear deformation of a solid can be described by specifying the change of the coordinates of every atom with respect to a reference equilibrium configuration:

$$R'_{I\alpha} = \sum_{\beta} (\delta_{\alpha\beta} + e_{\alpha\beta}) R_{I\beta}. \tag{6.6}$$

In this expression $R_{I\alpha}$ and $R'_{I\alpha}$ are the nuclear coordinates in the unstrained and strained solid, respectively, the quantity $e_{\alpha\beta}$ is a 3×3 matrix with very small elements ($e_{\alpha\beta} \ll 1$) and $\delta_{\alpha\beta}$ is Kronecker's delta. The Greek indices run over the Cartesian coordinates, in such a way that $\alpha = 1, 2, 3$ correspond to x, y and z, respectively. In the following we will use interchangeably 1, 2, 3 or x, y, z whenever one form or the other is more convenient. In the remainder of this chapter we will also represent sums like the one appearing in eqn 6.6 using the 'Einstein convention'. According to this convention we omit the summation symbol, and we follow the rule that whenever the same index appears in both terms of a product we perform a sum over that index. The Einstein convention is very convenient when dealing with elasticity in solids since it leads to a lighter notation. For example eqn 6.6 can be rewritten as:

$$R'_{I\alpha} = (\delta_{\alpha\beta} + e_{\alpha\beta}) R_{I\beta}. \tag{6.7}$$

As an illustration of this transformation, in Figure 6.3 we consider the deformation of a square lattice of side a in two dimensions. With reference to the figure we can see that the coordinates of atoms A_1 and A_3 transform as follows:

$$\mathbf{R}'_1 = \mathbf{R}_1 + a(e_{xx}\mathbf{u}_x + e_{yx}\mathbf{u}_y),$$

$$\mathbf{R}'_3 = \mathbf{R}_3 + a(e_{xy}\mathbf{u}_x + e_{yy}\mathbf{u}_y).$$

By looking at the figure we also realize that this deformation implicitly includes a rigid rotation of the two atoms around the origin. In fact the bond A_0–A_1 is rotating counterclockwise by an angle γ such that $e_{yx}a = \tan\gamma \, a$. If the matrix $e_{\alpha\beta}$ is very small, then $\tan\gamma \simeq \gamma$ and the rotation angle is precisely e_{yx}. Similarly, the bond A_0–A_3 is rotating clockwise by an angle e_{xy}. Therefore the average counterclockwise rotation of the square is $(e_{yx} - e_{xy})/2$. In order to describe the deformation of the solid irrespective of this global rotation, we need to subtract the rotation from the coordinate transformation. In the case of atom A_1 the deformation *without* rotation is:

$$\mathbf{R}''_1 = \mathbf{R}'_1 - \frac{a}{2}(e_{yx} - e_{xy})\mathbf{u}_y = \mathbf{R}_1 + a\left[e_{xx}\mathbf{u}_x + \frac{1}{2}(e_{xy} + e_{yx})\mathbf{u}_y\right].$$

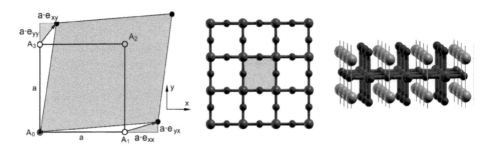

Fig. 6.3 Elastic deformation of a two-dimensional square lattice. The panel on the left shows the displacements of atoms A_0–A_3 according to the deformation $e_{\alpha\beta}$ of eqn 6.7. The components e_{xx} and e_{yy} of the deformation are referred to as 'normal strain' and are associated with dilations of the solid. The components e_{xy} and e_{yx} are referred to as 'shear strain' as they involve the sliding of the two opposite faces of the square. The displacements are highly exaggerated in the figure for clarity, but in reality they are very small: of the order of $0.01a$. An example of a two-dimensional square lattice is the basal CuO_2 plane of the copper oxide La_2CuO_4, where Cu atoms lie at the vertices of a square and the O atoms lie in the middle of its sides (light blue area in the middle panel). In the crystallographic unit cell the CuO_2 planes are stacked on top of each other and are separated by LaO layers (right panel). Copper oxides such as La_2CuO_4 have been investigated in great detail since they exhibit high-temperature superconductivity upon p-type doping (Pickett, 1989).

By repeating the exercise for atom A_3 we find that a similar expression holds for R_3''. These observations are of general validity and indicate that the pure elastic deformation of a solid is given by the *symmetric* part of $e_{\alpha\beta}$:

$$\epsilon_{\alpha\beta} = \frac{1}{2}(e_{\alpha\beta} + e_{\beta\alpha}), \tag{6.8}$$

while the remainder of the matrix (i.e. its antisymmetric part) describes the rotation. The quantity $\epsilon_{\alpha\beta}$ in eqn 6.8 is called the *strain tensor* and is a generalization of the scalar strain, ϵ, introduced in Section 6.2. The use of the name 'tensor' as opposed to the familiar term 'matrix' reflects the fact that $\epsilon_{\alpha\beta}$ is not just any matrix, but must be such that the physics of linear elasticity does not change if we change the system of coordinates. The reader interested in the properties and applications of tensor algebra will find a comprehensive introduction in the book by Itskov (2007).

Now that we have defined a proper mathematical framework to describe strain in solids, we can proceed to write the change in the total energy of a system following a deformation. In the most general case eqn 6.4 takes the following form:

$$\frac{\Delta U}{\Omega} = \frac{1}{2}C_{\alpha\beta\gamma\delta}\,\epsilon_{\alpha\beta}\,\epsilon_{\gamma\delta}, \tag{6.9}$$

where Ω is (as above) the volume of the crystalline unit cell (or the computational cell if we are considering a non-crystalline system), and $C_{\alpha\beta\gamma\delta}$ is a tensor comprising 3^4=81 elastic constants. The quadratic dependence of the energy on the strain in eqn 6.9 is a direct consequence of the fact that we are considering small deviations from the equilibrium structure, which itself must correspond to a parabolic minimum. The atomistic origin of eqn 6.9 can be traced back to the concept of equilibrium bond–length illustrated in Figure 5.1 and Figure 5.2.

The extension of our heuristic result in eqn 6.5 to the general case reads:

$$\sigma_{\alpha\beta} = \frac{1}{\Omega}\frac{\partial U}{\partial \epsilon_{\alpha\beta}}, \tag{6.10}$$

where we have introduced the *stress tensor* as the set of partial derivatives of the total energy with respect to all the elements of the strain tensor.

By combining eqn 6.9 and eqn 6.10 we find the general relation between strain and stress in the elastic regime:

$$\sigma_{\alpha\beta} = C_{\alpha\beta\gamma\delta}\,\epsilon_{\gamma\delta}. \tag{6.11}$$

This relation is the well known *Hooke's law* in the theory of elasticity (Timoshenko, 1951).

Exercise 6.2 ▶Derive eqn 6.11 starting from eqns 6.9 and 6.10. For this derivation it is important to keep in mind the Einstein convention for the repeated indices, and it might be easier to think of the 3×3 tensors as vectors with 9 elements. The following properties should be noted:

$$\frac{\partial \epsilon_{\alpha\beta}}{\partial \epsilon_{\gamma\delta}} = \delta_{\alpha\gamma}\delta_{\beta\delta},$$

and

$$C_{\alpha\beta\gamma\delta} = C_{\gamma\delta\alpha\beta}. \tag{6.12}$$

The latter relation arises from the fact that eqn 6.9 is invariant with respect to the exchange of the pairs of indices $\alpha\beta$ and $\gamma\delta$ in the elastic tensor.

From eqn 6.8 and eqn 6.10 it follows immediately that the stress tensor is also symmetric, i.e. $\sigma_{\alpha\beta} = \sigma_{\beta\alpha}$. In fact the symmetry of the strain tensor allows us to swap the indices α and β on the right-hand side of eqn 6.10 without altering the result. Physically the symmetry of the stress tensor reflects our choice of focusing on the pure deformation of a solid by ignoring eventual rotations. In fact if the strain tensor applied to the square lattice of Figure 6.3 were not symmetric, then the solid would spin around an axis perpendicular to the paper.

The symmetric nature of both the strain and the stress tensors implies that each of them only has six independent components out of the 3×3 matrix elements. It is common practice to choose as independent components the elements of the upper triangular part of these tensors, and rearrange them into a vector according to a counterclockwise cyclic order:

$$\epsilon_{\alpha\beta} = \begin{pmatrix} \epsilon_{xx} & \epsilon_{xy} & \epsilon_{xz} \\ & \epsilon_{yy} & \epsilon_{yz} \\ & & \epsilon_{zz} \end{pmatrix} \rightarrow \epsilon_i = \begin{pmatrix} \epsilon_1 & \epsilon_6 & \epsilon_5 \\ & \epsilon_2 & \epsilon_4 \\ & & \epsilon_3 \end{pmatrix}. \tag{6.13}$$

This convention is referred to as *Voigt notation*, and also applies to the strain tensor. Since strain and stress only have six independent components each, the most general linear relation between them will involve 6×6 coefficients; therefore only 36 out of the 81 elements of the elastic tensor in eqn 6.11 are inequivalent. The number of independent elements is further reduced to 21 due to the symmetry expressed by eqn 6.12. Taking into account the symmetry reduction of strain, stress and the elastic tensor we can restate Hooke's law in Voigt notation as follows:

$$\sigma_i = C_{ij}\,\epsilon_j. \tag{6.14}$$

The rule for obtaining C_{ij} from $C_{\alpha\beta\gamma\delta}$ is the same as in eqn 6.13 for each pair of indices $\alpha\beta$ and $\gamma\delta$. At this point we are ready to rewrite the relation in eqn 6.9 between the total energy of the solid and the associated deformation in the contracted Voigt notation (see Exercise 6.3 below):

$$\frac{\Delta U}{\Omega} = \frac{1}{2}C_{ij}u_i u_j \qquad \text{with} \qquad u_i = \begin{cases} \epsilon_i & \text{if } i = 1,2,3 \\ 2\epsilon_i & \text{if } i = 4,5,6. \end{cases} \tag{6.15}$$

The modified vectors of deformations u_i have been introduced in order to maintain a formal similarity between eqn 6.15 and eqn 6.9, and are referred to as *engineering strain*. Had we used instead the original strain vector, ϵ_i, eqn 6.15 would look slightly more complicated since the off-diagonal components, ϵ_4, ϵ_5 and ϵ_6, appear twice in the sum. For example, ϵ_4 appears once as ϵ_{yz} and once as ϵ_{zy}.

Exercise 6.3 ▶Derive eqn 6.15 starting from the tensorial relation in eqn 6.9 and using the rules for Voigt notation in eqn 6.13. This derivation is rather tedious but otherwise should not pose any difficulties.

The relation between the total energy of a solid and its deformation state in eqn 6.15 is very important for practical purposes and constitutes the basis for DFT calculations of the elastic constants of solids. In the next section we will see how to use this relation and we will discuss some illustrative examples.

6.4 Calculating elastic constants using the DFT total energy

The elastic constants, C_{ij}, can be calculated starting from eqn 6.15 by evaluating the DFT total energy of the solid for different values of the modified strain, u_i (Mehl *et al.*, 1990). In principle we need as many total energy evaluations as the number of independent elastic constants, plus one for the ground state.

Let us consider initially a crystal with a cubic lattice, such as silicon in the diamond structure or bcc tungsten (bcc stands for body-centred cubic). In this case the symmetries of the lattice imply that only three elastic constants are non-zero, C_{11}, C_{12} and C_{44}. The elastic tensor for a cubic lattice in Voigt notation reads (Nye, 1985):

$$C_{ij} = \begin{pmatrix} C_{11} & C_{12} & C_{12} & & & \\ C_{12} & C_{11} & C_{12} & & & \\ C_{12} & C_{12} & C_{11} & & & \\ & & & C_{44} & & \\ & & & & C_{44} & \\ & & & & & C_{44} \end{pmatrix}, \tag{6.16}$$

where all the empty entries correspond to $C_{ij} = 0$. In this case the change of total energy upon deformation from eqn 6.15 simplifies to:

$$\frac{\Delta U}{\Omega} = \frac{1}{2}C_{11}(u_1^2 + u_2^2 + u_3^2) + C_{12}(u_1u_2 + u_1u_3 + u_2u_3) + \frac{1}{2}C_{44}(u_4^2 + u_5^2 + u_6^2). \quad (6.17)$$

In order to determine C_{11}, C_{12} and C_{44} we need three equations. One equation can be obtained by considering a uniform isotropic expansion or compression of the unit cell. This choice can be implemented by setting the normal strain equal to a given parameter η for all directions and the shear strain equal to zero. Under these conditions eqn 6.17 becomes:

$$u_1 = u_2 = u_3 = \eta, \; u_4 = u_5 = u_6 = 0, \qquad \text{(isotropic deformation)}$$
$$\Delta U/\Omega = \frac{3}{2}(C_{11} + 2C_{12})\eta^2. \qquad\qquad (6.18)$$

This result means that we can determine the sum $C_{11} + 2C_{12}$ by calculating the total energy, U, in the unstrained structure and in the structure with the new lattice parameter $a' = (1 + \eta)a$.

Another relation for the elastic constants can be obtained by considering a shearless tetragonal deformation whereby the basal plane is compressed and the height of the cell increases while keeping the volume unchanged (to first order in the deformation). With this choice eqn 6.17 becomes:

$$u_1 = u_2 = -\eta, \; u_3 = 2\eta, \; u_4 = u_5 = u_6 = 0, \qquad \text{(tetragonal deformation)}$$
$$\Delta U/\Omega = 3(C_{11} - C_{12})\eta^2. \qquad\qquad (6.19)$$

In this case the calculation of the total energy, U, must be performed for the new tetragonal lattice parameters, $a' = (1 - \eta)a$ and $c' = (1 + 2\eta)a$.

Finally, a third relation for the elastic constants can be written by considering a pure shear deformation leading to a trigonal distortion in the xy plane:

$$u_1 = u_2 = u_3 = 0, \; u_4 = u_5 = 0, \; u_6 = \eta, \qquad \text{(trigonal deformation)}$$
$$\Delta U/\Omega = \frac{1}{2}C_{44}\eta^2. \qquad\qquad (6.20)$$

The effect of this last deformation is seen most clearly by writing explicitly the change of any lattice vector, say $\mathbf{a}_1$ for definiteness: $a'_{1x} = a_{1x} + \eta a_{1y}/2$, $a'_{1y} = a_{1y} + \eta a_{1x}/2$, and $a'_{1z} = a_{1z}$. In the simplest case where $\mathbf{a}_1$, $\mathbf{a}_2$ and $\mathbf{a}_3$ define a cube, this transformation leaves the z-axis unchanged and deforms the basal plane from a square to a rhombus. A schematic summary of the three deformations considered in eqns 6.18–6.20 is given in Figure 6.4.

The intuitive meaning of the elastic constants C_{11}, C_{12}, and C_{44} can be understood by taking the derivatives of the stress in eqn 6.14 with respect to the strain. Since $C_{11} = \partial\sigma_1/\partial\epsilon_1$, this constant describes the relation between the tensile stress and the ensuing elongation in the same direction. Similarly, C_{12} describes how a solid shrinks in the direction transverse to an applied tensile strain, and C_{44} characterizes the extent of a sliding deformation along the direction of a shear strain.

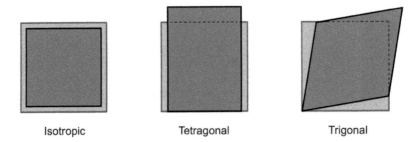

Isotropic   Tetragonal   Trigonal

Fig. 6.4 Structural deformations used to determine the elastic constants C_{11}, C_{12}, and C_{44} of a cubic crystal using DFT: isotropic (left, eqn 6.18), tetragonal (middle, eqn 6.19) and trigonal distortion (right, eqn 6.20). The deformations are schematically shown in two dimensions and are exaggerated for clarity.

6.5 Examples of calculations of elastic constants

In Figure 6.1 we have seen two examples of cubic crystals: silicon in the diamond structure and bcc tungsten. In both cases we can calculate the three independent elastic constants, C_{11}, C_{12} and C_{44}, by following the procedure described in Section 6.4. Single-crystal silicon is not only ubiquitous in electronics, but is also the most widely employed material in the fabrication of MEMS, i.e. micro-electromechanical systems (Petersen, 1982). MEMS are miniaturized electromechanical devices whose components have a characteristic size of 1–100 μm, and are used for instance in pressure and acceleration sensors, microphones and inkjet printer nozzles.

The design of silicon MEMS requires the precise knowledge of the elastic constants C_{11}, C_{12} and C_{44}. These constants can be determined very accurately by measuring the propagation of acoustic waves in the solid, since the speed of sound scales with the square root of the constants (Timoshenko, 1951). Details of typical experimental measurement are given for instance by Hall (1967). The measured elastic constants of single-crystal silicon at room temperature are $C_{11} = 165.6$ GPa, $C_{12} = 63.9$ GPa, $C_{44} = 79.5$ GPa (Hall, 1967; Hopcroft *et al.*, 2010). In Exercise 6.4 we calculate the elastic constant of silicon using the theory developed in Section 6.4, and we compare our results with the measured values.

Exercise 6.4 The following table reports the variation of the total energy of silicon (diamond structure) under the elastic deformations described in Section 6.4, calculated as a function of the distortion amplitude. The values in the table correspond to the energy per unit volume, $\Delta U/\Omega$, in meV/Å^3. The isotropic, tetragonal and trigonal distortions were obtained by using the strain vectors defined in eqn 6.18, eqn 6.19 and eqn 6.20 in terms of the dimensionless parameter η. The calculations were performed as described in relation to Figure 5.2.

η	-0.006	-0.004	-0.002	0.000	0.002	0.004	0.006
Isotropic	0.10786	0.04909	0.01377	0.00000	0.00807	0.03643	0.08437
Tetragonal	0.06567	0.02946	0.00754	0.00000	0.00765	0.02993	0.06746
Trigonal	0.01093	0.00536	0.00195	0.00000	0.00108	0.00330	0.00732

▶By fitting the data in this table to the expressions given in eqns 6.18–6.20, show that the DFT/LDA elastic constants of silicon are $C_{11} = 161$ GPa, $C_{12} = 62$ GPa and $C_{44} = 78$ GPa

(1 GPa = 6.242 meV/Å³). ▶Establish the error bar of this methodology by comparing the values of the elastic constants obtained by using three, five, and seven data-points in each fit. ▶Repeat the calculation by using fitting functions of the form $a(\eta - b)^2 + c$, with a, b and c fitting parameters, and check again the sensitivity of this procedure with respect to the number of data-points included in the fit. ▶Establish which one of these two fitting procedures is more robust and explain why. ▶Compare your results with the measurements by Hall (1967) and determine the relative deviation between calculations and experiment.

Tungsten is a prototypical transition metal which crystallizes in the bcc lattice (bcc W), with a lattice parameter of 3.16 Å at room temperature. Tungsten exhibits the highest melting point among the elemental metals (3,683 K), in fact one of the highest melting points of all materials. Since bcc W also exhibits very good mechanical properties up to at least 2,000 K (Škoro *et al.*, 2011), it is ideally suited for applications in filaments for light bulbs, arc welding and rocket nozzles. Tungsten is also used as the proton target in neutron spallation sources. A comprehensive review of the properties and uses of tungsten is provided by Lassner and Schubert (1999). In Exercise 6.5 we study the elastic properties of this metal using the theory of Section 6.4.

Exercise 6.5 In addition to Voigt's elastic constants, C_{ij}, it is often useful to introduce a derived quantity called the *bulk modulus*. In terms of the standard elastic constants the bulk modulus is defined as $B = (C_{11} + 2C_{12})/3$. Intuitively the bulk modulus describes the resistance of a material to compression under a hydrostatic load (e.g. when immersed in a liquid). ▶By using the definition above and eqn 6.18 show that the bulk modulus can be calculated as:

$$B = \Omega \frac{\partial^2 U}{\partial \Omega^2},$$

with the change in energy resulting from an isotropic compression or expansion. Note that we can express the volume Ω in terms of η as $\Omega = \Omega_0(1 + \eta)^3$, with Ω_0 the value at equilibrium. ☐The following table reports the total energy per atom of bcc tungsten, calculated for various lattice parameters around equilibrium using DFT/LDA. The W atom has 74 electrons in the configuration $[Xe]6s^2d^4$; however, it is possible to freeze the [Xe] core electrons as we did in Chapter 5 for N and for Si (Appendix E). This procedure allows us to perform calculations using only the six valence $6s^2d^4$ electrons. The total energy below is given with respect to the energy at equilibrium.

a (Å)	3.0976	3.1038	3.1100	3.1163	3.1225	3.1287	3.1349
U (meV/atom)	5.3408	2.4194	0.5935	0.0000	0.4654	1.9636	4.6004

▶Calculate the bulk modulus of bcc W. ▶Compare your result with the value of 331 GPa calculated by Roundy *et al.* (2001) and with the value of 314 GPa measured by Featherston and Neighbours (1963). ▶Calculate the bulk modulus of silicon in the diamond structure using the results of Exercise 6.4 and compare your calculated bulk moduli of Si and W.

By comparing the calculations in Exercises 6.4 and 6.5 with experiment we have found that DFT/LDA is able to predict the elastic constants of silicon and tungsten to within 5%. This is indeed a rather general trend, and as a rule of thumb we can remember that, typically, elastic constants obtained from DFT are accurate to within about 10%. This performance is quite remarkable when we think that the calculations do not involve any empirical parameters.

It is important to stress that in the theory developed in this chapter the temperature does not appear anywhere, i.e. all the calculations refer to $T = 0$. In reality the experimental measurements are often performed at room temperature, and the lattice constants always exhibit a moderate dependence on temperature. For example Hall (1967) and Featherston and Neighbours (1963) report the temperature dependence between 0 and 300 K of the elastic constants of silicon (diamond structure) and bcc tungsten, respectively. From these measurements we know that thermal effects yield changes in the elastic constants of the order of a few percent. General methods for including the effects of temperature in the calculations will be discussed in Section 8.5.

The two exercises presented in this section are based on DFT/LDA calculations. However, in Section 5.1 we mentioned that it is possible to perform DFT calculations using more elaborate exchange and correlation functionals in place of the LDA functional of eqns 3.20 and 3.22, e.g. the GGA. Since LDA and GGA yield slightly different equilibrium lattice parameters, it is natural to expect differences in the calculated elastic constants. The differences between LDA and GGA are well understood today and strategies to make the calculated elastic constants less sensitive to the choice of the exchange and correlation functional have been developed (Kunc and Syassen, 2010).

6.6 The stress theorem

In the case of complex crystalline solids exhibiting only a few symmetries, the calculation of the elastic constants using eqn 6.15 and the procedures outlined in Exercises 6.4–6.5 can become very complicated. In such cases it is possible to save some effort by replacing the total energy of eqns 4.9 and 3.10 inside eqn 6.10 and evaluating the derivative analytically. This operation leads to an explicit expression for the stress tensor, $\sigma_{\alpha\beta}$, which depends only on the electron density, $n(\mathbf{r})$, and Kohn–Sham wavefunctions, $\phi_i(\mathbf{r})$, inside the crystalline unit cell. As we can expect by examining eqns 4.9 and 3.10, this expression will comprise contributions arising from the classical electrostatic interactions ($\sigma_{\alpha\beta}^{\mathrm{M}}$), from the kinetic energy ($\sigma_{\alpha\beta}^{\mathrm{kin}}$), and from the exchange and correlation energy ($\sigma_{\alpha\beta}^{xc}$):

$$\sigma_{\alpha\beta} = \sigma_{\alpha\beta}^{\mathrm{M}} + \sigma_{\alpha\beta}^{\mathrm{kin}} + \sigma_{\alpha\beta}^{xc}. \tag{6.21}$$

Complete expressions for these components are given by Dal Corso and Resta (1994). In order to take a closer look we report explicitly the contribution $\sigma_{\alpha\beta}^{\mathrm{M}}$ to the stress tensor taken from Dal Corso and Resta:

$$\sigma_{\alpha\beta}^{\mathrm{M}} = -\frac{1}{2} \int d\mathbf{r} \int d\mathbf{r}' n(\mathbf{r}) n(\mathbf{r}') \frac{(r_\alpha - r'_\alpha)(r_\beta - r'_\beta)}{|\mathbf{r} - \mathbf{r}'|^3} - \frac{1}{2} \sum_{I,J} Z_I Z_J \frac{(R_{I\alpha} - R_{J\alpha})(R_{I\beta} - R_{J\beta})}{|\mathbf{R}_I - \mathbf{R}_J|^3}$$
$$+ \sum_I Z_I \int d\mathbf{r}\, n(\mathbf{r}) \frac{(r_\alpha - R_{I\alpha})(r_\beta - R_{I\beta})}{|\mathbf{r} - \mathbf{R}_I|^3}. \tag{6.22}$$

In this expression the integrals need to be evaluated within one unit cell of the crystal. While eqn 6.22 appears rather complicated, it is possible to recognize in the first line the terms arising from interactions among electrons or among nuclei, and in the second line the term coming from the electron–nucleus interactions. The contribution to the stress given by eqn 6.22 arises solely from the classical electrostatic interactions between electrons and nuclei. This quantity is well known in classical electrodynamics and is referred to as the 'Maxwell stress tensor' (Jackson, 1998). The remaining contributions on the right-hand side of eqn 6.21 take into account the quantum nature of the electrons (not included in the Maxwell stress) within the framework of DFT.

The explicit expression of the quantum-mechanical stress in terms of the electron density, the Kohn–Sham wavefunctions and the nuclear coordinates discussed in this section is generally referred to as the *stress theorem*. A complete derivation of the stress theorem is rather involved, and the interested reader is referred to Nielsen and Martin (1985) for a comprehensive presentation. Here we only point out that the stress theorem is connected with the Hellmann–Feynman theorem for atomic forces that was discussed in Section 4.3.

The direct calculation of the stress tensor using the stress theorem (i.e. without evaluating total energies as in Section 6.4) is particularly useful for studying materials under an external applied stress. In Exercise 5.2 we already have seen how pressure enters a DFT calculation of structural stability. When a material is subject to anisotropic stresses, the enthalpy $h = U + pV$ (introduced in Exercise 5.2) can be generalized to include a stress tensor in place of the hydrostatic pressure. The resulting formulation is very powerful when specific load conditions need to be investigated. The theory of calculations under hydrostatic pressure was developed by Andersen (1980), and its generalization to the most general loading conditions was made by Parrinello and Rahman (1980).

6.7 DFT predictions for materials under extreme conditions

Throughout this chapter we have discussed methods for calculating the elastic properties of solids using DFT. In Section 6.5 we have seen that in two simple cases the calculated elastic constants are found to be in good agreement (within about 10%) with experimental measurements, and we stated that this is indeed a rather general trend. Such a good agreement provides a basis for using DFT calculations to make *predictions* of elastic properties in those cases where performing experiments is particularly challenging.

One prototypical example where measuring elastic constants can be challenging is the study of the interior of planets, where materials are subject to extreme pressures. For example, the pressure in the Earth's core can be as high as 365 GPa (Oganov *et al.*, 2005). The standard experimental tool to study materials at high pressure is the diamond anvil cell, where two diamonds are used to exert pressure on a sample (Jayaraman, 1983). Although the record static pressures demonstrated with diamond anvil cells is currently 640 GPa (Dubrovinsky *et al.*, 2012), experiments are very difficult above 250 GPa due to the plastic deformations occurring in the diamond anvil. In these cases the use of DFT can provide very useful insights into the elastic and structural properties of materials at high pressure.

A well-known example where DFT calculations of elastic constants were instrumental to understanding the structure and properties of materials at high pressure is the

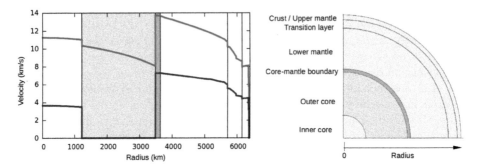

Fig. 6.5 Velocity of seismic waves in the Earth's interior (left) and schematic model of the Earth (right). The compressional (red) and shear (blue) velocities of sound waves are shown as a function of radius, starting from the centre of the planet. The velocities were obtained by Dziewonski and Anderson (1981) using seismic tomography, i.e. by inverting a large dataset of seismological observations. The D″ layer is the region between 3,480 km and 3,630 km highlighted in orange, and marks the boundary between the outer core and the lower mantle. The yellow region indicates the outer core. We can see that this region does not allow shear waves to propagate and therefore it must be in a liquid state.

case of magnesium silicate, $MgSiO_3$. Figure 6.5 shows a schematic representation of the Earth's interior, together with the measured velocities of sound waves as a function of depth. The lower mantle is mostly composed of $MgSiO_3$, a perovskite mineral consisting of a three-dimensional network of corner-sharing SiO_6 octahedra intermixed with a cubic lattice of Mg atoms (Figure 6.6).

The analysis of seismic tomography data reveals anomalous elastic properties in a narrow shell called the D″ layer near the mantle–core boundary (Figure 6.5). Sidorin *et al.* (1999) observed that in this layer the velocity of shear waves increases discontinuously by about 1%. In order to explain this and other similar anomalies these authors proposed that $MgSiO_3$ may undergo a phase transition in the D″ layer. A clarification of these aspects is important for our current understanding of geodynamics and seismicity.

Unambiguous evidence for the proposed structural phase transition was provided later by a set of experimental and theoretical studies (Murakami *et al.*, 2004; Oganov and Ono, 2004; Iitaka *et al.*, 2004). These studies demonstrated that, at the pressures existing in the D″ layer (around 130 GPa), a new phase of $MgSiO_3$ called 'post-perovskite' is more stable than the perovskite structure found higher up in the mantle. In this high-pressure phase of $MgSiO_3$, shown in Figure 6.6, the octahedra are tilted in such a way as to form flat sheets, leading to a slightly increased elastic anisotropy. In this context DFT calculations successfully predicted a structural phase transition of $MgSiO_3$ between 84 and 99 GPa (Oganov and Ono, 2004; Iitaka *et al.*, 2004). In addition, the calculated elastic constants indicated that the velocity of shear waves in the post-perovskite phase should be 1.9% higher than in the perovskite structure, in agreement with seismic data (see Table 6.1). The calculations of the elastic constants by Oganov and Ono (2004) and Iitaka *et al.* (2004) were performed precisely with the methods discussed in this chapter.

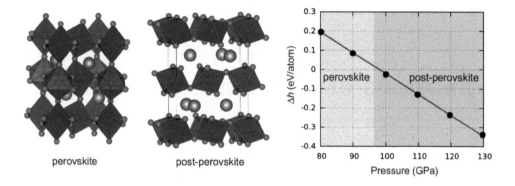

Fig. 6.6 Ball-stick-octahedra models of MgSiO$_3$ in the perovskite and post-perovskite phases, and enthalpy difference between the two phases as a function of pressure. The orange atoms represent Mg, O atoms are red, and Si atoms can be seen inside the octahedra. The enthalpy difference was calculated by Iitaka *et al.* (2004) as $\Delta h = h(\text{PP}) - h(\text{Pv})$, with the enthalpies $h(\text{PP})$ and $h(\text{Pv})$ corresponding to the post-perovskite (PP) and to the perovskite (Pv) phases, respectively. When $\Delta h > 0$ we have $h(\text{Pv}) < h(\text{PP})$ and the stable phase corresponds to the perovskite structure. When $\Delta h < 0$ the post-perovskite structure becomes energetically favourable. The phase transition between the two phases is marked by the pressure where $\Delta h = 0$, i.e. 98 GPa in the figure.

Table 6.1 Elastic constants calculated for perovskite and post-perovskite MgSiO$_3$ by Oganov and Ono (2004) using DFT (in GPa), together with the calculated velocity of shear waves, v_S (in km/s). The calculations were performed with a set pressure of 120 GPa, similar to the conditions at the core–mantle boundary. The shear velocity was calculated as the ratio $\sqrt{G/\rho}$, where ρ is the mass density and G an 'effective shear modulus'. G can be calculated from C_{ij} by using a standard averaging procedure described by Hill (1952).

	C_{11}	C_{22}	C_{33}	C_{12}	C_{13}	C_{23}	C_{44}	C_{55}	C_{66}	v_S
Perovskite	907	1157	1104	513	406	431	364	271	333	7.636
Post-perov.	1252	929	1233	414	325	478	277	266	408	7.783

The identification of the post-perovskite phase of MgSiO$_3$ and the explanation of the seismic anomalies of the D″ layer is only one among many examples where DFT calculations played an important role in the study of materials under extreme conditions. For additional examples of the uses of DFT in geophysics and planetary materials a recent review can be found in the book by Wentzcovitch and Stixrude (2010).

7
Vibrations of molecules and solids

In Chapters 4 and 5 we saw how to calculate the equilibrium structures of materials using DFT. In Chapter 6 we discussed how to study *static* deformations. It is natural at this point to take the theory one step further and consider *dynamic* deformations taking place at very small scales, i.e. atomic vibrations. Vibrations play an important role in many material properties, for example:

- o thermal conductivity,
- o thermal expansion,
- o electrical resistivity,
- o heat capacity,
- o optical absorption,
- o superconductivity,
- o thermopower, and
- o structural phase transitions.

In this chapter we discuss how to improve the theory of Chapter 4 by acknowledging that in real situations the nuclei are not clamped to their equilibrium positions. Most of the discussion will focus on solids at temperatures well below the melting point. Under these conditions the nuclei remain relatively close to their equilibrium positions, and the 'clamped nuclei approximation' of Section 2.4 can be used as the starting point of our discussion. Throughout this chapter we will rely on the approximation made in Section 4.2 that the nuclei can be described as *classical particles*. In Chapter 8 we will see how this picture changes when we take into account the quantum nature of the nuclei.

7.1   Heuristic notion of atomic vibrations

Before developing the formal theory of vibrations in molecules and solids (Section 7.2), it is useful to go through some key concepts using the simplest possible model system as an example. For consistency with Chapter 5 we consider once again the nitrogen molecule in its diamagnetic ground state.

In the gas phase each N_2 molecule has six degrees of freedom associated with the Cartesian coordinates of its two N atoms. These degrees of freedom can conveniently be separated into the motion of the centre of mass (three degrees), the azimuthal and polar rotations around a fixed reference axis (two degrees), and a motion along the N–N bond direction corresponding to the compression or elongation of the molecule (Goldstein, 1950). Here we want to focus on this last degree of freedom; therefore we choose our reference frame with its centre in the middle of the N–N bond and with the $\mathbf{u}_x$ axis along the bond, as already done in Exercises 4.4 and 4.5 for H_2^+ and H_2, respectively. Following the convention of Figure 7.1, the instantaneous positions of the two N atoms are denoted by x_1 and x_2 and the corresponding bond length is $d = x_2 - x_1$. By using the adiabatic approximation of Section 4.1, and the approximation of Section 4.2 that the nuclei can be considered as classical particles, the dynamics of the N–N distance, d, in the N_2 dimer is described by Newton's equations (eqn 4.12):

$$M_I \frac{d^2 \mathbf{R}_I}{dt^2} = -\frac{\partial U}{\partial \mathbf{R}_I}. \tag{7.1}$$

In order to avoid conflicting notation in this section we will use df/dt or $\dot{f}$ in order to denote the time derivative of a function f. When we specify eqn 7.1 for the N_2 under consideration we find:

$$M_N \frac{d^2 x_1}{dt^2} = -\frac{\partial U}{\partial x_1},$$

$$M_N \frac{d^2 x_2}{dt^2} = -\frac{\partial U}{\partial x_2},$$

with M_N the nitrogen mass. The law for the time evolution of the bond length, d, is obtained by taking the difference between the previous two equations:

$$M_N \ddot{d} = M_N \frac{d^2}{dt^2}(x_2 - x_1) = -\left(\frac{\partial U}{\partial x_2} - \frac{\partial U}{\partial x_1}\right) = -2\frac{\partial U}{\partial d},$$

where the partial derivatives have been expressed in terms of $\pm\partial U/\partial d$ using the chain rule. This result indicates that the time evolution of the N–N bond length corresponds to the classical motion of a fictitious particle of mass $M_N/2$ (i.e. the reduced mass of the N_2 molecule) in the potential $U(d)$:

$$\frac{M_N}{2}\ddot{d} = -\frac{\partial U}{\partial d}. \tag{7.2}$$

The potential U is given by eqn 4.9:

$$U(d) = 2E_N + E_b(d), \tag{7.3}$$

and can be extracted directly from Figure 5.1. This potential energy surface is shown in Figure 7.1.

The important observation that we make now is that, while the potential energy curve of Figure 7.1 spans several tens of eV, at room temperature our fictitious particle remains confined very near the bottom of the well. In fact, according to the *equipartition theorem* (Kittel, 1958), the average energy associated with the stretching degree of freedom of N_2 is of the order of $k_B T$, i.e. 26 meV at 300 K. As a consequence, when measured from the minimum of $U(d)$, the average 'elevation' of this particle along the vertical axis is very small on the scale of the left plot, and the time evolution of d will result in only very small changes of the bond length with respect to its equilibrium value, of the order of a few percent. This is illustrated in the close-up in Figure 7.1.

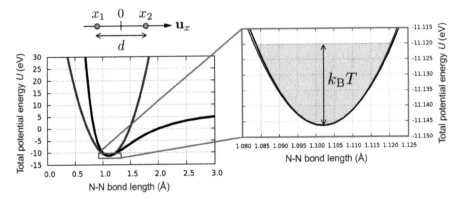

Fig. 7.1 Total potential energy, U, of the N_2 molecule as a function of N–N separation, d (black lines). The data have been obtained using eqn 7.3 and the binding energies calculated within DFT/LDA in Figure 5.1. The panel on the right is a close-up of the plot on the left. We note that the vertical scale in the left plot is of the order of tens of eV, while the scale on the right plot is of the order of tens of meV. The region in orange indicates the energies and bond lengths accessible at room temperature, as estimated from the equipartition theorem. The convention used for the reference frame is shown on the top left. The blue curves correspond to a parabolic approximation of the potential well near the bottom, from eqn 7.4.

Exercise 7.1 The following table reports the binding energy, E_b, of the N_2 molecule as a function of the N–N separation, d, as calculated in Figure 5.1 within DFT/LDA.

| d | E_b | d | E_b | d | E_b |
Å	meV	Å	meV	Å	meV
1.0800	-11107.571	1.0950	-11142.636	1.1100	-11141.409
1.0825	-11116.123	1.0975	-11144.840	1.1125	-11137.981
1.0850	-11123.549	1.1000	-11146.045	1.1150	-11133.639
1.0875	-11129.899	1.1025	-11146.289	1.1175	-11128.462
1.0900	-11135.186	1.1050	-11145.588	1.1200	-11122.443
1.0925	-11139.423	1.1075	-11143.955	1.1225	-11115.586

▶Determine the optimal fourth-order approximation of these data using the expression below:

$$E_b(d) = E_0 + \frac{K}{2}(d - d_0)^2 + \frac{K_3}{6}(d - d_0)^3 + \frac{K_4}{24}(d - d_0)^4.$$

The fitting parameters are K, K_3 and K_4, while d_0 and E_0 can be read directly from Figure 5.1. The fitting can be performed for example using the command 'fit' of the program 'Gnuplot'. As a sanity check, the quadratic constant should be $1.52 \cdot 10^5$ meV/Å^2. ▶By using the quartic expansion just obtained determine the smallest and the largest N–N bond lengths which can be accessed by the molecule at room temperature. We can assume that the stretching motion of the molecule is associated with a thermal energy $k_B T$. ▶Calculate the maximum relative deviation from the equilibrium bond length of N_2 at room temperature. ▶Show that, when the molecule is stretched by 1.6%, the quadratic, cubic and quartic components of the binding energy are approximately in the ratios $1000:46:3$. ▶Propose a simple approximation to the E_b vs. d curve of N_2 for studying properties at room temperature.

While Figure 7.1 and Exercise 7.1 are specific to the case of N_2, the main concepts hold true quite generally. Indeed, it turns out that in many cases of practical interest it is possible to replace the potential energy surface of a molecule or a solid with its parabolic approximation around the equilibrium configuration. Within this approximation, the potential energy, U, of Figure 7.1 can be replaced by the following simple expression:

$$U(d) = U_0 + \frac{1}{2}K(d - d_0)^2, \qquad (7.4)$$

where $U_0 = U(d_0)$ represents the bottom of the potential well. The replacement of the complete potential energy surface by its parabolic approximation is referred to as the *harmonic approximation* (Ashcroft and Mermin, 1976). By combining eqns 7.2 and 7.4 we obtain the equation governing the time evolution of the N–N bond length in the harmonic approximation:

$$\ddot{d} = -\frac{K}{M_N/2}(d - d_0). \qquad (7.5)$$

We can recognize in eqn 7.5 the equation of motion of a spring or mass $M_N/2$ and stiffness constant K. From eqn 7.4 the stiffness constant is formally obtained as:

$$K = \frac{\partial^2 U}{\partial d^2}. \qquad (7.6)$$

Exercise 7.2 ▶Verify that

$$d(t) = d_0 + [d(0) - d_0]\cos(\omega_{N_2}t) + \frac{\dot{d}(0)}{\omega_{N_2}}\sin(\omega_{N_2}t)$$

is a solution of eqn 7.5, with the resonant frequency given by $\omega_{N_2} = (2K/M_N)^{1/2}$. ▶Calculate the frequency ω_{N_2} of oscillation of the N_2 bond length and compare your result with the experimental value $\hbar\omega_{N_2} =288.9$ meV reported by Bendtsen (1974). ☐Experimental techniques for measuring vibrational frequencies will be discussed in Chapter 8. ▶Calculate the period of oscillation of the bond length of N_2.

In Exercise 7.2 we have seen that the vibrational frequency of the N_2 molecule calculated using DFT/LDA agrees with the experimental value to within 4%. Extensive tests show that DFT is able to predict molecular vibrational frequencies with good accuracy, typically within 5% (Sinnokrot and Sherrill, 2001).

At this point we may wonder what would happen if instead of the parabolic potential of eqn 7.4 we used a more refined approximation, for example by retaining the cubic term $K_3(d-d_0)^3/6$ of Exercise 7.1. Terms like this one and higher powers of $(d-d_0)$ are referred to as *anharmonic* contributions. Typically anharmonic contributions are very small and therefore are neglected in a first approximation. However, from Figure 7.1 we can see that, as we increase the temperature, the energy window accessible to the molecule becomes wider, hence the shape of the potential effectively probed by the molecule may deviate substantially from its parabolic approximation. In these cases it is important to carry out a careful analysis of anharmonicity. This aspect will be investigated in Exercise 8.2.

As a last point we note that in this section we completely ignored the rotation of the N_2 molecule. In reality it is intuitive that at room temperature the N_2 molecules in a gas will rotate as a consequence of random collisions. This rotation introduces a centrifugal force which can be taken into account by adding a term $L^2/(M_N d^2)$ to the effective one-dimensional potential, $U(d)$, of eqn 7.4, with L denoting the angular momentum mentioned in Section 5.1. In Exercise 7.3 it is shown that this effect is very small, leading to a change of the stretching frequency of less than 0.05%. This observation provides an a *posteriori* justification to our choice of ignoring rotational effects throughout this section. A detailed and elegant theoretical analysis of the motion of diatomic molecules within classical mechanics can be found in the book by Goldstein (1950), and the corresponding extension to include the quantum nature of the nuclei can be found in the book by Bransden and Joachain (1983).

Exercise 7.3 A complete description of the dynamics of diatomic molecules requires taking into account rotational degrees of freedom in addition to the stretching motion along the molecular bond. In absence of external forces the total angular momentum, L, of the two N atoms is conserved, and the atoms rotate in a plane perpendicular to this momentum. Under these conditions Goldstein (1950) shows that the one-dimensional potential of eqn 7.4 must be modified in order to take into account the centrifugal force, as follows:

$$U'(d) = U(d) + \frac{L^2}{M_N d^2}.$$

▶Using a Taylor expansion around $d = d_0$ rewrite this effective potential as $U_0' + K'(d-d_0')^2/2$ and show that the modified stiffness constant is:

$$K' = K + \frac{6L^2}{M_N d_0^4}.$$

▶From the equipartition theorem we know that the energy $L^2/2I$ associated with the rotational degree of freedom is of the order of $k_B T$. Using the expression for the moment of inertia of N_2 given in Section 5.1 show that:

$$\frac{K'}{K} = 1 + 6\frac{k_B T}{K d_0^2}.$$

▶Calculate the new stretching frequency which is obtained using the stiffness constant K' when the molecules are at room temperature, and compare your result with the value obtained in Exercise 7.2.

7.2   Formal theory of vibrations for classical nuclei

In the previous section we have seen that atomic vibrations involve very small displacements from the equilibrium positions of the nuclei. In the most general case of a solid we can formalise this observation by expressing the time-dependent position of each nucleus, $\mathbf{R}_I(t)$, as the sum of the equilibrium position, $\mathbf{R}_I^0$, which does not depend on time, and the displacement, $\mathbf{u}_I(t)$:

$$\mathbf{R}_I(t) = \mathbf{R}_I^0 + \mathbf{u}_I(t). \tag{7.7}$$

This decomposition is shown schematically in Figure 7.2. Following the same steps as in Section 7.1 we make use of the adiabatic approximation (Section 4.1) and the

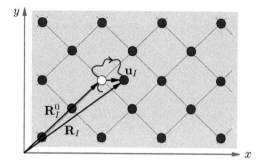

Fig. 7.2 Schematic representation of a crystalline solid (in two dimensions for clarity), with the atoms indicated as dark discs. At any given time the position of the nucleus I can be written as $\mathbf{R}_I = \mathbf{R}_I^0 + \mathbf{u}_I$, where $\mathbf{R}_I^0$ is the equilibrium coordinate and is independent of time.

classical approximation for the nuclei (Section 4.2). These choices allow us to write the equation of motion for each atomic displacement using Newton's equations (eqn 4.12):

$$M_I \ddot{\mathbf{u}}_I = -\frac{\partial U}{\partial \mathbf{u}_I}, \tag{7.8}$$

where all the partial derivatives are evaluated at the equilibrium positions, i.e. $\mathbf{u}_1 = \mathbf{u}_2 = \cdots = \mathbf{u}_M = 0$. In the most general case the total potential energy U is given by eqn 4.9, and can be calculated using DFT as discussed in Chapters 3 and 4.

In analogy with eqn 7.4, we can perform a Taylor expansion of the total energy, U, up to the second order in the atomic displacements, $\mathbf{u}_I$:

$$U(\mathbf{R}_1, \ldots, \mathbf{R}_M) = U(\mathbf{R}_1^0, \ldots, \mathbf{R}_M^0) + u_{I\alpha} \frac{\partial U}{\partial R_{I\alpha}} + \frac{1}{2} \frac{\partial^2 U}{\partial R_{I\alpha} \partial R_{J\beta}} u_{I\alpha} u_{J\beta}. \tag{7.9}$$

Here $R_{I\alpha}$ denotes the coordinate of the I-th nucleus along Cartesian direction α. In this expansion we used the Einstein convention for the summation over repeated indices (see Chapter 6), and all the derivatives are evaluated at the equilibrium coordinates, $\mathbf{R}_1^0, \ldots, \mathbf{R}_M^0$. In order to simplify the notation we can call U_0 the total energy corresponding to the nuclei in their equilibrium positions, and express the first derivatives of U in terms of the atomic forces using eqn 4.13:

$$U(\mathbf{R}_1, \ldots, \mathbf{R}_M) = U_0 - F_{I\alpha} u_{I\alpha} + \frac{1}{2} \frac{\partial^2 U}{\partial R_{I\alpha} \partial R_{J\beta}} u_{I\alpha} u_{J\beta}. \tag{7.10}$$

The second partial derivatives of the total energies with respect to the nuclear coordinates, evaluated at the equilibrium positions, are referred to as the *Born–von Karman force constants* (Born and Huang, 1954; Pick *et al.*, 1970):

$$K_{I\alpha, J\beta} = \frac{\partial^2 U}{\partial R_{I\alpha} \partial R_{J\beta}}. \tag{7.11}$$

As implied by their name, the force constants are independent of time, since they are evaluated at the equilibrium positions of the nuclei. Now we observe that the equilibrium configuration corresponds to the situation where the force on each nucleus vanishes, i.e. $\mathbf{F}_1 = \cdots = \mathbf{F}_M = 0$; therefore we can combine eqns 7.10 and 7.11 to obtain:

$$U = U_0 + \frac{1}{2} K_{I\alpha,J\beta}\, u_{I\alpha}\, u_{J\beta}. \tag{7.12}$$

Apart from the slightly more complicated notation, we can recognize in this expression an extension of eqn 7.4 to the case of systems with many atoms. Equation 7.12 defines the *harmonic approximation* and represents the starting point of any discussions of vibrations in solids, from steels to plastics.

At this point we can continue the analogy with Section 7.1 and derive the equation of motion for the atomic displacements. In Exercise 7.4 it is shown how to combine eqns 7.8 and 7.12 in order to obtain Newton's equations for the nuclei in the harmonic approximation:

$$M_I \ddot{u}_{I\alpha} = -K_{I\alpha,J\beta}\, u_{J\beta}. \tag{7.13}$$

Exercise 7.4 In this exercise we want to derive eqn 7.13 starting from eqns 7.8 and 7.12. The algebra is similar to that used in Exercise 6.2. ▶Show that, within the harmonic approximation, the partial derivative of the total energy, U, with respect to the displacement variable $u_{I'\alpha'}$ is given by:

$$\frac{\partial U}{\partial u_{I'\alpha'}} = \frac{1}{2} K_{I\alpha,J\beta} \left(\frac{\partial u_{I\alpha}}{\partial u_{I'\alpha'}} u_{J\beta} + u_{I\alpha} \frac{\partial u_{J\beta}}{\partial u_{I'\alpha'}} \right).$$

▶After expressing the partial derivatives within the parenthesis in terms of Kronecker deltas, show that:

$$\frac{\partial U}{\partial u_{I'\alpha'}} = \frac{1}{2} \left(K_{I'\alpha',J\beta} + K_{J\beta,I'\alpha'} \right) u_{J\beta}.$$

▶Using the definition in eqn 7.11, prove that the matrix of Born–von Karman force constants is symmetric with respect to the exchange of its indices $(I\alpha)$ and $(J\beta)$:

$$K_{I\alpha,J\beta} = K_{J\beta,I\alpha}.$$

▶Combine the previous two results with eqn 7.8 in order to obtain eqn 7.13.

In order to derive a formal solution of eqn 7.13 it is convenient to push the similarity with eqn 7.5 a bit further. To this end, we divide both sides of eqn 7.13 by $M_I^{1/2}$, and we introduce the factors $M_J^{1/2}$ and $M_J^{-1/2}$ for each term of the sum on the right-hand side:

$$M_I^{\frac{1}{2}} \ddot{u}_{I\alpha} = -\frac{K_{I\alpha,J\beta}}{(M_I M_J)^{\frac{1}{2}}} M_J^{\frac{1}{2}} u_{J\beta}.$$

By defining the *mass-weighted displacements* $v_{I\alpha}$ as:

$$v_{I\alpha} = M_I^{\frac{1}{2}} u_{I\alpha}, \tag{7.14}$$

the previous expression becomes:

$$\ddot{v}_{I\alpha} = -\frac{K_{I\alpha,J\beta}}{(M_I M_J)^{\frac{1}{2}}} v_{J\beta}. \tag{7.15}$$

Now the analogy with eqn 7.5 should be evident. This equation describes the motion of a set of $3M$ springs interconnected by the Born–von Karman force constants. The notation has become more involved than in the previous section only because eqn 7.15 describes M atoms instead of a diatomic molecule. The matrix appearing in eqn 7.15,

$$D_{I\alpha,J\beta} = \frac{K_{I\alpha,J\beta}}{(M_I M_J)^{\frac{1}{2}}}, \tag{7.16}$$

is referred to as the *dynamical matrix*. We note that in the case of crystalline solids the term 'dynamical matrix' carries a slightly different meaning, as will be discussed in Section 7.4. Using this matrix the equation of motion for the mass-weighted displacements acquires the simple form:

$$\ddot{v}_{I\alpha} = -D_{I\alpha,J\beta}\, v_{J\beta}. \tag{7.17}$$

The dynamical matrix $\mathbb{D} = D_{I\alpha,J\beta}$ has $3M \times 3M$ entries in the combined indices $I\alpha$ and $J\beta$. Since these combined indices identify $3M$ degrees of freedom (i.e. three Cartesian coordinates for each of the M nuclei), it is sometimes convenient to rearrange them into a one-dimensional index, ν, as follows:

	nucleus 1			nucleus 2				nucleus M			
$I\alpha$	1,1	1,2	1,3	2,1	2,2	2,3	...	M,1	M,2	M,3	(7.18)
ν	1	2	3	4	5	6	...	$3M-2$	$3M-1$	$3M$.	

In the following we will make use of $I\alpha$ or ν interchangeably according to this convention.

Equation 7.17 represents a coupled linear system of second-order differential equations, and its solution can be found by moving to a new set of coordinates where the dynamical matrix has a diagonal form. To this end, we first rewrite eqn 7.17 using the more concise matrix notation:

$$\frac{d^2\mathbf{v}}{dt^2} = -\mathbb{D}\,\mathbf{v}, \tag{7.19}$$

where $\mathbf{v}$ is the $(3M)$-dimensional vector with components $v_{I\alpha}$ or equivalently v_ν. Second, we diagonalize the dynamical matrix by solving the secular equation:

$$\det\left(\mathbb{D} - \omega^2 \mathbb{1}\right) = 0, \tag{7.20}$$

where $\mathbb{1}$ indicates the identity matrix of size $3M \times 3M$. Since by construction $\mathbb{D}$ is a real and symmetric matrix (see Exercise 7.4), it follows that (i) it can be diagonalized by an orthogonal matrix, i.e. a matrix $\mathbb{E}$ such that $\mathbb{E}^{\mathsf{T}}\mathbb{E} = \mathbb{1}$ (T denoting the matrix transpose), and (ii) its eigenvalues are real numbers (Chow, 2000). The columns of the matrix $\mathbb{E}$ are referred to as the *mass-weighted normal modes of vibration*, or simply vibrational eigenmodes. The corresponding eigenvalues are traditionally indicated by ω_ν^2 and their square roots, ω_ν, are called the *normal frequencies of vibration* or vibrational eigenfrequencies. If we indicate by Ω the diagonal matrix having the eigenfrequencies on its diagonal, our eigenvalue problem can be written in compact form as follows:

$$\mathbb{D}\mathbb{E} = \mathbb{E}\Omega^2. \tag{7.21}$$

Using this concise notation we can express $\mathbb{D}$ as $\mathbb{E}\Omega^2 \mathbb{E}^\top$ and substitute inside eqn 7.19 to find:

$$\frac{d^2}{dt^2}\left(\mathbb{E}^\top \mathbf{v}\right) = -\Omega^2\left(\mathbb{E}^\top \mathbf{v}\right). \tag{7.22}$$

Finally, we define the $(3M)$-dimensional vector, $\mathbf{w}$, as:

$$\mathbf{w} = \mathbb{E}^\top \mathbf{v}, \tag{7.23}$$

so that eqn 7.22 simplifies to:

$$\ddot{\mathbf{w}} = -\Omega^2\,\mathbf{w}, \quad \text{or equivalently:} \quad \ddot{w}_\nu = -\omega_\nu^2\,w_\nu \text{ with } \nu = 1, \ldots, 3M. \tag{7.24}$$

This last result means that the coupled linear system in eqn 7.17 has been split into a set of $3M$ completely independent equations. From eqn 7.24 we see that each coordinate $w_\nu(t)$ behaves as a simple harmonic oscillator; therefore, the solution of each differential equation will be the sum of terms $\cos(\omega_\nu t)$ and $\sin(\omega_\nu t)$, precisely as in Exercise 7.2.

The price to pay for this simplification is that the new coordinates, $w_\nu(t)$, do not represent the displacements of individual nuclei, but correspond to some kind of *collective displacements* of all the nuclei. The relation between these collective displacements and the original nuclear displacements, $u_{I\alpha}(t)$, in eqn 7.13 is found by combining together eqns 7.14 and 7.23:

$$u_{I\alpha}(t) = M_I^{-\frac{1}{2}}\sum_\nu E_{I\alpha,\nu}[A_\nu \cos(\omega_\nu t) + B_\nu \sin(\omega_\nu t)], \tag{7.25}$$

with $E_{I\alpha,\nu}$ the matrix elements of $\mathbb{E}$, and the constants A_ν, B_ν to be determined using the initial conditions on the nuclear positions and velocities.

Exercise 7.5 ▶By evaluating the displacements as well as their first derivative in eqn 7.25 for $t = 0$, show that the constants A_ν and B_ν are related to the initial nuclear displacements, $u_{J\beta}(0)$, and velocities, $\dot{u}_{J\beta}(0)$, as follows:

$$A_\nu = \sum_{J\beta} M_J^{\frac{1}{2}} E_{\nu,J\beta}^\top u_{J\beta}(0), \qquad B_\nu = \omega_\nu^{-1}\sum_{J\beta} M_J^{\frac{1}{2}} E_{\nu,J\beta}^\top \dot{u}_{J\beta}(0).$$

▶Combine this result with eqn 7.25 in order to show that a complete formal solution of eqn 7.13 inclusive of initial conditions is given by:

$$u_{I\alpha}(t) = \sum_{J\beta}\sqrt{\frac{M_J}{M_I}}\sum_\nu E_{I\alpha,\nu}E_{\nu,J\beta}^\top\left[u_{J\beta}(0)\cos(\omega_\nu t) + \frac{\dot{u}_{J\beta}(0)}{\omega_\nu}\sin(\omega_\nu t)\right].$$

This result shows that the shape of the solution of Newton's equations for the nuclei in the harmonic approximation is similar to what we already found for the N_2 molecule in Exercise 7.2. Here the situation is more complicated since the motion of each nucleus 'contains' many harmonic oscillators at once.

7.3   Calculations of vibrational eigenmodes and eigenfrequencies

In the previous section we established the general theoretical framework for studying vibrations in molecules and solids; however, nothing was said about how to proceed with a real calculation. The starting point of any atomistic calculation of vibrations is the numerical evaluation of the dynamical matrix in eqn 7.16. For convenience we rewrite this matrix by showing explicitly the derivatives of the total energy (see eqn 7.11):

$$D_{I\alpha,J\beta} = (M_I M_J)^{-\frac{1}{2}} \frac{\partial^2 U}{\partial R_{I\alpha} \partial R_{J\beta}}. \tag{7.26}$$

Since it is possible to compute the total potential energy, U, of eqn 4.9 using DFT at *any given* nuclear configuration (cf. Chapters 3 and 4), it is natural to try and evaluate the partial derivatives in eqn 7.26 using finite-difference formulas. Similarly to what we already have done for eqn 4.44 of Chapter 4 we can write (Kiusalaas, 2005):

$$\frac{\partial^2 U}{\partial R_{J\beta}^2} = \frac{U(R_{J\beta}^0 + u) - 2U_0 + U(R_{J\beta}^0 - u)}{u^2} + \mathcal{O}(u^2), \tag{7.27}$$

where $U(R_{J\beta}^0 \pm u)$ is a shorthand for the total energy calculated when all the atoms are in their equilibrium positions, $\mathbf{R}_I^0$, except for atom J, which has been displaced along the direction β by a small amount, $\pm u$. The term $\mathcal{O}(u^2)$ indicates that the difference between the exact derivative on the left side and its approximation by finite differences on the right is of the order of u^2.

The finite-differences expression in eqn 7.27 shows that the second derivative of U with respect to the same variable requires the evaluation of the total energy at two different nuclear configurations, plus one calculation of the energy at equilibrium, U_0.

Exercise 7.6 We want to establish how many evaluations of the total energy, U, are required in order to calculate each one of the mixed partial derivatives appearing in the dynamical matrix of eqn 7.26. In order to simplify the notation we consider a general function $f(x,y)$ and its mixed second-order partial derivative, $\partial^2 f/\partial x \partial y$, around $x=y=0$. ▶By performing two successive Taylor expansions of $f(u,0)$ and $f(u,v)$ for small u and v show that:

$$f(u,v) = f(0,0) + \left(\frac{\partial f}{\partial x}u + \frac{\partial f}{\partial y}v\right) + \frac{1}{2}\left(u^2\frac{\partial^2 f}{\partial x^2} + 2uv\frac{\partial^2 f}{\partial x \partial y} + v^2\frac{\partial^2 f}{\partial y^2}\right) + \mathcal{O}(u^3 + u^2v + uv^2 + v^3)$$

(note that this derivation is rather lengthy). □We now consider the values of function f at the four corners $(\pm u, \pm u)$ of a square of size $2u$ centred at $x=y=0$. ▶Using the previous result show that:

$$f(u,u) + f(-u,-u) - f(u,-u) - f(-u,u) = 4u^2\frac{\partial^2 f}{\partial x \partial y} + \mathcal{O}(u^4).$$

This equation indicates that the mixed second-order partial derivative of $f(x,y)$ can be obtained by finite differences as follows:

$$\frac{\partial^2 f}{\partial x \partial y} = \frac{f(u,u) + f(-u,-u) - f(u,-u) - f(-u,u)}{4u^2},$$

with an error of the order $\mathcal{O}(u^2)$. Therefore the calculation of each mixed derivative in eqn 7.26 requires the evaluation of the total energy, U, at four different nuclear configurations.

From Exercise 7.6 we know that each of the $3M(3M-1)/2$ inequivalent mixed partial derivatives appearing in eqn 7.26 requires the evaluation of the total energy for four different nuclear configurations. Equation 7.27 tells us that the $3M$ second derivatives with respect to the same variable require each two evaluations of U, plus one calculation of the energy at equilibrium, U_0. By adding up these figures we discover that the calculation of the complete dynamical matrix requires the evaluation of the total energy for a total of $18M^2+1$ configurations.

In order to understand what this estimate means in practice, we consider the model silicon surface in Figure 5.4. This model contains 60 atoms, and one evaluation of the total energy, U, requires approximately 1,000 seconds on a parallel computing cluster with 32 cores. When we multiply this figure by $18 \cdot 60^2 + 1$ we obtain a total execution time of 750 days. Obviously this kind of calculation would be highly impractical.

The solution to this impasse is provided by the Hellmann–Feynman theorem discussed in Section 4.3. In fact, by rewriting the second derivative of the total energy appearing in eqn 7.26 using the expression for the atomic forces in eqn 4.13 we find immediately:

$$D_{I\alpha,J\beta} = -(M_I M_J)^{-\frac{1}{2}} \frac{\partial F_{I\alpha}}{\partial R_{J\beta}}.$$

The first-order derivatives of the forces can be evaluated numerically using finite differences as follows:

$$D_{I\alpha,J\beta} = -(M_I M_J)^{-\frac{1}{2}} \frac{F_{I\alpha}(R^0_{J\beta}+u) - F_{I\alpha}(R^0_{J\beta}-u)}{2u} + \mathcal{O}(u^2). \qquad (7.28)$$

Similarly to eqn 7.27, here $F_{I\alpha}(R^0_{J\beta} \pm u)$ is a shorthand for the force on nucleus I along the direction α when all the nuclei are in their equilibrium positions except for J, which has been displaced along the direction β by a small amount, $\pm u$. Since for a given nuclear configuration all the forces can be calculated in one go using the Hellmann–Feynman theorem expressed by eqn 4.36, this new formulation requires DFT calculations for only $2 \times 3M$ nuclear configurations. Going back to the above example of the silicon surface, the use of eqn 7.28 allows us to calculate the entire dynamical matrix on 32 cores in four days, thereby affording us a tremendous saving in computer time.

Exercise 7.7 In this exercise we calculate the dynamical matrix of a very simple system, namely one molecule of water (H_2O). This choice allows us to put into practice all the concepts presented in Sections 7.2 and 7.3 while keeping the numerical calculations to a minimum. Any other system can be studied using the same procedure outlined here, the only difference being that in most DFT calculations there will be tens or hundreds of atoms.

Water is essentially everywhere on the Earth's surface, represents approximately two-thirds of the human body mass and is key to biological processes. Despite the molecule's simplicity, within the context of first-principles calculations the study of water is still an active field of research (Grossman *et al.*, 2004; Prendergast and Galli, 2006; Morrone and Car, 2008). A DFT/LDA calculation of the water molecule yields an equilibrium geometry where the O–H bond length is $d = 0.97$ Å and the H–O–H angle is $2\gamma = 105.5°$, in agreement with experiment (Darling and Dennison, 1940). The calculated atomic coordinates at equilibrium are O: $(0,0,0)$; H(1): $d(\sin\gamma/\sqrt{2}, \sin\gamma/\sqrt{2}, \cos\gamma)$; H(2): $d(-\sin\gamma/\sqrt{2}, -\sin\gamma/\sqrt{2}, \cos\gamma)$. The following table reports the forces, F_x, F_y and F_z, on the three nuclei calculated for 18 different configurations. In each configuration two nuclei are in their equilibrium positions, and one nucleus has been displaced from equilibrium along a given direction. For example, the label 'O dis. by $+0.01d\mathbf{u}_x$' means that in this configuration the O atom has been displaced from its equilibrium position by $+0.01d$ along the x axis. All the forces are in units of meV/Å.

	F_x	F_y	F_z	F_x	F_y	F_z	F_x	F_y	F_z
	O dis. by $+0.01d\mathbf{u}_x$			O dis. by $+0.01d\mathbf{u}_y$			O dis. by $+0.01d\mathbf{u}_z$		
O	-307.1	-307.2	-7.7	-307.2	-307.1	-7.7	0.0	0.0	-390.1
H(1)	154.2	157.0	169.2	157.0	154.2	169.2	131.0	131.0	195.1
H(2)	152.8	150.2	-161.5	150.2	152.8	-161.5	-131.0	-131.0	195.1
	H(1) dis. by $+0.01d\mathbf{u}_x$			H(1) dis. by $+0.01d\mathbf{u}_y$			H(1) dis. by $+0.01d\mathbf{u}_z$		
O	153.3	150.9	122.0	150.9	153.3	122.0	161.9	161.9	189.3
H(1)	-165.7	-163.1	-142.2	-163.1	-165.7	-142.2	-142.7	-142.7	-183.1
H(2)	12.4	12.1	20.2	12.1	12.4	20.2	-19.3	-19.3	-6.2
	H(2) dis. by $+0.01d\mathbf{u}_x$			H(2) dis. by $+0.01d\mathbf{u}_y$			H(2) dis. by $+0.01d\mathbf{u}_z$		
O	153.8	156.3	-132.0	156.3	153.8	-132.0	-161.9	-161.9	189.3
H(1)	13.4	13.6	-18.1	13.6	13.4	-18.1	19.3	19.3	-6.2
H(2)	-167.1	-169.9	150.1	-169.9	-167.1	150.1	142.7	142.7	-183.1
	O dis. by $-0.01d\mathbf{u}_x$			O dis. by $-0.01d\mathbf{u}_y$			O dis. by $-0.01d\mathbf{u}_z$		
O	307.1	307.2	-7.7	307.2	307.1	-7.7	0.0	0.0	380.8
H(1)	-152.8	-150.2	-161.5	-150.2	-152.8	-161.5	-123.8	-123.8	-190.4
H(2)	-154.2	-157.0	169.2	-157.0	-154.2	169.2	123.8	123.8	-190.4
	H(1) dis. by $-0.01d\mathbf{u}_x$			H(1) dis. by $-0.01d\mathbf{u}_y$			H(1) dis. by $-0.01d\mathbf{u}_z$		
O	-153.8	-156.3	-132.0	-156.3	-153.8	-132.0	-168.7	-168.7	-196.0
H(1)	167.1	169.9	150.1	169.9	167.1	150.1	150.1	150.1	187.5
H(2)	-13.4	-13.6	-18.1	-13.6	-13.4	-18.1	18.7	18.7	8.5
	H(2) dis. by $-0.01d\mathbf{u}_x$			H(2) dis. by $-0.01d\mathbf{u}_y$			H(2) dis. by $-0.01d\mathbf{u}_z$		
O	-153.3	-150.9	122.0	-150.9	-153.3	122.0	168.7	168.7	-196.0
H(1)	-12.4	-12.1	20.2	-12.1	-12.4	20.2	-18.7	-18.7	8.5
H(2)	165.7	163.1	-142.2	163.1	165.7	-142.2	-150.1	-150.1	187.5

► Calculate the matrix $K_{I\alpha,J\beta}$ of Born–von Karman force constants for the water molecule, using eqns 7.11, 7.28 and the index convention set by eqn 7.18. □ Owing to numerical errors in the calculation, it is typically the case that the calculated matrix of force constants is not symmetric, as we would expect on theoretical grounds (see Exercise 7.4). In these

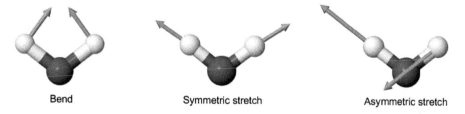

Bend   Symmetric stretch   Asymmetric stretch

Fig. 7.3 Ball-and-stick representation of the displacement patterns corresponding to the vibrational eigenmodes of the water molecule. The O and H atoms are in grey and white, respectively, and the arrows indicate the directions and relative amplitude of the displacements. The amplitude of the displacements has been exaggerated for clarity: in reality the displacements are of the order of 0.01 Å. The displacements of the O atom are much smaller than those of the H atoms and are not shown.

cases it is convenient to 'symmetrize' the force constants in order to reduce numerical noise. ▶Symmetrize the matrix of force constants by replacing it with $(K + K^\top)/2$. ▢More generally it is possible to further reduce numerical noise by imposing additional constraints, as discussed by Ackland *et al.* (1997). ▶Using the force constants just determined, calculate the dynamical matrix of the water molecule. Show that this matrix is given by (only the upper triangular part is shown owing to the symmetry):

$$\hbar^2 \mathbb{D} = \begin{pmatrix} 82 & 82 & 0 & -164 & -164 & -176 & -164 & -164 & 176 \\ & 82 & 0 & -164 & -164 & -176 & -164 & -164 & 176 \\ & & 103 & -136 & -136 & -206 & 136 & 136 & -206 \\ & & & 710 & 711 & 625 & -55 & -55 & -81 \\ & & & & 710 & 625 & -55 & -55 & -81 \\ & & & & & 791 & 81 & 81 & 31 \\ & & & & & & 710 & 711 & -625 \\ & & & & & & & 710 & -625 \\ & & & & & & & & 791 \end{pmatrix} \cdot 100\,\text{meV}^2 .$$

Exercise 7.8 ▶Solve the secular equation (eqn 7.20) for the dynamical matrix of water obtained in Exercise 7.7 and determine the vibrational eigenfrequencies.▢The lowest six eigenvalues of the dynamical matrix are to be ignored since they correspond to the six degrees of freedom associated with the rigid translations and the rigid rotations of the molecule (i.e. they are not vibrations). The proper vibrational eigenmodes correspond to the three largest eigenvalues. ▶Compare the vibrational eigenfrequencies of water calculated above with the experimental measurements by Darling and Dennison (1940) yielding $\hbar\omega$=198 meV, 453 meV and 466 meV, respectively. Determine the relative deviation between calculated and measured eigenfrequencies. ▶Calculate the period of oscillation of the water molecule for each of these eigenmodes. ▶Calculate the vibrational eigenmodes for the three highest eigenfrequencies and sketch the corresponding atomic displacements. Bear in mind that the eigenmodes contain mass factors as discussed in relation to eqn 7.14. ▶Sketch the atomic displacements for these eigenmodes and label each vibrational eigenfrequency by comparing your sketch with the displacements shown in Figure 7.3.

From Exercise 7.8 we see that the eigenfrequencies calculated entirely from first principles using DFT fall within 5% of the experimental measurements. This turns out to be a rather general trend and can be taken as an estimate of the error bar of DFT/LDA calculations for vibrations.

The procedure carried out in the two exercises above is very general and is not limited to simple molecules. It is standard practice to use the same procedure for calculating the vibrational eigenfrequencies and eigenmodes of materials ranging from large molecules to polymers and amorphous solids. The main difference between the calculations performed in Exercises 7.7 and 7.8 and those for more complex systems is in the number of atoms. When we deal with many atoms it becomes necessary to write small programs (e.g. shell scripts in Unix-type environments) to organize the various operations into an automated procedure.

The procedure described in this section is also valid for crystalline solids; however, in the case of crystals it is often more convenient to take advantage of the periodicity of the structure. This aspect is discussed in the following section.

7.4   Vibrations of crystalline solids

In the case of crystalline solids the theory developed in Sections 7.2 and 7.3 remains unchanged, but it is very advantageous to take into account the fact that the atoms are arranged in a periodic lattice. The classic textbooks by Kittel (1976) and Ashcroft and Mermin (1976) provide comprehensive introductions to the concepts of lattice vibrations in crystals, and should be consulted by readers who are not familiar with this topic. Here we take a point of view slightly different from the standard textbooks, in order to build on what we have learned in the previous sections.

In order to illustrate how the notion of periodicity modifies eqns 7.13 and 7.26 in the case of crystalline solids, we consider the ideal situation shown in Figure 7.4. In this situation the crystal is a one-dimensional chain of identical atoms. These atoms are equally spaced at equilibrium, and are allowed to move only along the direction of the chain, u_x. This is the simplest possible example for introducing vibrations in crystals (Kittel, 1976). Obviously this model is highly idealized; however, there are a few systems which fit this description, such as atomic wires of molybdenum or europium inside carbon nanotubes (Muramatsu et al., 2008; Kitaura et al., 2009). For this ideal system, since there is only one type of atom and all atoms move only along the chain, we indicate the nuclear masses by M_0 and we omit the Cartesian indices in the displacements and force constants. Equation 7.13 simplifies as follows:

$$M_0 \ddot{u}_I = - \sum_J K_{I,J} u_J, \qquad (7.29)$$

where we have suspended the Einstein convention in order to avoid ambiguity in the following.

We are all familiar with the intuitive notion that *sound waves* in solids are related to atomic vibrations. Bearing this in mind, it is natural to check whether eqn 7.29 admits solutions in the form of waves. The general equation describing a sound wave, $p(x, t)$, in one dimension is:

$$\frac{1}{v_s^2} \frac{\partial p}{\partial t^2} = \frac{\partial p}{\partial x^2}, \qquad (7.30)$$

with v_s the speed of sound. As can be verified immediately, this equation admits solutions of the form:

$$p(x, t) = p_0 \, e^{i(qx - \omega t)}, \qquad (7.31)$$

where p_0 is a constant, and the frequency, ω, is related to the wavevector, q, by $\omega = v_s q$. This solution represents a plane wave propagating at velocity v_s, as can be seen by rewriting it as $p(x, t) = p_0 \exp[iq(x - v_s t)]$. For a general discussion of the wave equation and its solution the reader is referred to the textbook by Chow (2000). The example of the sound wave prompts us to look for solutions of eqn 7.29 which take the form:

$$u_I(t) = u_0 \, e^{i(qR_I - \omega t)}, \qquad (7.32)$$

with u_0 a constant. In the remainder of this section we temporarily use R_I instead of R_I^0 to denote the equilibrium positions of the nuclei. This small change is only to

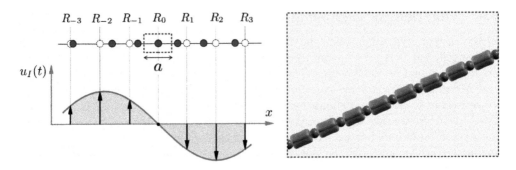

Fig. 7.4 Left: A one-dimensional chain of equally spaced atoms, i.e. the simplest model for studying vibrations in crystals. In this model the atoms (purple discs) are allowed to move only along the chain (x axis). There is one atom in each unit cell (indicated by the blue rectangle), and the unit cell size is a. It is convenient to label atoms according to their equilibrium position along the chain: $R_I = Ia$. For example, at equilibrium, the atom with label $I = -3$ is located at $x = -3a$. For simplicity we can assume that the chain is infinite; therefore I goes from $-\infty$ to $+\infty$. The diagram at the bottom shows the instantaneous displacements *along the chain* from the equilibrium positions (white discs).

Right: Hypothetical infinitely long cumulene. Cumulene is a long linear molecule of carbon atoms which is found within molecular clouds in interstellar space (Thaddeus and McCarthy, 2001). Small cumulene chains including up to eight C atoms have been produced in the laboratory using laser evaporation of graphite (Heath *et al.*, 1987). While infinitely long cumulene can be useful for visualizing the model crystal discussed in this section, this and similar one-dimensional systems with equally spaced atoms are unstable towards the formation of chains with short and long bonds in alternation (Peierls, 1955). The figure shows the carbon atoms in grey and the DFT electron density is shown as an isosurface. We can see that the electrons mostly concentrate in between adjacent C atoms.

maintain a reasonably light notation. By replacing the *trial solution* of eqn 7.32 inside eqn 7.29 we obtain:

$$M_0\omega^2 = \sum_J e^{-iq(R_I - R_J)} K_{I,J}, \tag{7.33}$$

where the sum over J runs from $-\infty$ to $+\infty$. If the nuclei are labelled sequentially using the convention indicated in Figure 7.4, then the *periodicity* of the crystal structure implies that the force constants are insensitive to rigid translations of the chain by integer multiples of the lattice parameter:

$$K_{I,J} = K_{I+L,J+L} \quad \text{for any integer } L. \tag{7.34}$$

This property allows us to simplify eqn 7.33 as follows:

$$M_0\omega^2 = \sum_J e^{-iqR_J} K_{0,J}.$$

On the right-hand side of this equation we can recognize the Fourier transform of the Born–von Karman force constants. We define:

$$D(q) = \frac{1}{M_0} \sum_J e^{-iqR_J} \frac{\partial^2 U}{\partial R_0 \partial R_J},$$

(7.35)

to obtain:

$$\omega^2 = D(q).$$

(7.36)

This result shows that for a given wavevector q, the trial function in eqn 7.32 is indeed a solution of the equation of motion (eqn 7.29), with a q-dependent frequency $\pm D(q)^{1/2}$. Since eqn 7.29 is a homogeneous linear differential equation, any linear combination of functions as in eqn 7.32 will also be a valid solution. However, many solutions are simply duplicates. In fact, if we write the quantity $D(q)$ from eqn 7.35 at $q + 2\pi L/a$ (with L an integer) we find:

$$D\left(q + \frac{2\pi}{a}L\right) = \frac{1}{M_0} \sum_J e^{-iqR_J} \frac{\partial^2 U}{\partial R_0 \partial R_J} e^{-i2\pi JL} = D(q).$$

Therefore, from eqn 7.36, the vibrational frequencies for wavevectors q and $q + 2\pi L/a$ do coincide. In addition, also the term $\exp(iqR_I)$ in eqn 7.32 remains unchanged if we use $q + 2\pi L/a$ instead of q. In practice only the solutions with wavevectors in the range $[-\pi/a, \pi/a]$ are inequivalent. In the language of solid-state physics this portion of the wavevector space is referred to as the first *Brillouin zone* (Kittel, 1976). A brief summary of the basic concepts relating to wavevectors and Brillouin zones in crystals is given in Appendix D. Taking into account this restriction we can now write the most general solution of eqn 7.29 as:

$$u_I(t) = \int_{-\pi/a}^{+\pi/a} dq \left\{ A_q e^{i[qR_I - \omega(q)t]} + B_q e^{i[qR_I + \omega(q)t]} \right\} \quad \text{with } \omega(q) = [D(q)]^{\frac{1}{2}}.$$

(7.37)

The constants A_q and B_q need to be determined using the initial conditions for the nuclear displacements and velocities, as we already have seen for eqn 7.25 of Section 7.2.

Exercise 7.9 ▶In the same spirit of Exercise 7.5, determine the constants A_q and B_q in eqn 7.37 using the initial conditions $u_I(0)$ and $\dot{u}_I(0)$. This derivation is rather lengthy and requires the use of an inverse Fourier transform. As a reference, the answers are:

$$A_q = \frac{a}{4\pi} \sum_I \left[u_I(0) + i\frac{\dot{u}_I(0)}{\omega(q)} \right] e^{-iqR_I} \quad \text{and} \quad B_q = \frac{a}{4\pi} \sum_I \left[u_I(0) - i\frac{\dot{u}_I(0)}{\omega(q)} \right] e^{-iqR_I}.$$

▶Show that the complete solutions of the equation of motion (eqn 7.29) including the initial conditions above are given by:

$$u_I(t) = \sum_J \int_0^{\pi/a} \frac{dq}{\pi/a} \left\{ u_J(0) \cos[\omega(q)t] + \frac{\dot{u}_J(0)}{\omega(q)} \sin[\omega(q)t] \right\} \cos[q(R_I - R_J)].$$

Note the similarity between these solutions and those obtained in Exercise 7.5.

Exercise 7.10 We want to determine the vibrational eigenfrequencies, $w(q)$, of the monoatomic linear chain using the simplest possible approximation. We start by combining eqns 7.35 and 7.36:

$$w^2(q) = \frac{1}{M_0} \sum_J e^{-iqR_J} \frac{\partial^2 U}{\partial R_0 \partial R_J}.$$

▶Using the translational invariance of the chain and eqn 4.13, show that this equation can be written more compactly as:

$$w^2(q) = -\frac{1}{M_0} \left[\frac{\partial F_0}{\partial R_0} + 2 \sum_{J>0} \frac{\partial F_J}{\partial R_0} \cos(qR_J) \right].$$

▶Using again translational invariance, and the fact that the centre of mass of the isolated chain is immobile, show that the following 'sum rule' must be satisfied:

$$\frac{\partial F_0}{\partial R_0} + 2 \sum_{J>0} \frac{\partial F_J}{\partial R_0} = 0. \qquad (7.38)$$

▶Combine the previous two equations to show that:

$$w^2(q) = \frac{2}{M_0} \sum_{J>0} \frac{\partial F_J}{\partial R_0} [1 - \cos(Jqa)]. \qquad (7.39)$$

☐The derivatives on the right-hand side can be expressed using forward finite differences, similarly to eqn 7.28:

$$w^2(q) \simeq \frac{2}{M_0 u_0} \sum_{J>0} F_J(u_0)[1 - \cos(Jqa)].$$

Here $F_J(u_0)$ indicates the force on nucleus J when all the nuclei are in their equilibrium positions, except nucleus $I = 0$, which is displaced by a very small amount, u_0, along the chain. ☐Intuitively we can expect that the force on the J-th nucleus, arising when the nucleus $I = 0$ is displaced from equilibrium, becomes smaller as the distance $|R_J - R_0|$ increases. We now make the *extreme* approximation that the forces $F_J(u_0)$ are non-negligible only for the nuclei closest to $I = 0$, i.e. the nearest-neighbours $J = \pm 1$. In this approximation eqn 7.38 yields, after replacing the derivatives by finite differences: $F_0(u_0) + 2F_1(u_0) = 0$. As a result the expression for $w^2(q)$ above simplifies as follows:

$$w(q) \simeq \Omega \left[1 - \cos(qa)\right]^{\frac{1}{2}}, \qquad \text{with} \qquad \Omega^2 = -\frac{F_0(u_0)}{M_0 u_0}. \qquad (7.40)$$

☐A DFT/LDA calculation of the equilibrium structure of a long cumulene chain yields a C–C bond length $d = 1.275$ Å. Since there is one atom per unit cell, the lattice parameter is $a = d$. When atom $I = 0$ at the centre of the reference frame is displaced from its equilibrium position by $0.01d$, the restoring force acting on it is calculated to be 1,174 meV. The actual calculation has been performed by simulating an infinite chain using 20 C atoms in the simulation box and periodic boundary conditions, similarly to Figure 5.4. ▶Estimate the highest vibrational eigenfrequency using eqn 7.40. ▶Plot the frequency, w, vs. the wavevector, q, for this chain within the first Brillouin zone using eqn 7.40, and compare your result with Figure 7.5. ☐The curve w vs. q is referred to as *frequency–wavevector dispersion relation*. The model dispersion provided by eqn 7.40 is identical to the standard spring–model discussed in the classic textbooks by Kittel (1976) and Ashcroft and Mermin (1976). While this model is useful for understanding the concept of vibrations in crystals and for deriving a simple dispersion

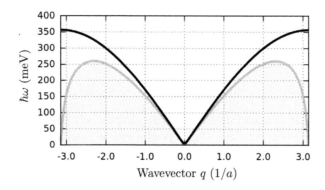

Fig. 7.5 Frequency (ω) vs. wavevector (q) dispersion relation in the one-dimensional monoatomic chain with atoms equally spaced. The black line corresponds to the model defined by eqn 7.40, where all the forces beyond the first neighbours of the displaced atom have been set to zero. The force on the displaced atom has been calculated for an infinite cumulene chain using DFT/LDA, as described in Exercise 7.10. For comparison the dispersion relation calculated using DFT/LDA using the complete expression in eqn 7.39 is shown in light grey. The latter calculation was performed using the techniques developed by Baroni *et al.* (2001). We see that the approximation of neglecting the forces beyond the first neighbours is not accurate here, and is meaningful only as a pedagogical expedient. As a curiosity, the dips of the grey curve at $q = \pm\pi/a$ can be interpreted as the signatures of a structural instability. This instability drives the formation of alternate short and long C–C bonds in cumulene (Cahangirov *et al.*, 2010).

relation, it is important to realize that in many cases the forces are *not* negligible beyond the first neighbours as we assumed in this exercise. For example, in the case of the infinite cumulene chain the forces are significant even very far away from the displaced atom, and the actual dispersion will be different from eqn 7.40. The grey line in Figure 7.5 shows the result of a DFT/LDA calculation using the complete solution of eqn 7.39.

In the general case of a solid in three dimensions all the concepts introduced in this section remain valid, the only difference being that the equations are more complicated since we have to deal with three Cartesian directions and possibly many atoms in the crystalline unit cell. Since the complete formalism does not add anything fundamentally new to what we have discussed until this point, in the following it will be assumed that the reader will be able to perform the explicit derivations autonomously. In the standard literature on vibrations in crystals it is common practice to write the nuclear positions as follows:

$$\mathbf{R}_I = \mathbf{R}_l + \boldsymbol{\tau}_s, \tag{7.41}$$

where $\mathbf{R}_l$ indicates one of the infinitely many unit cells of the crystals (i.e. it is a direct lattice vector, see Appendix D) and $\boldsymbol{\tau}_s$ indicates the position of the nucleus *within* the unit cell. Equation 7.41 is simply a way to relabel the nuclei using composite indices $I = (l, s)$ and $J = (l', s')$. Using this new convention, for a crystalline solid the matrix of Born–von Karman force constants defined in eqn 7.11 takes the form:

$$K_{ls\alpha,l's'\beta} = \frac{\partial^2 U}{\partial(R_{l\alpha} + \tau_{s\alpha})\partial(R_{l'\beta} + \tau_{s'\beta})}, \tag{7.42}$$

and the equation of motion, eqn 7.13, becomes:

$$M_s \ddot{u}_{ls\alpha} = -K_{ls\alpha,l's'\beta}\, u_{l's'\beta}. \tag{7.43}$$

Here $u_{ls\alpha}(t)$ is the instantaneous displacement of atom s in unit cell l along Cartesian direction α. The periodicity of the crystal implies that the force constants are insensitive to translations by a lattice vector, similarly to eqn 7.34. In order to look for solutions of the equation of motion in the form of propagating waves, as in eqn 7.32, we define:

$$u_{ls\alpha}(t) = u_{s\alpha}^0\, e^{i[\mathbf{q}\cdot(\mathbf{R}_l + \tau_s) - \omega t]}, \tag{7.44}$$

with $u_{s\alpha}^0$ a constant. By replacing this expression inside eqn 7.43 and using the translational invariance of the matrix of force constants we obtain:

$$\sum_{s'\beta} D_{s\alpha,s'\beta}(\mathbf{q})\, v_{s'\beta}^0 = \omega^2 v_{s\alpha}^0, \tag{7.45}$$

with

$$v_{s\alpha}^0 = M_s^{\frac{1}{2}} u_{s\alpha}^0, \tag{7.46}$$

and

$$D_{s\alpha,s'\beta}(\mathbf{q}) = \frac{1}{(M_s M_{s'})^{\frac{1}{2}}} \sum_l e^{i\mathbf{q}\cdot\mathbf{R}_l} e^{i\mathbf{q}\cdot(\tau_{s'} - \tau_s)} K_{0s\alpha,ls'\beta}. \tag{7.47}$$

The matrix $\mathbb{D}(\mathbf{q})$ whose matrix elements are given by $D_{s\alpha,s'\beta}(\mathbf{q})$ above is referred to as the *dynamical matrix for lattice vibrations*. If there are M atoms in the crystalline unit cell, $\mathbb{D}(\mathbf{q})$ is a $3M \times 3M$ matrix whose entries depend on the wavevector $\mathbf{q}$.

By comparing the equations above with their counterpart in Section 7.2, we see that also in the case of crystalline solids one needs to solve an eigenvalue problem. The key difference with respect to Section 7.2 is that now the dynamical matrix depends on the wavevector, $\mathbf{q}$. We can also note a close similarity between eqn 7.47 and its simplified version, eqn 7.35. The main difference is the extra exponential factor $\exp[i\mathbf{q}\cdot(\tau_{s'} - \tau_s)]$ appearing in eqn 7.47, which reflects the presence of several atoms inside each unit cell.

The first step in the study of vibrations in three-dimensional crystals using DFT is the evaluation of the dynamical matrix. This operation can be performed in one of two ways. In the first approach one computes the matrix of force constants in eqn 7.42 using the same methods discussed in Section 7.2 for non-crystalline solids. The dynamical matrix at each wavevector, $\mathbf{q}$, is then obtained using a Fourier transform according to eqn 7.47. In this case the periodic simulation box must be large enough to accommodate the wavelength $\lambda = 2\pi/|\mathbf{q}|$, i.e. it is necessary to use a 'supercell' that covers several crystalline unit cells. The reader interested in the details of this procedure can consult, among others, Parlinski *et al.* (1997), Yin and Cohen (1982*a*) and Ackland *et al.* (1997). A second possibility is to directly compute the dynamical matrix at each wavevector $\mathbf{q}$. This latter approach, which goes under the name of 'density functional perturbation theory' or 'linear-response theory' (Baroni et al., 1987; Gonze and Lee, 1997; Savrasov, 1996) has the advantage that the simulation box can be restricted to a single unit cell of the crystal. A detailed review of this approach can be found in Baroni et al. (2001).

The next step after the evaluation of the dynamical matrix, $\mathbb{D}(\mathbf{q})$, is the calculation of its eigenfrequencies and eigenmodes by diagonalization. Since $\mathbb{D}(\mathbf{q})$ is a $3M \times 3M$ matrix, for each wavevector $\mathbf{q}$ we will find $3M$ eigenfrequencies, $\omega_{\mathbf{q}\nu}$, and eigenmodes, $E_{s\alpha,\nu}(\mathbf{q})$ (with $\nu = 1, \ldots, 3M$). At this point, in complete analogy with eqn 7.25 and eqn 7.37, it will be possible to construct the most general solution of the equation of motion (eqn 7.43) as follows:

$$ u_{lsa}(t) = M_s^{-\frac{1}{2}} \sum_\nu \int_{BZ} d\mathbf{q}\, E_{s\alpha,\nu}(\mathbf{q})\, e^{i\mathbf{q}\cdot(\mathbf{R}_l + \tau_s)} \left[A_{s\alpha}(\mathbf{q})\, e^{-i\omega_{\mathbf{q}\nu} t} + B_{s\alpha}(\mathbf{q})\, e^{i\omega_{\mathbf{q}\nu} t} \right]. $$

Here the integral is restricted to the first Brillouin zone (BZ), similarly to what was done in eqn 7.37 (see Appendix D). Furthermore, the unknown constants $A_{s\alpha}(\mathbf{q})$ and $B_{s\alpha}(\mathbf{q})$ can be determined using the initial conditions, as it was done in Exercise 7.9. It is important to keep in mind that, even though for ease of notation the displacements $u_{lsa}(t)$ are expressed in terms of complex exponentials, they are real-valued quantities.

Exercise 7.11 We want to determine the frequency vs. wavevector dispersion relations for diamond. Diamond is a fascinating allotrope of carbon, and has the highest hardness of any known bulk material. In diamond, carbon crystallizes in a face-centred cubic lattice with two atoms in the unit cell. The crystal structure is the same as that of silicon shown in Figure 5.2, whereby each atom is fourfold coordinated to its neighbours in a tetrahedral configuration. The measured lattice parameter of diamond is 3.56 Å (Kittel, 1976), while the calculated lattice parameter at equilibrium is $a = 3.532$ Å. A DFT/LDA calculation of diamond based on density functional perturbation theory (Baroni et al., 2001) yields the following dynamical matrices for the wavevectors: $\mathbf{q}_1 = 0$, $\mathbf{q}_2 = \pi/a\mathbf{u}_x$ and $\mathbf{q}_3 = 2\pi/a\mathbf{u}_x$. The matrices are given in units of 100 meV^2 and have 6×6 entries, since there are two atoms per unit cell. The convention for the indices of the matrix elements is the same as in eqn 7.18.

$$ \hbar^2 \mathbb{D}(\mathbf{q}_1) = \begin{pmatrix} 132.4 & 0.0 & 0.0 & -132.4 & 0.0 & 0.0 \\ 0.0 & 132.4 & 0.0 & 0.0 & -132.4 & 0.0 \\ 0.0 & 0.0 & 132.4 & 0.0 & 0.0 & -132.4 \\ -132.4 & 0.0 & 0.0 & 132.4 & 0.0 & 0.0 \\ 0.0 & -132.4 & 0.0 & 0.0 & 132.4 & 0.0 \\ 0.0 & 0.0 & -132.4 & 0.0 & 0.0 & 132.4 \end{pmatrix}, $$

$$ \hbar^2 \mathbb{D}(\mathbf{q}_2) = \begin{pmatrix} 181.1 & 0.0 & 0.0 & -(1+i)65.4 & 0.0 & 0.0 \\ 0.0 & 135.4 & 0.0 & 0.0 & -(1+i)62.7 & -(1-i)14.4 \\ 0.0 & 0.0 & 135.4 & 0.0 & -(1+i)14.4 & -(1+i)62.7 \\ -(1-i)65.4 & 0.0 & 0.0 & 181.1 & 0.0 & 0.0 \\ 0.0 & -(1-i)62.7 & -(1+i)14.4 & 0.0 & 135.4 & 0.0 \\ 0.0 & -(1+i)14.4 & -(1-i)62.7 & 0.0 & 0.0 & 135.4 \end{pmatrix}, $$

$$\hbar^2\mathbb{D}(\mathbf{q}_3) = \begin{pmatrix} 226.4 & 0.0 & 0.0 & 0.0 & 0.0 & 0.0 \\ 0.0 & 137.7 & 0.0 & 0.0 & 0.0 & -41.1 \\ 0.0 & 0.0 & 137.7 & 0.0 & -41.1 & 0.0 \\ 0.0 & 0.0 & 0.0 & 226.4 & 0.0 & 0.0 \\ 0.0 & 0.0 & -41.1 & 0.0 & 137.7 & 0.0 \\ 0.0 & -41.1 & 0.0 & 0.0 & 0.0 & 137.7 \end{pmatrix}.$$

▶Calculate the vibrational eigenfrequencies of diamond at the wavevectors $\mathbf{q}_1$, $\mathbf{q}_2$ and $\mathbf{q}_3$. Small imaginary components in the eigenvalues of the dynamical matrix can be ignored, as they would disappear by including more significant digits in the matrices above. ▶Sketch the frequency vs. wavevector dispersion relations for diamond along the line Δ of the fcc Brillouin zone. The Δ line is the segment in the Brillouin zone connecting the wavevector Γ ($\mathbf{q} = 0$) with the wavevector X ($\mathbf{q} = 2\pi/a\mathbf{u}_x$). ▶Compare the highest vibrational frequency at the centre of the Brillouin zone (Γ) with the value $\hbar\omega = 163$ meV measured by Schwoerer-Böhning *et al.* (1998). ▶Calculate the period of oscillation of the carbon atoms of diamond in the highest vibrational eigenmode at Γ.

Exercise 7.12 ▶Calculate the vibrational eigenmodes of diamond at the centre of the Brillouin zone (Γ point, i.e. $\mathbf{q} = 0$) using the corresponding dynamical matrix provided in Exercise 7.11. ▶Sketch the six vibrational eigenmodes for the two C atoms in the diamond unit cell. The atomic displacements can be represented by arrows, as in Figure 7.3.□You should find that the three vibrational eigenmodes corresponding to $\hbar\omega = 0$ are pure translations of the entire crystal. These modes are not proper vibrations, hence should be ignored. For the three vibrational eigenmodes corresponding to $\hbar\omega = 162.7$ meV, you should find that the displacements of the two C atoms are in the same direction but have opposite signs. These eigenmodes are referred to as 'bond-stretching' modes.

In this chapter we have learned how to study the vibrational properties of molecules and solids within the harmonic approximation. At this point we know how to obtain the vibrational eigenfrequencies and eigenmodes, and how to address the dynamics of classical nuclei within the framework of DFT. However, we did not answer yet the questions on (i) how vibrational frequencies and frequency vs. wavevector dispersion relations can be measured, and (ii) how DFT calculations compare with experiment. These two questions will be addressed in Chapter 8.

8

Phonons, vibrational spectroscopy and thermodynamics

In Chapter 7 we introduced the theoretical framework for understanding and computing vibrations of materials at the atomic scale, from simple molecules to crystalline solids. What is still missing in this picture is the connection between such a theoretical framework and experimental measurements of vibrations. In this chapter we fill this gap by discussing how it is possible to measure vibrational frequencies in molecules and solids using *vibrational spectroscopy*. As will be shown below, in order to understand experimental techniques it is necessary to take a step beyond the approximation which we adopted in Chapter 4 (from eqn 4.10 onwards) whereby we considered nuclei as classical particles. This extra step will naturally lead to the concepts of quanta of vibrations, which are referred to as *vibrons* (in molecules) or *phonons* (in solids). As a byproduct, in the final part of this chapter we will discuss the connection between the vibrations of solids and their *phase diagrams*.

8.1 Basics of Raman and neutron scattering spectroscopy

Vibrational spectroscopy refers to a set of experimental techniques for measuring the vibrational properties of materials, ranging from molecules to solids. The vibrational frequencies introduced in Chapter 7 constitute unique and reliable fingerprints of the structure and composition of solids, surfaces, nanostructures and molecules. For this reason the measurement of such frequencies through vibrational spectroscopy is an extremely useful tool for obtaining structural and compositional information on materials.

Among the most common techniques we find Raman spectroscopy, inelastic neutron scattering, infrared spectroscopy (Stuart, 2004), inelastic X-ray scattering (Burkel *et al.*, 1987; Burkel, 2000), electron energy loss spectroscopy (Ibach, 1970) and inelastic electron tunnelling (Jaklevic and Lambe, 1966; Stipe *et al.*, 1998).

Since we are mainly interested in understanding how DFT calculations compare with experiment, in the following we will only touch upon the two most popular techniques, namely Raman and inelastic neutron scattering spectroscopy. This choice is merely due to space constraints; one should keep in mind that the concepts discussed in this chapter are completely general and apply to all forms of vibrational spectroscopy.

8.1.1 Raman scattering spectroscopy

When a monochromatic beam of light passes through a material, the frequency spectrum of the outgoing radiation shows a peak corresponding to the original frequency, ω, of the incoming radiation, as well as satellites around this peak. The main peak corresponds to the elastic (i.e. energy-conserving) scattering of light, and is referred to as the *Rayleigh* peak. The satellites are arranged symmetrically around the principal peak, at frequencies $\omega \pm \omega_0$, and their intensity is several orders of magnitude smaller than the main peak. In units of energy, the separation $\hbar\omega_0$ between the satellites and the Rayleigh peak is of the order of 1–300 meV. The appearance of these satellites goes under the name of the *Raman effect* (Raman and Krishnan, 1928; Raman, 1928). Since $\hbar\omega_0$ falls precisely within the range of vibrational energies discussed in Chapter 7, we expect the Raman effect to be connected with the vibrations of the nuclei in a material.

In order to see this connection let us consider the simplest possible example, namely the Raman effect in a gas of molecules, e.g. N_2. If the incoming monochromatic light is $\mathbf{E}(t) = E_0 \cos(\omega t)\mathbf{u}_z$, and the electric polarizability of the molecule is α, then the induced electric dipole of each molecule will be $\mathbf{p}(t) = \alpha\mathbf{E}(t)$. In this case the difference between the local electric field on the molecule and the average electric field over the gas (Kittel, 1976) can safely be ignored. For simplicity we take the polarizability, α, to be a scalar quantity, although in reality it is a tensor in the Cartesian coordinates. If the molecule in question is at the centre of our reference frame, then from classical electrostatics we know that the induced dipole generates an electric field at point $\mathbf{r}$ far away from the molecule given by (Jackson, 1998):

$$\mathbf{E}_{\text{dip}}(\mathbf{r}, t) = \frac{3[\mathbf{p}(t) \cdot \hat{\mathbf{r}}]\,\hat{\mathbf{r}} - \mathbf{p}(t)}{4\pi\epsilon_0 \, r^3}, \tag{8.1}$$

with $r = |\mathbf{r}|$ and $\hat{\mathbf{r}} = \mathbf{r}/r$. By expressing the dipole in terms of the electric field, $\mathbf{E}(t)$, we obtain $\mathbf{E}_{\text{dip}}(\mathbf{r}, t) = \{E_0[3(\mathbf{u}_z \cdot \hat{\mathbf{r}})\,\hat{\mathbf{r}} - \mathbf{u}_z]/(4\pi\epsilon_0 \, r^3)\}\,\alpha\cos(\omega t)$. If we now consider the contributions from all the molecules of the gas, the term within braces will be replaced by some average $\mathbf{E}_{\text{ave}}$ depending on the position of each molecule, while the time dependence in the second term is the same for every molecule and can be factored out. We can therefore write the total field re-radiated by the gas as:

$$\mathbf{E}_{\text{dip}}^{\text{gas}}(t) = \mathbf{E}_{\text{ave}}\,\alpha\cos(\omega t). \tag{8.2}$$

At this point we can take a closer look at the polarizability, α. This quantity describes the charge fluctuations inside the molecule in response to the external field, $\mathbf{E}(t)$. In Raman spectroscopy this field typically corresponds to visible radiation, for example the green light from a Nd:YAG laser (wavelength $\lambda = 532$ nm). Therefore the typical period of oscillation of the electric field is $\lambda/c = 1.8$ fs, with c denoting the speed of light. In Exercises 7.2 and 7.11 we saw that the typical period of atomic oscillations is > 10 fs. As a consequence, during one period of oscillation of the electric field, the nuclei are almost immobile, and the polarizability, α, reflects the rearrangement of the electronic charge when the nuclei are at the instantaneous distance d. We can describe this effect by acknowledging that the polarizability depends on the bond length, $\alpha = \alpha(d)$, and use a Taylor expansion to first order in d:

$$\alpha(d) = \alpha(d_0) + \frac{\partial\alpha}{\partial d}(d - d_0). \tag{8.3}$$

The bond length in turn oscillates according to the law determined in Exercise 7.2:

$$d(t) = d_0 + [d(0) - d_0]\cos(\omega_0 t), \tag{8.4}$$

where for simplicity we have set the initial velocity $\dot{d}(0) = 0$. By combining eqns 8.2–8.4 we obtain:

$$\mathbf{E}_{\text{dip}}^{\text{gas}}(t) = \alpha(d_0)\cos(\omega t)\mathbf{E}_{\text{ave}} + \frac{1}{2}\frac{\partial\alpha}{\partial d}[d(0) - d_0][\cos(\omega + \omega_0)t + \cos(\omega - \omega_0)t]\mathbf{E}_{\text{ave}}.$$

Since the intensity of the radiation field is proportional to the time average of $|\mathbf{E}_{\text{dip}}^{\text{gas}}(t)|^2$ (Jackson, 1998), based on the last expression we see that a detector (say, a charge-coupled device, CCD) will measure three peaks. One peak will be found at the frequency ω and corresponds to the elastically scattered Rayleigh radiation. The other two peaks will be found at the frequencies $\omega \pm \omega_0$, with intensities proportional to:

$$I^{\pm} = \left|\frac{\partial\alpha}{\partial d}\right|^2 [d(0) - d_0]^2. \tag{8.5}$$

These are precisely the Raman peaks, and are referred to as the Stokes peak $(\omega - \omega_0)$ and anti-Stokes peak $(\omega + \omega_0)$, respectively. The Rayleigh peak and the Raman Stokes and anti-Stokes peaks are shown schematically in Figure 8.1.

The frequency shifts between the Rayleigh peak and the Raman peaks are $\pm\omega_0$, i.e. they correspond precisely to the frequency of oscillation of the nuclei. Therefore the measurement of Raman shifts yields directly and unambiguously the vibrational frequencies of the nuclei discussed in Chapter 7. This observation makes the Raman effect a very powerful tool for probing the vibrational properties of materials.

While the heuristic derivation given here started by considering a simple diatomic molecule in a gas, the same principles apply when studying the Raman effect in larger systems, such as polyatomic molecules, nanostructures and solids. In these more complicated situations there could be as many Raman peaks as the number of vibrational eigenmodes. In practice, however, the nuclear displacements along certain eigenmodes may not result in changes of the polarizability as was assumed in eqn 8.3. In these cases terms like $\partial\alpha/\partial d$ in eqn 8.5 would vanish, thereby suppressing the Raman activity of the mode.

A comprehensive introduction to Raman spectroscopy, with a focus on chemistry, solid-state chemistry and biological applications can be found in the textbook by Ferraro *et al.* (2003).

In the case of solids our heuristic derivation needs to be modified in two ways. First, the molecular polarizability, α, of eqn 8.3 needs to be replaced by the macroscopic dielectric function. This new quantity will be considered in Chapter 10. Second, as we have seen in Chapter 7, vibrations in solids propagate as waves, hence the wavevector **q** appearing in eqn 7.44 must also be taken into account along with the wavevector **k** of the incoming light beam. By carrying out a derivation similar to eqns 8.2–8.5 it is possible to show that the Raman radiation from solids carries wavevectors $\mathbf{k} \pm \mathbf{q}$. Since the frequency of the incoming radiation and that of the Raman radiation are very similar (as $\omega_0/\omega \sim 0.001$–$0.1$), it follows that also the wavevectors must be very

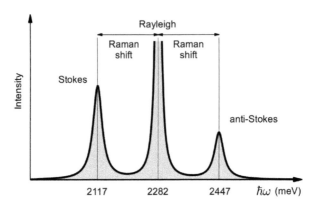

Fig. 8.1 Schematic representation of a Raman spectrum. In this example the incident radiation has a wavelength of 532 nm (green), i.e. $\hbar\omega = 2.282$ eV. The Rayleigh peak corresponds to elastically scattered light and is found at the same frequency as the incoming radiation. The two satellites in the figure are the Stokes (left) and anti-Stokes (right) peaks, and are shifted with respect to the Rayleigh peak by plus or minus the frequency of a vibrational eigenmode. In this example the numbers correspond to the C–C stretching mode at the centre of the Brillouin zone in diamond (Solin and Ramdas, 1970; Prawer and Nemanich, 2004). In general, Raman spectra are considerably more complicated than in this illustration, since there can be many 'active' vibrational modes, as well as higher-order effects not captured by our simple heuristic derivation in eqns 8.2–8.5.

similar owing to the relations $c = \omega/|\mathbf{k}|$ and $c = (\omega \pm \omega_0)/|\mathbf{k} \pm \mathbf{q}|$. As a result, the only possibility is that $\mathbf{q} \simeq 0$, i.e. the Raman radiation only comes from vibrational eigenmodes at or very near the centre of the Brillouin zone, $\mathbf{q} = 0$ (Appendix D).

Strictly speaking the above considerations apply only to 'first–order' Raman scattering. In fact, if we were to consider a Taylor expansion to second order in eqn 8.3, we would find additional very small Raman peaks at $\pm 2\omega_0$. These peaks are referred to as the 'second–order' Raman radiation. In these cases the restriction $\mathbf{q} \simeq 0$ for solids does not apply any more. The interested reader will find a careful discussion of the Raman effect in solids and of the underlying theory in the textbook by Yu and Cardona (2010). Table 8.1 reports a comparison between the vibrational frequencies calculated in Chapter 7 for N_2, H_2O and diamond, and the corresponding measured Raman shifts. As anticipated in Chapter 7 the frequencies calculated using DFT/LDA are in good agreement with experiment. The deviation between calculated and measured frequencies ranges between 1% and 5%. This result is quite remarkable if we remember that the density functional formalism does not make use of any *ad hoc* assumptions, and that the only external parameters in the theory are the atomic numbers of the N, H, O and C atoms. More generally, extensive comparisons reported systematically since the mid-1980s indicate that DFT predicts vibrational frequencies consistently well across very different systems, from the smallest molecules to crystalline solids, as we can see in Table 8.1.

Table 8.1 Comparison between calculations and measurements of the Raman vibrational frequencies of molecular nitrogen, water vapour and diamond. All the frequencies are given in units of energy as $\hbar\omega$ and reported in meV. The calculated frequencies are from Exercise 7.2 for N_2, from Exercise 7.7 for H_2O and from Exercise 7.11 for diamond. The vibrational frequency of diamond corresponds to the Γ point of the Brillouin zone ($\mathbf{q} = 0$). The measured frequencies of molecular nitrogen and water vapour are from the Raman spectra taken by Bendtsen (1974) and Murphy (1977–1978), respectively. The C–C stretching frequency in diamond was measured by Solin and Ramdas (1970). A detailed discussion of Raman experiments in diamond and diamond-like carbon materials can be found in (Prawer and Nemanich, 2004) and (Ferrari and Robertson, 2004), respectively.

	N_2	H_2O			diamond
DFT/LDA calculations	300.4	189.1	458.1	472.8	162.7
Experiment	288.9	197.7	453.4	465.5	165.1

The ability to describe with good accuracy materials across the periodic table makes DFT an increasingly popular tool in the area of vibrational spectroscopy. In fact, measured Raman spectra can be considerably more complicated than the example given in Figure 8.1, and in these cases DFT can be very helpful for assigning Raman lines to specific structural motifs. For example, by comparing their DFT calculations with experiments Umari *et al.* (2003) showed that the Raman lines observed in amorphous silica (a-SiO_2) for shifts between 55 and 60 meV can be used to quantify the porosity of the glass at the nanoscale.

8.1.2 Inelastic neutron scattering spectroscopy

Inelastic neutron scattering (INS) is possibly the most widely used technique for measuring vibrational frequency vs. wavevector dispersion relations in crystals (Brockhouse and Stewart, 1958). Neutron scattering experiments consist of directing a neutron beam towards a sample, and measuring the energetic distribution of the deflected neutrons as a function of the scattering angle. In these experiments neutrons are obtained by 'spallation'. This procedure consists of firing high-energy protons from a particle accelerator onto a Pb target. The Pb nuclei hit by protons release their energy by ejecting fast neutrons. The neutrons generated in this reaction are decelerated by passage through heavy water, and subsequently enter a monochromator, which retains only those in a narrow range of kinetic energies. Typical energies are of the order of 10–100 meV, and in these cases one speaks of 'thermal neutrons'. The monochromatic neutron beam is finally directed towards the sample, and the scattered neutrons are counted using a nuclear scintillator coupled to a photodetector.

The physical principle underlying INS is that neutrons interact only with other nucleons (i.e. neutrons or protons) through the nuclear strong force, which is non-negligible only when two nucleons are closer than $\sim 10^{-4}$ Å. As a consequence neutrons passing through the sample directly probe the position and dynamics of individual nuclei. Neutrons also possess a magnetic moment and so they interact with electron and nuclear spins; however we will ignore this aspect in our short presentation. A comprehensive introduction to the field of neutron scattering can be found in the classic textbook by Squires (1978).

In what follows we discuss briefly the connection between INS experiments and the frequency vs. wavevector dispersion relations of vibrations in solids. A complete formal derivation would require considerably more room than is warranted by a simple introduction; therefore we will proceed with semi-heuristic arguments. Formal presentations can be found in the original work by Van Hove (1954), and also in the appendices of the textbooks by Ashcroft and Mermin (1976) and by Grimvall (1981). A central quantity in the interpretation of neutron scattering experiments is the *dynamic structure factor*, $S(\mathbf{q}, E)$. This quantity is a generalization of the concept of *static* structure factor used in Chapter 5 in the context of X-ray crystallography. If the incoming neutron beam propagates along direction $\mathbf{k}_i$ with energy E_i, then $S(\mathbf{q}, E)$ is proportional to the number of neutrons scattered along $\mathbf{k}_f = \mathbf{k}_i + \mathbf{q}$ with energy $E_f = E_i + E$. This is a result of the quantum theory of scattering (Merzbacher, 1998) and was derived by Van Hove (1954). Here we want to understand why a measurement of $S(\mathbf{q}, E)$ provides information on the vibrational frequencies of crystalline solids.

Van Hove (1954) has shown that the dynamic structure factor can be expressed in terms of the instantaneous positions, $\mathbf{R}_I(t)$, of the M nuclei in the solid as follows:

$$S(\mathbf{q}, E) = \frac{1}{M} \int_{-\infty}^{+\infty} \frac{dt}{2\pi} e^{-\frac{i}{\hbar} Et} \sum_{I,J} \left\langle e^{-i\mathbf{q}\cdot\mathbf{R}_I(0)} e^{i\mathbf{q}\cdot\mathbf{R}_J(t)} \right\rangle_T,$$

where $\langle \cdots \rangle_T$ indicates a 'thermal average', i.e. an average over all the possible initial conditions for the nuclear positions and velocities yielding the same temperature, T. In order to simplify the notation we now customize this expression to the case of a one-dimensional chain of identical atoms, as was done in Section 7.4. By expressing the time-dependent positions of the nuclei in terms of their displacements from equilibrium (see eqn 7.7) we find:

$$S(q, E) = \frac{1}{M} \int_{-\infty}^{+\infty} \frac{dt}{2\pi} e^{-\frac{i}{\hbar} Et} \sum_{I,J} e^{iq(R_J^0 - R_I^0)} \left\langle e^{iq[u_J(t) - u_I(0)]} \right\rangle_T. \tag{8.6}$$

This expression can be simplified considerably by noting that the displacements are very small quantities (remember Figure 7.1); therefore it makes sense to perform a Taylor expansion of the last exponential as $1 + iq[u_J(t) - u_I(0)] - q^2[u_J(t) - u_I(0)]^2/2 + \mathcal{O}\{[u_J(t) - u_I(0)]^3\}$. In this expansion the first constant term leads to the *static structure factor*, $S(q)$:

$$\frac{1}{M} \int_{-\infty}^{+\infty} \frac{dt}{2\pi} e^{-\frac{i}{\hbar} Et} \sum_{I,J} e^{iq(R_J^0 - R_I^0)} = \delta(E) \sum_I e^{iqR_I^0} = S(q)\delta(E),$$

δ being the Dirac delta function. The linear terms $iqu_J(t)$ and $iqu_I(0)$ in the Taylor expansion vanish when performing the thermal average, since at equilibrium there will be an equal number of atoms displaced along the positive or negative x axis. The thermal average of the quadratic terms $u_J^2(t)$ and $u_I^2(0)$ yields in both cases the mean square displacement of the nuclei, which we will call $\langle u^2 \rangle_T$. Finally the thermal average of the cross terms $\langle u_J(t)u_I(0)\rangle_T$ is slightly more complicated and needs to be derived explicitly. Putting together these observations we can rewrite eqn 8.6 as follows:

$$S(q,E) = \underbrace{\left[1-q^2\langle u^2\rangle_T\right]S(q)\delta(E)}_{\text{Neutron diffraction}} + \underbrace{\frac{q^2}{M}\int_{-\infty}^{+\infty}\frac{dt}{2\pi}e^{-\frac{i}{\hbar}Et}\sum_{I,J}e^{iq(R_J^0-R_I^0)}\langle u_J(t)u_I(0)\rangle_T}_{\text{Inelastic neutron scattering}}.$$

(8.7)

The first term in eqn 8.7 describes the diffraction of neutrons, and involves no change in energy. In fact, owing to the delta function, this term is non-zero only when the energy gained by the neutrons is $E = 0$, which corresponds to the case of elastic collisions. This first term contains information about the crystal structure, and is the analogue of the XRD structure factor introduced in eqn 5.4.

The second term in eqn 8.7 describes 'correlations' between the motion of the nuclei in the system. The thermal averages, $\langle u_J(t)u_I(0)\rangle_T$, can be evaluated formally by expressing the time dependence of the nuclear displacements in terms of normal modes of vibrations. This is accomplished by replacing eqn 7.37 for $u_J(t)$ and $u_I(0)$ in eqn 8.7. After a rather long series of steps, and using the results of Exercise 7.9, we find the elegant result:

$$S(q,E) = \left[1-q^2\langle u^2\rangle_T\right]S(q)\delta(E)+q^2\langle u^2\rangle_T\delta[E+\hbar\omega(q)]+q^2\langle u^2\rangle_T\delta[E-\hbar\omega(q)]. \quad (8.8)$$

This last equation is telling us that, in addition to the diffraction peak, the intensity of the scattered beam of neutrons will display additional sharp signals for energy transfers $E = \pm\hbar\omega(q)$. These signals correspond to neutrons whose energy increased or decreased by the amount $\hbar\omega(q)$, $\omega(q)$ being the vibrational frequency of the system for the wavevector q. Since the mean square displacement, $\langle u^2\rangle_T$, is very small, eqn 8.8 is telling us that the two additional peaks will appear as small satellites around the main diffraction peak.

While the derivation of eqn 8.8 is based on simple considerations, the final result is essentially the same (up to the third order in the displacements) as that obtained by Ashcroft and Mermin (1976) using more sophisticated conceptual tools based on second quantization (see Appendix N of their textbook). In particular, while we restricted our derivation to the case of a simple linear chain of atoms, the result for three-dimensional crystals is formally identical. The main difference is that for a given wavevector, **q**, there will be two inelastic peaks for each vibrational eigenfrequency, $\omega_{q\nu}$.

At this point it should be clear that, if we were to carry on with the Taylor expansion of the exponential in eqn 8.6, we would find additional peaks of even smaller intensity for energy transfers corresponding to sums and differences of $\hbar\omega(q)$, e.g. $E = 2\hbar\omega(q_1)$ or $E = \hbar\omega(q_1) + 2\hbar\omega(q_2)$ and so on. We can therefore imagine that an actual neutron scattering spectrum will consist of a diffraction peak, some sharp satellites, and a continuous background arising from all these additional terms. A schematic representation of such spectrum is given in Figure 8.2.

An example of INS measurement of vibrational frequencies in solids is shown in Figure 8.3. The experimental data-points in the figure (blue dots) are for diamond, and have been measured by Warren *et al.* (1967) on a unique sample named the

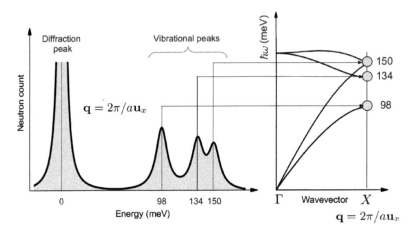

Fig. 8.2 Measurement of the vibrational frequency vs. wavevector dispersion relations in solids using inelastic neutron scattering (INS). The system considered in this hypothetical measurement is diamond. In order to determine the dispersion relations (right panel) one performs a scan of the wavevectors, $\mathbf{q}$, by recording INS spectra at various scattering angles. For each wavevector the energy distribution of the scattered neutrons will display a diffraction peak, as well as vibrational peaks corresponding to neutrons which have lost energy $\hbar\omega_{\mathbf{q}\nu}$ (left panel). In this example the energies of the three peaks provide three data-points for the wavevector $\mathbf{q} = 2\pi/a\mathbf{u}_y$, i.e. at the X point of the diamond Brillouin zone (see Figure 8.3). From this example it should become apparent that INS can access wavevectors throughout the Brillouin zone, while in the case of Raman scattering, discussed in Section 8.1.1, we have seen that only a few selected wavevectors can be probed.

'Oppenheimer diamond'. This is among the largest diamonds in the world (about 2 cm in diameter) and is preserved in the Smithsonian's Museum. The vibrational frequencies have been measured for wavevectors $\mathbf{q}$ along a few selected directions of the diamond Brillouin zone, as shown in the left panel. The black lines in Figure 8.3 represent DFT/LDA calculations performed exactly as in Exercise 7.11. The only difference between Exercise 7.11 and Figure 8.3 is that this time the dynamical matrix, $\mathbb{D}(\mathbf{q})$, has been evaluated for 60 different wavevectors in order to obtain smooth curves. It is apparent that DFT/LDA calculations are able to predict rather well the measured dispersion relations. A careful examination of this figure shows that the deviation between calculations and experiment is between 1% and 5% throughout the Brillouin zone.

In general, DFT calculations based on the formalism discussed in Chapter 7 yield vibrational frequencies in very good agreement with experiment, with typical deviations similar to those in Figure 8.3. As one may expect, the calculated frequencies also depend slightly on the choice of the exchange and correlation functional in eqn 3.10. For example the difference between DFT calculations based on LDA or GGA exchange and correlation functionals is of the order of a few percent (Favot and Dal Corso, 1999). This difference partly arises from the slightly different structural parameters provided by LDA and GGA calculations (see Section 6.5).

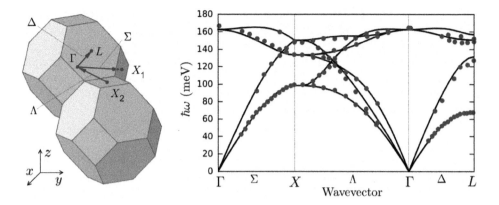

Fig. 8.3 Vibrational frequency vs. wavevector dispersion relations of diamond: comparison between DFT/LDA calculations and inelastic neutron scattering (INS) experiments. The truncated octahedra on the left are two adjacent Brillouin zones of diamond (and more generally of any fcc crystal). The centre of the Brillouin zone ($\mathbf{q} = 0$) is the Γ point; the other points labelled in the figure are X_1: $\mathbf{q} = (2\pi/a)\mathbf{u}_y$, X_2: $\mathbf{q} = (2\pi/a)(\mathbf{u}_x + \mathbf{u}_y)$, L: $\mathbf{q} = (\pi/a)(\mathbf{u}_x + \mathbf{u}_y + \mathbf{u}_z)$. The lines joining these points (in blue) are called $\Sigma : \Gamma \to X_1$, $\Lambda : \Gamma \to X_2$ and $\Delta : \Gamma \to L$. The panel on the right shows the vibrational eigenfrequencies of diamond as a function of wavevector $\mathbf{q}$, along the lines Σ, Λ and Δ indicated on the left. The blue discs are from the INS measurements of Warren *et al.* (1967). The black lines are DFT/LDA calculations. A comprehensive set of DFT/LDA calculations on the vibrational properties of diamond can be found in the work by Pavone *et al.* (1993).

It is worth noting that, at variance with Figure 7.5, here we have some *branches* of the dispersion relations for which the frequency, $\omega_{\mathbf{q}\nu}$, goes to zero linearly as we approach Γ, and some branches for which the frequencies remains non-zero at Γ. The first group of branches are referred to as *acoustic branches*, and are responsible for the propagation of sound in the material. The second group are called *optical branches* since the associated vibrations can exchange energy with an electromagnetic field, as we have seen for the Raman effect. The classification of vibrational modes into acoustic and optical branches is discussed in depth in the textbooks by Kittel (1976) and by Ashcroft and Mermin (1976).

8.2 Going beyond the classical approximation for nuclei

In the preceding section we have seen that DFT calculations can provide us with a description of the vibrational properties of materials in good agreement with experimental measurements such as Raman scattering or INS. However, when we consider once more Figures 8.1 and 8.2, we realize that our comparisons between theory and experiment referred only to the vibrational frequencies, and we never considered the *intensity* of Raman or INS peaks.

At first sight this does not seem to pose a problem, since eqns 8.8 and 8.5 already contain recipes for calculating the peak intensities. In fact, in the case of INS eqn 8.8 states that the peak intensities are proportional to the mean square displacement, $\langle u^2 \rangle_T$, of the nuclei from their equilibrium positions. In the case of Raman scattering the situation is analogous, as can be seen by performing a thermal average of the displacement in eqn 8.5. Now, from the equipartition theorem we know that the energy associated with the atomic displacements is proportional to the temperature: $\langle u^2 \rangle_T \propto k_B T$ (see Section 7.1). As a direct consequence of this observation, we expect the intensities of both INS and Raman spectra to vanish completely when a sample is cooled to close to absolute zero. The trouble is that this dependence of the intensities on temperature is *not* observed in the experimental spectra: Raman and INS spectra display inelastic peaks even at very low temperatures.

The inadequate description of spectral intensities is connected with our choice of describing nuclei as if they were classical particles. In fact, in Section 4.1 we replaced the quantum-mechanical momentum of the nuclei by its corresponding classical momentum (eqn 4.10) in order to obtain simple Newton's equations (eqn 4.12). While a classical description of nuclear dynamics represents a very good approximation for calculating vibrational eigenfrequencies, when it comes to spectral intensities (and more generally to *temperature-dependent properties*) we must proceed with a proper quantum description of the nuclear motion.

To this end we need to go back to the Schrödinger equation for nuclei derived in Section 4.1. For convenience we reproduce eqn 4.6 below:

$$\left[-\sum_I \frac{\nabla_I^2}{2M_I} + \frac{1}{2} \sum_{I \neq J} \frac{Z_I Z_J}{|\mathbf{R}_I - \mathbf{R}_J|} + E(\mathbf{R}_1, \ldots, \mathbf{R}_M) \right] \chi = E_{\text{tot}} \, \chi. \tag{8.9}$$

In order to calculate vibrational properties within a quantum theory for the nuclear motion we need to determine the nuclear wavefunctions, $\chi(\mathbf{R}_1, \ldots, \mathbf{R}_M)$, and eigenenergies, E_{tot}, for all possible quantum states. In Exercise 8.1 we are going to perform this operation for the usual simplest possible case, i.e. the N_2 molecule. Exercise 8.1 is rather demanding and it is advisable to jump directly to its conclusions during a first reading of the book.

Exercise 8.1 The Schrödinger equation for the nuclear wavefunction, $\chi(\mathbf{R}_1, \mathbf{R}_2)$, of the N_2 molecule in its ground state can be written as follows (see eqns 8.9 and 4.9):

$$\left[-\frac{\hbar^2 \nabla_1^2}{2M_N} - \frac{\hbar^2 \nabla_2^2}{2M_N} + U(\mathbf{R}_1, \mathbf{R}_2) \right] \chi(\mathbf{R}_1, \mathbf{R}_2) = E_{\text{tot}} \, \chi(\mathbf{R}_1, \mathbf{R}_2),$$

where the reduced Planck constant is shown explicitly in order to avoid confusion with the units. The total potential energy, U, depends only on the N–N distance, $d = |\mathbf{R}_1 - \mathbf{R}_2|$, and is shown in Figure 7.1. In order to make the calculations below easy to handle, the data of Figure 7.1 have been fitted with a simple *Morse potential* (Morse, 1929), yielding:

$$U(d) = U_0 + D \left[1 - e^{-(d - d_0)/\lambda} \right]^2, \tag{8.10}$$

with U_0 and d_0 as in Figure 5.1, $D = 13.848$ eV and $\lambda = 0.416$ Å. ▶Plot this Morse potential as a function of the interatomic distance, and verify that the fit reproduces the DFT/LDA calculations reported in Figure 7.1 within better than 5% for d between 0.9 and 1.2 Å.◻We now consider the nuclear wavefunctions, χ, describing states with zero angular momentum. In the centre-of-mass coordinates system, these wavefunctions depend only on the separation between the two N nuclei. By expressing the Laplace operator in spherical coordinates and making the substitution $\chi(d) = u(d)/d$, we can rewrite the Schrödinger equation above as an ordinary differential equation in the variable d:

$$-\frac{\hbar^2}{2(M_N/2)}\frac{\partial^2 u}{\partial d^2} + U(d)u(d) = E_{\text{tot}}\, u(d). \tag{8.11}$$

This equation describes the energetics of a particle of mass $M_N/2$ in the potential energy surface $U(d)$, and is the quantum-mechanical counterpart of eqn 7.2. A detailed derivation of eqn 8.11 can be found in the textbook by Bransden and Joachain (1983), or in Morse (1929). We now want to investigate the consequences of moving from the classical description of the nuclei in Exercises 7.1 and 7.2 to a quantum-mechanical description. For this purpose we carry out a numerical solution of eqn 8.11. We start by discretizing the distance, d, into a grid of points separated by the increment δ, so that $d_m = m\delta$, $m = 0, 1, \ldots, p+1$. Accordingly we will evaluate the potential energy and the wavefunction at each one of these points. ▶By expressing the second derivative via finite differences, and using the boundary conditions $u(d_0) = u(d_{p+1}) = 0$, show that eqn 8.11 can be written as:

$$\begin{cases} (2 + \mu_m)u_m - u_{m-1} - u_{m+1} = \epsilon\, u_m, & \text{if } m = 1, 2, \ldots, p \\ u_m = 0 & \text{if } m = 0, p+1 \end{cases} \tag{8.12}$$

with $u_m = u(d_m)$, $\mu_m = M_N(\delta/\hbar)^2 U(d_m)$ and $\epsilon = M_N(\delta/\hbar)^2 E_{\text{tot}}$. ▶Show that the above differences equations can be recast into a linear algebra eigenproblem as follows:

$$\begin{pmatrix} 2+\mu_1 & -1 & & & \\ -1 & 2+\mu_2 & -1 & & \\ & & \ddots & & \\ & & -1 & 2+\mu_{p-1} & -1 \\ & & & -1 & 2+\mu_p \end{pmatrix} \begin{pmatrix} u_1 \\ u_2 \\ \vdots \\ u_{p-1} \\ u_p \end{pmatrix} = \epsilon \begin{pmatrix} u_1 \\ u_2 \\ \vdots \\ u_{p-1} \\ u_p \end{pmatrix},$$

where all the empty entries stand for zeroes. ▶Solve this eigenvalue problem using $\delta = 0.01$ Å and $p = 200$, and verify that the lowest-energy eigenvalues are:

n	0	1	2	3	4
$(E_{\text{tot},n} - U_0)$ (meV)	153.6	458.0	759.0	1056.4	1350.3

In order to perform this calculation it is necessary to use a numerical computing environment such as 'Octave' (less than 10 lines of code should be sufficient). ▶Plot the square modulus of the nuclear wavefunctions for the lowest three eigenstates and compare your solutions with Figure 8.4. □When the DFT/LDA energy $U(d)$ in eqn 8.11 is replaced by the approximate Morse potential of eqn 8.10, the eigenvalues can be calculated exactly. ▶Compare the exact eigenvalues of eqn 8.10 as given by Morse (1929) with your numerical solutions.

Exercise 8.2 In this exercise we want to repeat the procedure carried out in Exercise 8.1, after having made the harmonic approximation for the potential energy surface. ▶Show that, within the harmonic approximation, the Morse potential of eqn 8.10 becomes:

$$U(d) = U_0 + \frac{D}{\lambda^2}(d - d_0)^2. \tag{8.13}$$

▶Repeat the same procedure of Exercise 8.1 in order to evaluate numerically the eigenvalues of eqn 8.11 within the harmonic approximation. Compare your results with the following table:

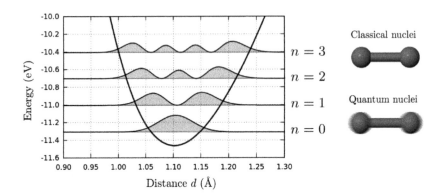

Fig. 8.4 DFT/LDA calculation of the quantum nuclear wavefunctions corresponding to the electronic ground state of N_2. The thick line is the total potential energy, U, from Figure 7.1 vs. the N–N distance, d. The thin lines and shaded areas indicate the square moduli of the nuclear wavefunctions, $|u(d)|^2$, corresponding to the four lowest eigenvalues, $n = 0, \ldots, 3$. These wavefunctions were obtained by solving eqn 8.12 as indicated in Exercise 8.1. Each curve has been translated vertically in such a way that the zero of u matches the associated eigenvalue (thin horizontal lines). From this plot we see that in the lowest-energy state the distance between the two N nuclei is not sharply defined. This effect can be visualized intuitively using the ball-and-stick representation on the right: when we consider a quantum-mechanical description of the nuclei the plot becomes slightly blurred owing to the fuzziness of the bond length in the lowest-energy state (Patrick and Giustino, 2013).

n	0	1	2	3	4
$(E_{\text{tot},n} - U_0)$ (meV)	153.8	461.4	769.0	1076.6	1384.2

☐You will notice that these eigenvalues follow the progression $E_0, 3E_0, 5E_0, 7E_0, 9E_0$ with $E_0 = 153.8$ meV, or equivalently $E = (2n + 1)E_0$ with $n = 0, \ldots, 4$. As expected, this progression is precisely the ladder of eigenvalues of the simple harmonic oscillator (Merzbacher, 1998). ▶By comparing eqn 8.13 with eqn 7.4 show that $E_0 \simeq \hbar\omega_{N_2}/2$, ω_{N_2} being the classical vibrational frequency of the nuclei determined in Exercise 7.2. ▶Calculate the difference between the energy eigenvalues obtained in this exercise using the harmonic approximation and those calculated in Exercise 8.1. The differences between the two sets of results constitute the *anharmonic* corrections to the vibrational eigenenergies. ▶Show that the anharmonic corrections to the lowest two eigenvalues are smaller than 1%.

In Exercise 8.1 we have learned how to solve the Schrödinger equation for nuclei (eqn 4.6) in a very simple situation. The main outcome of the exercise is that, even in their lowest-energy state, nuclei do not occupy fixed positions in space, but keep performing very small oscillations around their equilibrium positions. These oscillations are referred to as *zero-point motion* and are schematically represented by the blurred ball-and-stick model in Figure 8.4. The extent of the nuclear displacements in the lowest-energy state is about 0.1 Å in this example. This value is comparable to what was found in Figure 4.1 using heuristic arguments.

Another important result of Exercise 8.1 is that the total energy of the system at equilibrium, E_{tot}, does not correspond to the bottom of the potential energy surface, $U(d)$ (i.e. the bottom of the thick black line in Figure 8.4), but is slightly higher than that. The difference between the lowest-energy eigenvalue and the bottom of the potential well is referred to as *zero-point energy*.

In Exercise 8.2 we have learned that the replacement of the DFT potential energy surface, $U(d)$, by its harmonic approximation yields rather accurate eigenvalues, and that anharmonic corrections are very small in the case of N_2, at least for the lowest eigenstates. In addition we have found that the zero-point energy is related to the classical vibrational frequency of the molecule, ω_{N_2}.

The above observations can be rationalized using the standard quantum theory of the simple harmonic oscillator (Merzbacher, 1998). In fact, within the harmonic approximation defined by eqn 7.4, the nuclear Schrödinger equation for N_2 (eqn 8.11) takes the form:

$$-\frac{\hbar^2}{2(M_N/2)}\frac{\partial^2 u}{\partial d^2} + \frac{1}{2}K(d-d_0)^2 u(d) = (E_{tot} - U_0)\,u(d).$$

If we define the characteristic length, b, as:

$$b = \left[\frac{\hbar}{(M_N/2)\,\omega_{N_2}}\right]^{\frac{1}{2}},$$

and we simplify the notation using $x = (d-d_0)/b$, $f(x) = \sqrt{b}\,u(d_0+bx)$ and $\epsilon = 2(E_{tot}-U_0)/(\hbar\omega_{N_2})$, then the equation above becomes:

$$\frac{d^2 f}{dx^2} + (\epsilon - x^2)f = 0, \tag{8.14}$$

where x, f and ϵ are all dimensionless quantities. The solutions to this second-order differential equation with the boundary conditions $f(\pm\infty) = 0$ are well known; see for instance Abramowitz and Stegun (1965). The equation admits infinitely many eigenvalues, one for each non-negative integer n, given by $\epsilon_n = 2n + 1$. The corresponding eigenfunctions are:

$$f_n(x) = (2^n n!\sqrt{\pi})^{-\frac{1}{2}}e^{-\frac{x^2}{2}}h_n(x), \tag{8.15}$$

where $h_n(x)$ is the n-th Hermite polynomial. The first few Hermite polynomials are deceptively simple, for example $h_0(x) = 1$, $h_1(x) = 2x$, $h_2(x) = 4x^2 - 2$. Figure 8.5 shows the lowest eigenfunctions, f_n, and the corresponding probability distributions, $|f_n|^2$. It is apparent that the functions in Figure 8.5 are remarkably similar to the numerical solutions presented in Figure 8.4. This was to be expected as we already noted that the harmonic approximation provides a good description of the lowest-lying nuclear quantum states in N_2.

If we now express the eigenvalues of eqn 8.14 in our original units we find:

$$E_{tot,n} = U_0 + (2n+1)\frac{\hbar\omega_{N_2}}{2} \qquad \text{with } n = 0,1,\ldots, \tag{8.16}$$

which is precisely the same progression obtained in Exercise 8.2 if we identify E_0 with $\hbar\omega_{N_2}/2$. This expression indicates that the *total energy* of the molecule is quantized,

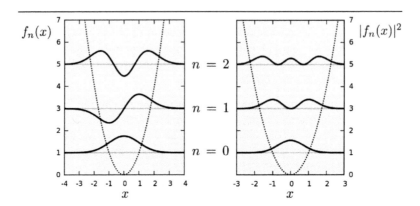

Fig. 8.5 The first three eigenstates of eqn 8.14, as given explicitly in eqn 8.15. The left panel shows the functions $f_n(x)$ for $n = 0, 1, 2$, and the right panels shows the associated probability distributions, $|f_n(x)|^2$. The normalization is such that $\int_{-\infty}^{+\infty} |f_n(x)|^2 dx = 1$. Similarly to Figure 8.4, each curve has been translated vertically so that the level $f_n = 0$ matches the corresponding eigenvalue $\epsilon_n = 2n + 1$, marked by a horizontal line.

i.e. this energy can change only by discrete amounts which are multiples of $\hbar\omega_{N_2}$. Since the smallest amount of energy which can be exchanged with external fields (e.g. the electromagnetic radiation in Raman experiments) or particles (e.g. neutrons in INS experiments) is $\hbar\omega_{N_2}$, this quantity is referred to as a 'quantum of vibrational energy'. At this stage it should be clear why in Chapter 7 we insisted on quoting vibrational frequencies in units of energy.

In the case of molecules, nanostructures and defects in solids, the quanta of vibrational energy are referred to as *vibrons*, while in solids they are referred to as *phonons*.

As the energy separation of any stationary quantum state from the zero-point state is an integer multiple of a vibration quantum, i.e. $E_{\text{tot},n} - E_{\text{tot},0} = n\,\hbar\omega_{N_2}$, we can say that in the state of energy $E_{\text{tot},n}$ we have n quanta, or equivalently that 'there are n phonons (or vibrons)'.

At this point it is worth spending a few more words on the lowest-energy state of the system. In our original units the probability of finding the N nuclei at a distance d in the zero-point state reads:

$$|u_0(d)|^2 = \frac{1}{\sqrt{\pi}\,b} \exp\left[-\frac{(d - d_0)^2}{b^2}\right],$$

i.e. it is a Gaussian with full width at half maximum $2b\sqrt{\log 2} \simeq 0.07$ Å.

Exercise 8.3 ▶Show that the full width at half maximum of the zero-point state of a diatomic molecule scales as $M^{-1/4}$, M being the reduced mass of the molecule. To this end you can combine the expression for the characteristic length b given on page 135 with the dependence of the vibrational frequency on the reduced mass obtained in Exercise 7.2. ▶Plot the full width at half maximum for a range of nuclear masses spanning the periodic table from H to Pb, as a ratio to the width of the N_2 molecule, assuming that the force constants, K, are the same in all cases. Note that this exercise is similar to Exercises 4.1 and 4.2.

Exercise 8.4 The expectation value of the mean square displacement of the nuclei of N_2 from their equilibrium distance in the quantum vibrational state $u(d)$ is given by $\langle u|(d - d_0)^2|u\rangle$. ▶Show that, in the zero-point state, this expectation value is equal to $b^2/2$. For this exercise it is convenient to first convert the integral in the dimensionless units defined for eqn 8.14.

Despite its apparent simplicity, the result obtained in Exercise 8.4 is very important. In fact this result is precisely the missing link required to understand the intensities of vibrational spectroscopy experiments. At the beginning of this section it was said that the intensities of Raman and INS spectra are proportional to the mean square vibrational amplitude, $\langle u^2\rangle_T$, and that in a classical theory of nuclear dynamics this amplitude vanishes when T goes to zero. Now from Exercise 8.4 we learn that a proper quantum theory of nuclear dynamics yields a vibrational amplitude which is *finite* even at $T = 0$. In fact at this temperature the molecule is in its lowest-energy state, i.e. the zero-point state, and therefore $\langle u^2\rangle_T = b^2/2$.

Generally speaking the occurrence of zero-point motion is precisely the reason why we can observe vibrational spectra even at low temperature, not only in the case of small molecules, but for systems of any size and shape, including bulk solids.

8.3 Vibrons and phonons

In this section we generalize the concepts presented in Section 8.2 to the case of systems with more than two atoms, either finite (molecules) or extended (solids).

Once again, within the Born–Oppenheimer approximation the stationary quantum states of the nuclei are obtained by solving the following equation (see eqn 4.6):

$$\left[-\sum_I \frac{\nabla_I^2}{2M_I} + U(\mathbf{R}_1, \ldots, \mathbf{R}_M)\right]\chi(\mathbf{R}_1, \ldots, \mathbf{R}_M) = E_{\text{tot}}\,\chi(\mathbf{R}_1, \ldots, \mathbf{R}_M), \quad (8.17)$$

where the potential energy, U, is the sum of the nuclear–nuclear repulsion and the ground-state energy of the electrons, $E(\mathbf{R}_1, \ldots, \mathbf{R}_M)$, for nuclei immobile in the positions $\mathbf{R}_1, \ldots, \mathbf{R}_M$ (eqn 4.9):

$$U(\mathbf{R}_1, \ldots, \mathbf{R}_M) = \frac{1}{2}\sum_{I\neq J} \frac{Z_I Z_J}{|\mathbf{R}_I - \mathbf{R}_J|} + E(\mathbf{R}_1, \ldots, \mathbf{R}_M). \quad (8.18)$$

In eqns 8.17 and 8.18 we are using Hartree atomic units for consistency with Chapter 4. Now we acknowledge that at temperatures well below the melting point in solids or the dissociation temperature in molecules, the nuclei move only slightly away from their equilibrium positions. We can therefore use the harmonic approximation, as we already did in eqns 7.9 and 7.12:

$$U = U_0 + \frac{1}{2}\sum_{I\alpha}\sum_{J\beta} K_{I\alpha,J\beta}\, u_{I\alpha} u_{J\beta}, \quad (8.19)$$

where the quantities $u_{I\alpha}$ and $u_{J\beta}$ are the nuclear displacements and $K_{I\alpha,J\beta}$ is the matrix of Born–von Karman force constants defined in eqn 7.11. For the sake of clarity in this equation we suspended the Einstein convention and indicated explicitly the sums over nuclei and Cartesian directions. By replacing eqn 8.19 inside eqn 8.17 we find immediately:

$$\left[-\sum_I \frac{1}{2M_I} \nabla_I^2 + \frac{1}{2} \sum_{I\alpha} \sum_{J\beta} K_{I\alpha,J\beta} u_{I\alpha} u_{J\beta} \right] \chi = (E_{\text{tot}} - U_0)\chi. \tag{8.20}$$

The next step is to simplify this equation by moving to the set of collective coordinates w_ν introduced in Section 7.2 (see eqn 7.24). This can be achieved by applying in sequence the coordinate transformations of eqn 7.14 and 7.23:

$$u_{I\alpha} = \sum_\nu M_I^{-\frac{1}{2}} E_{I\alpha,\nu} w_\nu. \tag{8.21}$$

Here the matrix $\mathbb{E}$ whose matrix elements are given by $E_{I\alpha,\nu}$ contains the eigenvectors of the dynamical matrix, as given by eqns 7.21 and 7.16. This transformation allows us to define a new wavefunction, Ξ, in terms of the normal modes coordinates, w_ν, in analogy with was done for the N_2 molecule on page 135:

$$\Xi(w_1, \ldots, w_{3M}) = \chi \left(\ldots, R_{I\alpha}^0 + \sum_\nu M_I^{-\frac{1}{2}} E_{I\alpha,\nu} w_\nu, \ldots \right). \tag{8.22}$$

By replacing the last two equations inside eqn 8.20 we obtain the nuclear Schrödinger equation in the normal modes coordinates:

$$\sum_\nu \left(-\frac{1}{2} \frac{\partial^2}{\partial w_\nu^2} + \frac{1}{2} \omega_\nu^2 w_\nu^2 \right) \Xi = (E_{\text{tot}} - U_0) \Xi, \tag{8.23}$$

where the quantities ω_ν are the vibrational eigenfrequencies from eqn 7.20. The transformation from eqn 8.20 to eqn 8.23 is rather lengthy but otherwise should not pose any difficulties. The key ingredients are the relation between the Born–von Karman force constants and the dynamical matrix (eqn 7.16), the relation between the dynamical matrix and the normal modes (eqn 7.21), and the fact that the matrix of normal modes is orthogonal. In addition we must transform the derivatives with respect to $u_{I\alpha}$ into derivatives with respect to w_ν using the chain rule.

The left-hand side of eqn 8.23 contains the Hamiltonian of a set of *non-interacting* harmonic oscillators. This situation is reminiscent of the independent electron approximation discussed in Section 2.5; therefore it must be possible to obtain a solution in terms of a product of wavefunctions, as in eqn 2.35:

$$\Xi(w_1, \ldots, w_{3M}) = \Phi_1(w_1)\Phi_2(w_2)\cdots. \tag{8.24}$$

In complete analogy with eqn 2.36 of Section 2.5 we can choose each Φ_ν to be the solution of the following single-particle equation in the variable w_ν:

$$\left[-\frac{1}{2} \frac{\partial^2}{\partial w_\nu^2} + \frac{1}{2} \omega_\nu^2 w_\nu^2 \right] \Phi_\nu = E_\nu \Phi_\nu. \tag{8.25}$$

The solutions of this equation are precisely the Hermite functions given in eqn 8.15 if we make the substitutions $\epsilon = 2E_\nu/\omega_\nu$ and $x = w_\nu \sqrt{\omega_\nu}$:

$$\Phi_\nu(\omega_\nu) = f_n (w_\nu \sqrt{\omega_\nu}), \qquad E_\nu = (2n+1)\frac{\omega_\nu}{2}.$$

Since for each vibrational eigenmode of frequency ω_ν we have infinitely many solutions $f_0, f_1, \ldots$, it is convenient to label these solutions by the integer n_ν in order to distinguish between the various eigenmodes. The complete wavefunction, $\Xi(w_1, \ldots, w_{3M})$, will be a product of Hermite functions, and the final wavefunction, χ, in the original coordinates, $\mathbf{R}_1, \ldots, \mathbf{R}_M$, will be obtained by using the inverse transform of eqn 8.22:

$$w_\nu = \sum_{J\beta} E_{\nu,J\beta}^\top M_J^{\frac{1}{2}} (R_{J\beta} - R_{J\beta}^0).$$

The total energy of the system can be found by replacing eqn 8.24 inside eqn 8.23 and using eqn 8.25. The procedure is identical to what was done in Section 2.5 and the result is analogous to eqn 2.38 of that section:

$$E_{\text{tot}} = U_0 + \sum_\nu (2n_\nu + 1)\frac{\hbar\omega_\nu}{2}. \tag{8.26}$$

Here we have re-introduced the Planck constant in order to be consistent with our discussion of the N_2 molecule (in other words, eqn 8.26 is now valid in any system of units).

Equation 8.26 is the generalization to the case of many atoms of eqn 8.16 obtained for N_2. The main difference is that now we can have an integer number, $n_\nu \geq 0$, of vibrons or phonons in each mode of frequency ω_ν. At this point any solution of the nuclear Schrödinger equation (eqn 8.20) is uniquely defined by specifying how many phonons or vibrons we have in each vibrational eigenmode: $E_{\text{tot}} = E_{\text{tot}}(n_1, n_2, \ldots, n_{3M})$.

In this case the zero-point state, i.e. the lowest-energy state of the system, is obtained by setting to zero the number of phonons or vibrons in each mode. If we set $n_\nu = 0$ for every mode in eqn 8.26 we find:

$$(E_{\text{tot}})_{\text{min}} = U_0 + \sum_\nu \frac{\hbar\omega_\nu}{2}. \tag{8.27}$$

This result is important insofar as it tells us that the ground-state energy of any atomic system (from molecules to solids) is slightly higher than the minimum of the potential energy surface, U_0. The additional energy, $\sum_\nu \hbar\omega_\nu/2$, is referred to as 'zero-point correction' and cannot be ignored in thermochemical and thermophysical calculations.

In the case of crystalline solids everything that we have discussed so far remains valid almost without any changes. The only differences are that (i) the vibrational eigenfrequencies depend on the wavevector, $\mathbf{q}$, as we have seen in Section 7.4, and

(ii) the sum in eqn 8.26 must also be performed for each wavevector inside the Brillouin zone of the crystal. The total energy per crystalline unit cell becomes:

$$E_{\text{tot}} = U_0 + \sum_{\nu} \int \frac{d\mathbf{q}}{\Omega_{\text{BZ}}} (2n_{\mathbf{q}\nu} + 1) \frac{\hbar \omega_{\mathbf{q}\nu}}{2}, \tag{8.28}$$

where Ω_{BZ} indicates the volume of the Brillouin zone (Appendix D). The integral is restricted to the first Brillouin zone in order to avoid duplicate vibrational modes, as we have seen already in Section 7.4. In this case the zero-point state is obtained by zeroing all the phonon numbers, $n_{\mathbf{q}\nu} = 0$.

At the end of this section it may be useful to mention that the quantization of the vibrational energy leads naturally to the intuitive notion that vibrons and phonons are somewhat 'particle' in nature and can be exchanged with the external environment. Accordingly, when in vibrational spectroscopy the electromagnetic field or the neutrons lose energy to the sample, it is common to say that the material has 'absorbed' a phonon. Similarly, when radiation or neutrons gain energy, we say that the material has 'emitted' a phonon. These events are generally referred to as one-phonon processes. When we consider exchanges of energy such as $2\,\hbar\omega_{\mathbf{q}\nu}$ (see Section 8.1.2) we talk instead of two-phonon processes and so on. The reader interested in the zoology of multi-phonon processes can find an introductory presentation in the textbook by Ashcroft and Mermin (1976), and a more advanced discussion in the book by Yu and Cardona (2010).

The reader will also have noted that we started this section with the assumption that the harmonic approximation provides a reliable description of nuclear dynamics. The vast majority of current DFT calculations is based indeed on the harmonic approximation. However, there are cases where more accurate information is needed, and anharmonic effects need to be considered. The theory of anharmonic effects in crystals can be found in Cowley (1968), and the papers by Lazzeri *et al.* (2003) and Monserrat *et al.* (2013) constitute examples of recent DFT calculations of anharmonic effects.

8.4 Phonon density of states

In the study of the vibrational properties of solids and nanostructures it is often useful to know how many normal modes of vibration exist in the neighbourhood of a given phonon energy. For example, this information is necessary when performing calculations of phase diagrams (Section 8.5), specific heat, thermal conductivity, electrical conductivity and also the critical temperature of superconductors (Grimvall, 1981).

In principle we could obtain this information by simply constructing a histogram of all the phonon energies, $\hbar\omega_{\mathbf{q}\nu}$, calculated for a certain material. However, for practical purposes it is more convenient to use a slightly more formal definition as follows:

$$g(E) = \frac{1}{M} \sum_{\nu} \int \frac{d\mathbf{q}}{\Omega_{\text{BZ}}} \delta(E - \hbar\omega_{\mathbf{q}\nu}), \tag{8.29}$$

where E is the phonon energy and M indicates, as usual, the number of nuclei. In this expression, for each vibrational mode the Dirac delta function contributes a very narrow peak normalized to 1, so that:

$$\int_0^\infty g(E)\,dE = \frac{1}{M}\sum_\nu \int \frac{d\mathbf{q}}{\Omega_{\text{BZ}}} \int_0^\infty \delta(E - \hbar\omega_{\mathbf{q}\nu})\,dE = \frac{1}{M}\sum_\nu \int \frac{d\mathbf{q}}{\Omega_{\text{BZ}}} = 3. \qquad (8.30)$$

The quantity $g(E)$ defined by eqn 8.29 is referred to as phonon *density of states* and is usually abbreviated as 'DOS'. In the literature it is common to find slightly different naming conventions. For instance, if the solid is not crystalline we talk about 'vibrational density of states'. Clearly, normalization choices different from that used in eqn 8.29 are also possible; for example, in the case of crystals it is common to see density of states per unit cell.

As an example of vibrational DOS, Figure 8.6 shows the comparison between the density of states of vitreous silica (a-SiO_2) calculated using DFT/LDA and that measured by inelastic neutron scattering. Vitreous silica is a disordered oxide whereby every Si atom forms covalent bonds with four O atoms, and each O atom is connected in turn to two Si atoms. The resulting structure is a continuous random network of SiO_4 tetrahedral units, as can be seen in the ball-and-stick model in Figure 8.6.

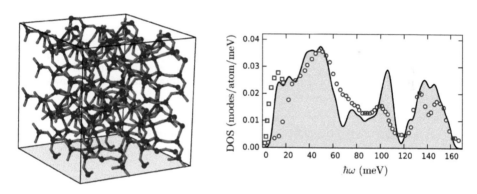

Fig. 8.6 Ball-and-stick model of a-SiO_2 (left) and corresponding vibrational density of states (right). Glassy materials do not have a crystalline structure with a periodically repeated unit cell; therefore the determination of atomistic structural models can be very challenging. In this example the model structure was generated using a first-principles molecular dynamics simulation (Car and Parrinello, 1985). In this simulation one first determines a model of liquid SiO_2, and then rapidly freezes the liquid in order to obtain a glassy structure (Sarnthein *et al.*, 1995). The model in the figure consists of 24 Si atoms and 48 O atoms in a periodic computational cell. The theoretical DOS was first calculated using DFT/LDA by Sarnthein *et al.* (1997). The black curve is taken from the subsequent calculations of Giustino and Pasquarello (2006). The experimental DOS is from the inelastic neutron scattering measurements by Carpenter and Price (1985, circles) and Buchenau *et al.* (1986, squares). Both the experimental and theoretical DOS are normalized as in eqn 8.30. The Dirac deltas of eqn 8.29 have been replaced by Gaussians of width 2.5 meV in order to have a smooth curve.

Vitreous silica is a ubiquitous material as it is the main component of window glass and optical fibres, and is also used in the electronics industry for the electrical insulation of silicon wafers. The DOS in the right panel of Figure 8.6 was obtained by calculating the vibrational eigenfrequencies of a structural model of a-SiO_2 containing 72 atoms (Sarnthein *et al.*, 1995). The calculation of the frequencies was performed exactly as in Exercise 7.7; the only difference is that in this case there are many more atoms in the computational cell and $72 \times 3 = 216$ eigenmodes.

Figure 8.6 shows that there is a rather good agreement between INS experiments and DFT calculations, especially when it comes to the energy of the main peaks. Some disagreement can be noted instead for the intensity of the peaks, in particular for the peak around 100 meV. These discrepancies result from two factors. The first is that the DOS measured in INS experiments is slightly more complicated that the simple expression given in eqn 8.29 (Price and Carpenter, 1987). When this aspect is taken into account the calculated peak around 100 meV becomes smaller and the agreement with experiment improves considerably (Pasquarello *et al.*, 1998). The second factor is that these DFT calculations can be performed only on relatively small atomistic models (e.g. up to 200–300 atoms), hence these models may not fully capture the complexity of the random network in the real sample. Since different atomistic models yield slightly different vibrational properties, as shown by Giacomazzi *et al.* (2009), the key challenge here is to build a model which can be representative of the 'average structure' of the real system.

Exercise 8.5 ▶Rewrite the expression for the vibrational density of states in eqn 8.29 for the case of non-crystalline systems. ▶Show that the zero-point correction to the ground-state energy (see eqn 8.27) can be calculated in terms of the vibrational DOS as follows:

$$\sum_{\nu} \frac{1}{2} \hbar \omega_{\nu} = 3M \frac{\hbar \langle \omega \rangle}{2},$$

where $\langle \omega \rangle$ is the mean vibrational frequency of the system, averaged over the density of states:

$$\hbar \langle \omega \rangle = \frac{\int_0^{\infty} g(E) E \, dE}{\int_0^{\infty} g(E) \, dE}.$$

▶By reading the calculated DOS of a-SiO_2 from Figure 8.6, provide a simple estimate for the zero-point correction per atom in vitreous silica. ☐As a sanity check, the numerical evaluation of the previous two equations using the data in Figure 8.6 yields a correction of 38 meV/atom.

8.5 Phonon DOS and pressure–temperature phase diagrams

As an example of the uses of the phonon DOS and the expression for the total energy in eqn 8.26, in this final section we illustrate the principles underlying the calculation of pressure–temperature phase diagrams using DFT.

8.5.1 Thermal average of total energy

In all the previous chapters and in the present one up to this point we have discussed the structural, mechanical and vibrational properties of materials in their ground state, i.e. in the lowest-energy state at temperature $T = 0$. In order to study pressure–temperature phase diagrams we need to extend our description to temperatures $T > 0$ (Wallace, 1998; Grimvall, 1999).

When $T > 0$, instead of studying one quantum state with a total energy given by eqn 8.26 for a given set of phonon numbers, n_ν, we need to consider an 'ensemble' of quantum states in thermal equilibrium with each other at temperature T. We can think of such an ensemble as a large number of replicas of the same system, where each replica is described by a well defined quantum state, but all the macroscopic measurements result from taking averages over all the replicas (e.g. the microcrystals of a polycrystalline solid).

Let us indicate a quantum state identified by the phonon numbers $\{n_\nu\}$ using the 'state index' i, and the corresponding total energy using $E_{\text{tot},i}$. The probability, P_i, that our system occupies the state of energy $E_{\text{tot},i}$ at equilibrium (i.e. the fraction of replicas having this energy) is given by the *Gibbs distribution law* (Landau and Lifshitz, 1969):

$$P_i = C \exp\left(-\frac{E_{\text{tot},i}}{k_{\text{B}}T}\right), \tag{8.31}$$

with C a constant to be determined. This result is among the fundamental relations of statistical mechanics, and can be proved using simple arguments. The reader will find a very clear derivation of this result in the book by Fermi (1966, pages 205–209). The constant appearing in eqn 8.31 is simply determined by requiring that the probability of finding the system in any state be one:

$$\sum_i P_i = 1 \rightarrow C \sum_i \exp\left(-\frac{E_{\text{tot},i}}{k_{\text{B}}T}\right) = 1.$$

In statistical mechanics the dimensionless sum on the right-hand side is called the *partition function* and is denoted by the symbol Z (Kittel, 1958; Landau and Lifshitz, 1969):

$$Z = \sum_i \exp\left(-\frac{E_{\text{tot},i}}{k_{\text{B}}T}\right). \tag{8.32}$$

Using this definition probability P_i can be written as:

$$P_i = \frac{1}{Z} \exp\left(-\frac{E_{\text{tot},i}}{k_{\text{B}}T}\right). \tag{8.33}$$

The knowledge of the probabilities allows us to calculate rigorously the *thermal average* of the quantum-mechanical total energy of the system at equilibrium. In fact it is intuitive that, if every state with energy $E_{\text{tot},i}$ occurs with a probability P_i, then the average energy of the system must be:

$$\langle E_{\text{tot}} \rangle_T = \sum_i P_i E_{\text{tot},i}. \tag{8.34}$$

By inserting eqns 8.33 and 8.32 inside 8.34 and carrying out some simple algebra we find:

$$\langle E_{\text{tot}} \rangle_T = k_{\text{B}}T^2 \frac{\partial}{\partial T} \log Z. \tag{8.35}$$

This equation relates the thermal average of the total energy of our system, at equilibrium at temperature T, with its partition function, Z.

In order to see how eqn 8.35 is connected with what was discussed in the previous sections, let us replace eqn 8.26 for the total energies $E_{\text{tot},i}$ appearing in the partition function. We have:

$$E_{\text{tot},i} = U_0 + \sum_\nu (2n_\nu + 1)\frac{\hbar\omega_\nu}{2}, \quad \text{where } i \text{ is a label for } \{n_1, n_2, \ldots, n_{3M}\},$$

therefore:

$$Z = \sum_{n_1=0}^{\infty} \sum_{n_2=0}^{\infty} \cdots \sum_{n_{3M}=0}^{\infty} \exp\left\{-\frac{1}{k_\text{B}T}\left[U_0 + \sum_{\nu=1}^{3M}(n_\nu + 1/2)\hbar\omega_\nu\right]\right\}.$$

In this last expression the sum over the states, i, has been replaced by the explicit sum over all possible combinations of phonon numbers with $n_\nu = 0, 1, \ldots, \infty$ for any vibrational mode, ν. The above expression is somewhat scary at first sight, but it is actually very easy to evaluate. To see this, let us factor out all the terms corresponding to integer multiples of $\hbar\omega_1$. We find:

$$Z = \sum_{n_1=0}^{\infty} \exp\left\{-\frac{\hbar\omega_1}{k_\text{B}T}n_1\right\} \sum_{n_2=0}^{\infty} \cdots \sum_{n_{3M}=0}^{\infty} \exp\left\{\ldots\right\}.$$

Now the sum over n_1 can be transformed in a well-known geometric series by performing the variable substitution $x = \exp(-\hbar\omega_1/k_\text{B}T)$:

$$\sum_{n_1=0}^{\infty} \exp\left\{-\frac{\hbar\omega_1}{k_\text{B}T}n_1\right\} = \sum_{n_1=0}^{\infty} x^{n_1} = \frac{1}{1-x} = \frac{1}{1 - \exp(-\hbar\omega_1/k_\text{B}T)}.$$

All the other terms containing integer multiples of $\hbar\omega_\nu$ can be dealt with in the same way, while the terms containing U_0 and those containing $\hbar\omega_\nu/2$ simply factor out. The final result is:

$$Z = \exp(-E_{\text{tot,min}}/k_\text{B}T) \prod_\nu \frac{1}{1 - \exp(-\hbar\omega_\nu/k_\text{B}T)}, \tag{8.36}$$

where $E_{\text{tot,min}}$ is the total energy in the ground state including the zero-point correction, as in eqn 8.27. The symbol Π indicates the product over all the vibrational modes of the system. At this point we can use this expression for the partition function inside eqn 8.35. After a few straightforward steps we obtain:

$$\langle E_{\text{tot}}\rangle_T = E_{\text{tot,min}} + \sum_\nu \frac{1}{\exp(\hbar\omega_\nu/k_\text{B}T) - 1}\hbar\omega_\nu.$$

In statistical mechanics the function defined by:

$$n_T(E) = \frac{1}{\exp(E/k_\text{B}T) - 1}, \tag{8.37}$$

is called the *Bose–Einstein* distribution (Kittel, 1958; Landau and Lifshitz, 1969). Using this function the previous result can be written compactly as:

$$\langle E_{\text{tot}} \rangle_T = E_{\text{tot,min}} + \sum_\nu n_T(\hbar\omega_\nu)\, \hbar\omega_\nu = U_0 + \sum_\nu \left[2n_T(\hbar\omega_\nu) + 1\right] \frac{\hbar\omega_\nu}{2}. \qquad (8.38)$$

This result is very important since it allows us to calculate the total energy of the system at any given temperature, starting from the DFT total energy and vibrational frequencies.

Strictly speaking, eqn 8.38 is valid only in a limited range of temperatures, since the vibrations are described within the harmonic approximation (Section 8.2); however, for simplicity we will ignore this limitation in the following.

An interesting aspect of eqn 8.38 is that it looks formally identical to our previous result, eqn 8.26. The only difference is that we have replaced the phonon numbers, n_ν, with the Bose–Einstein functions, $n_T(\hbar\omega_\nu)$. Given this similarity we may interpret $n_T(\hbar\omega_\nu)$ as the 'average number of phonons' in the mode with energy $\hbar\omega_\nu$ at temperature T.

Exercise 8.6 We consider the peaks at $\sim$40 meV and $\sim$140 meV in the density of vibrational states of vitreous silica shown in Figure 8.6. ▶Plot the Bose–Einstein distributions of eqn 8.37 corresponding to the two phonon energies $\hbar\omega_1 = 40$ meV and $\hbar\omega_2 = 140$ meV, for temperatures between 0 and 1,000 K. ▶Calculate the average number of phonons per atom in vitreous silica with the energy $\hbar\omega_1$ or $\hbar\omega_2$ at the following temperatures: 10 K, 300 K, 1,000 K. ▶Show that the temperature, T_1, at which there would be, on average, one phonon of energy $\hbar\omega$ per atom in the system is $T_1 = \hbar\omega/k_B \log 4$. ☐Particles which are described by the Bose–Einstein distribution are called 'bosons'. Therefore phonons can be considered as one particular kind of boson. This should be contrasted with the case of electrons, which follow the Fermi–Dirac distribution in eqn 5.16, and hence are called 'fermions', as mentioned in Section 2.6.

It is often useful to express the partition function, Z, from eqn 8.36 in terms of the vibrational density of states. This can be achieved simply by evaluating $\log Z$ instead of Z, as follows:

$$\log Z = -\frac{E_{\text{tot,min}}}{k_B T} - \sum_\nu \log[1 - \exp(-\hbar\omega_\nu/k_B T)]$$

$$= -\frac{E_{\text{tot,min}}}{k_B T} - \int_0^\infty dE \sum_\nu \log[1 - \exp(-E/k_B T)]\delta(E - \hbar\omega_\nu)$$

$$= -\frac{E_{\text{tot,min}}}{k_B T} + M \int_0^\infty dE\, \log[1 + n_T(E)]\, g(E). \qquad (8.39)$$

Here M is the number of atoms per unit cell, $g(E)$ is normalized as in eqn 8.30, and we used the definition of the Bose–Einstein distribution from eqn 8.37. In this form we clearly see that the partition function depends only on the ground-state energy of the system and the phonon DOS.

Since the total energy and the DOS are calculated in DFT for a given computational cell (as in Figure 5.2), the partition function is uniquely defined by the symmetry of the crystal lattice, the volume of the unit cell and the temperature.

When lattice vibrations are described within the harmonic approximation, but the dependence of the vibrational frequencies on the volume of the unit cell is taken into account (by calculating the DOS for different volumes), we say that we are using the *quasiharmonic approximation* (Carrier *et al.*, 2007; Baroni *et al.*, 2010).

8.5.2   The Gibbs energy

In order to calculate phase diagrams, the knowledge of the total energy alone, as given by eqn 8.38, is not sufficient. To see this, let us imagine a situation where we calculated the total energy of two crystalline phases of a solid for the same unit cell volume, as in Figure 5.2. Suppose that these calculations gave the same total energy for these systems, but different pressures, $p_1 > p_2$. In this case the work of the external forces required to establish the pressure in the two systems would be larger for the system at p_1. Therefore we should expect the system at p_2 to be more stable, even though the two systems have the same total energy.

These intuitive considerations can be laid on solid formal grounds by using a quantity which is closely related to the total energy, and is called the *Gibbs energy*, G. The discussion of the Gibbs energy requires some background knowledge of statistical mechanics and thermodynamics, hence the reader not already familiar with these subjects is invited to consult one of the many books on the subject (e.g. Callen, 1985; Landau and Lifshitz, 1969). In order to keep the chapter self-contained, here we give an extremely condensed summary of the key concepts, without any pretence of rigour.

Let us first see how to calculate two important quantities, the *entropy* and the *pressure* of the system. Using the probabilities P_i in eqn 8.33, Gibbs' definition of the entropy is as follows:

$$S = -k_B \sum_i P_i \log P_i. \tag{8.40}$$

From this definition it can be shown that the entropy provides a measure of how many microscopic states (i.e. the replicas of Section 8.5.1) correspond to the same macroscopic state. Intuitively we can understand that the macroscopic equilibrium state of a system, i.e. the most probable state, is the one which can be realized in the largest number of microscopic states. Therefore the thermodynamic equilibrium of a system can be identified with the configuration of maximum entropy.

As for the thermal average of the total energy in eqn 8.35, the entropy can be expressed in terms of the partition function. In fact, by combining eqn 8.40 with eqns 8.33–8.35 and after performing a few manipulations, we obtain:

$$S = k_B \left(1 + T\frac{\partial}{\partial T}\right) \log Z. \tag{8.41}$$

Since Z can be expressed in terms of the DFT total energy and the vibrational energies of the system (eqn 8.36), this last result provides us with a recipe for calculating the entropy.

In order to calculate the pressure, p, of the system using DFT we can use the first and second laws of thermodynamics. These laws can be stated compactly using the differential identity $d\langle E_{\text{tot}}\rangle_T = TdS - p\,d\Omega$ (Callen, 1985). If we keep the temperature, T, constant, we can replace TdS in this expression by $d(TS)$, thereby obtaining:

$$p = -\frac{\partial}{\partial\Omega}\left(\langle E_{\text{tot}}\rangle_T - TS\right) \quad \text{with} \quad T = \text{const.}$$

By inserting eqns 8.35 and 8.41 in this expression we obtain immediately:

$$p = k_B T \frac{\partial}{\partial\Omega} \log Z. \tag{8.42}$$

The total energy, the entropy and the pressure can be combined in a common function called the *Gibbs energy* and defined as follows:

$$G = \langle E_{\text{tot}}\rangle_T - TS + p\Omega. \tag{8.43}$$

At $T = 0$ the Gibbs energy reduces to the enthalpy discussed in Exercise 5.2, Section 6.6 and Figure 6.6. The Gibbs energy is extremely useful for studying thermodynamic equilibria at fixed temperature and pressure. In fact, as stated above, the equilibrium configuration of a solid is the one that maximizes the value of the entropy. Now it can be shown rigorously that the maximization of the entropy at fixed pressure and temperature is mathematically equivalent to finding the *minimum* of the Gibbs energy in eqn 8.43 (a proof of equivalence is given by Callen, 1985, pages 153–157). The connection between entropy and Gibbs energy implies that, in order to identify the equilibrium phase of a solid, we need to *determine the structure with the lowest Gibbs energy* for given temperature and pressure.

Since $\langle E_{\text{tot}}\rangle$, S and p are all functions of $\log Z$, so is the Gibbs energy. In fact, by combining eqns 8.43, 8.42, 8.35 and 8.41 we find immediately:

$$G = -k_B T \left(1 - \Omega\frac{\partial}{\partial\Omega}\right) \log Z. \tag{8.44}$$

In this form the Gibbs energy can be determined from DFT by calculating the total energy including the zero-point correction, $E_{\text{tot,min}}$, the vibrational density of states $g(E)$, as well as their numerical derivatives with respect to the volume Ω.

Exercise 8.7 We consider a hypothetical solid with a phonon DOS sharply peaked around the energy $\hbar\omega_0$: $g(E) = 3\,\delta(E - \hbar\omega_0)$, where δ indicates the Dirac delta function. The ground-state energy of this solid at clamped nuclei is U_0. ▶Using eqns 8.39 and 8.37 show that the partition function of this solid is given by:

$$\log Z = -\frac{U_0}{k_B T} - 3M\frac{1}{2}\frac{\hbar\omega_0}{k_B T} + 3M \log[1 + n_T(\hbar\omega_0)]. \tag{8.45}$$

▶Evaluate the thermal average of the total energy, $\langle E_{\text{tot}}\rangle_T$, using eqn 8.35 and show that:

$$\langle E_{\text{tot}}\rangle_T = U_0 + 3M\left[\frac{1}{2} + n_T(\hbar\omega_0)\right]\hbar\omega_0. \tag{8.46}$$

▶Next we consider the entropy. Using eqn 8.41 and eqn 8.45 show that:

$$S = 3Mk_{\mathrm{B}} \left\{ \log[1 + n_T(\hbar\omega_0)] + \frac{\hbar\omega_0}{k_{\mathrm{B}}T} n_T(\hbar\omega_0) \right\}. \tag{8.47}$$

☐The last remaining step is to determine the pressure. We define the dimensionless *Grüneisen parameter*, γ, as follows (Grimvall, 1999):

$$\gamma = -\frac{\Omega}{\omega_0} \frac{\partial\omega_0}{\partial\Omega}.$$

This parameter provides information on the change of the vibrational frequency of the solid with its volume. ▶Using the definition of the Grüneisen parameter and eqns 8.42 and 8.45, show that:

$$p = -\frac{\partial U_0}{\partial\Omega} + 3M \left[\frac{1}{2} + n_T(\hbar\omega_0) \right] \frac{\hbar\omega_0}{\Omega} \gamma. \tag{8.48}$$

☐The parameters of the solid under consideration are as follows: there is one atom in the unit cell, the peak phonon energy is $\hbar\omega_0 = 20$ meV and the Grüneisen parameter is $\gamma = 1.5$. The pressure term $-\partial U_0/\partial\Omega$ can be neglected, and the zero of the energy axis can be taken to coincide with U_0. ▶Plot $\langle E_{\mathrm{tot}}\rangle_T$, TS and $p\Omega$ for this system as a function of temperature, for T between 0 and 1,000 K. Express the energies in meV. ▶Using the definition of the Gibbs energy in eqn 8.43, plot the Gibbs energy vs. temperature, in the same temperature range as above. ☐From these plots we see that, as the temperature increases, the pressure also increases. In a three-dimensional plot of the Gibbs energy in the pT plane this situation would correspond to moving along a line which is not parallel to the temperature axis, but is bending towards higher pressures.

Practical calculations of phase diagrams involve the determination of $E_{\mathrm{tot,min}}$ and $g(E)$ for a set of volumes, and then the calculation of the Gibbs energy as a function of temperature using eqns 8.44 and 8.39. The equation for the pressure (eqn 8.42) is used to identify the (p, T) point in the phase diagram. In the presence of multiple possible phases, the 'stable phase' at each point will be that with the lowest Gibbs energy. This procedure is illustrated schematically in Figure 8.7 for a hypothetical solid that can crystallize in two phases.

Exercise 8.8 In this exercise we want to try a simplified version of the procedure for calculating the transition temperature between two phases of a hypothetical solid. The solid in question has one atom per unit cell ($M = 1$), and we will denote the two phases by α and β. We are interested in finding the stable phase at ambient pressure; therefore we are going to evaluate the Gibbs energy with the approximation that the pressure term, $p\Omega$, in eqn 8.43 can be neglected. The DFT total energy of the α phase in its ground state at clamped nuclei is higher than the ground-state energy of the β phase by 1.68 meV (note that this is a fictitious value: in actual DFT calculations it would be very difficult to resolve such a small energy difference). The phonon DOS of these two phases have been calculated and are reported in the following table for a small set of equally spaced vibrational energies.

$\hbar\omega$	g_α	g_β	$\hbar\omega$	g_α	g_β	$\hbar\omega$	g_α	g_β
meV	1/meV	1/meV	meV	1/meV	1/meV	meV	1/meV	1/meV
0	0.000	0.000	12	0.049	0.070	24	0.229	0.121
2	0.001	0.003	14	0.085	0.075	26	0.210	0.151
4	0.002	0.008	16	0.110	0.077	28	0.150	0.209
6	0.006	0.023	18	0.144	0.078	30	0.067	0.302
8	0.015	0.045	20	0.172	0.085	32	0.019	0.087
10	0.029	0.064	22	0.212	0.102	34	0.000	0.000

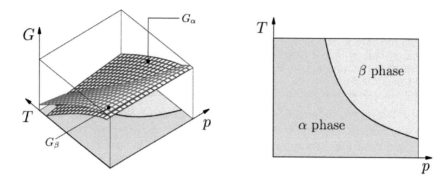

Fig. 8.7 Schematic representation of a calculation of phase diagrams using DFT. In this example a solid can crystallize in two phases, which we denote as the α phase and β phase. In order to determine the pT phase diagram it is necessary to perform DFT calculations of the total energy and of the vibrational frequencies for a set of unit cell volumes, for both the α and β phases. Using these data it is possible to calculate $\log Z$ for each volume, and also evaluate the numerical derivatives of $\partial \log Z / \partial \Omega$, which are needed in eqn 8.42 in order to identify the pressure. These quantities are then used in eqn 8.44 to obtain the Gibbs energy for a discrete set of (p, T) coordinates. The result will be two surfaces, $G_\alpha = G_\alpha(p, T)$ and $G_\beta = G_\beta(p, T)$, indicated by the blue and red surfaces on the left, respectively. At each (p, T) point the equilibrium phase is the one with the lowest Gibbs energy. The intersection between the two Gibbs surfaces marks the phase boundary (black line). The two-dimensional map thus extracted from the three-dimensional plot on the left defines the pT phase diagram on the right.

▶By using the trapezium rule verify that g_α and g_β in the table are correctly normalized to three modes per atom. This operation can be performed manually, or by using the command 'trapz' of the program 'Octave' for example. ▶Starting from the expression for the vibrational zero-point energy in eqn 8.27, and following similar steps as in eqn 8.39, express the zero-point energy per atom, $\Delta E_{\rm zp} = (E_{\rm tot,min} - U_0)/M$, in terms of the phonon DOS. Show that the result is:

$$\Delta E_{\rm zp} = \frac{1}{2} \int_0^\infty dE \, E \, g(E).$$

▶Using the trapezium rule, calculate the zero-point corrections to the ground-state energy of the two phases of our hypothetical solid, using the DOS given in the table. ▶By combining eqns 8.44, 8.43 and 8.42 show that, if we ignore the term $p\Omega$, the Gibbs energy per atom can be expressed in terms of the density of states as follows:

$$G/M = U_0/M + \Delta E_{\rm zp} - k_{\rm B}T \int_0^\infty dE \, \log[1 + n_T(E)] \, g(E)$$

Strictly speaking this quantity, which corresponds to $G - p\Omega$, is called the 'Helmholtz energy'. ▶Using this expression, evaluate the Gibbs function of the α and β phases at the temperatures $T = 1, 50, 100, 150, 200, 250$ and 300 K. Perform the integration using the trapezium rule. ▶Plot the Gibbs functions G_α and G_β and determine the range of temperatures at which each phase is the most stable. ▶Show that the transition temperature obtained from this graphical analysis is approximately $T_c \simeq 150$ K.

☐For the sake of simplicity in this exercise we neglected the term $p\Omega$ in the Gibbs energy. In reality this approximation is not allowed. In fact, by increasing the temperature at fixed volume, the pressure p increases as well, as discussed in Exercise 8.7. Another point worth mentioning is that, in actual calculations, it is not meaningful to sample the DOS at very few energies as in the table above. In order to minimize integration errors the DOS is usually described using fine energy grids, e.g. at 0.1 meV intervals.

Exercises 8.7 and 8.8 show that the calculation of the Gibbs energy and the study of phase diagrams is already very laborious even in the simplest possible cases. On top of the computational complexity, the phase boundaries are also rather sensitive to the numerical accuracy of the calculations, owing to the fact that typically the Gibbs energies of competing phases have similar magnitude and slopes in the (p, T) plane.

An example of DFT calculations of phase diagrams is shown in Figure 8.8 for magnesium. Mg is among the most abundant elements on Earth, and is commonly used in structural components, especially in those applications where weight matters. The electronic configuration of the Mg atom is $[\mathrm{Ne}]3s^2$, and at ambient conditions this element is metallic and crystallizes in a hexagonal close-packed (hcp) lattice. At room temperature and at pressures around 50 ± 6 GPa, Mg undergoes a phase transition to a body-centred cubic (bcc) structure (Olijnyk and Holzapfel, 1985). The left panel of Figure 8.8 shows DFT calculations of the solid–solid hcp–bcc phase transition in Mg, by Mehta *et al.* (2006). In this work the authors used methods which are conceptually identical to the theory discussed in this section. However, in order to control the numerical accuracy of their results, Mehta *et al.*

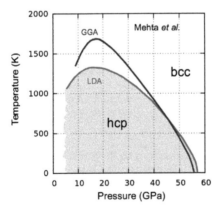
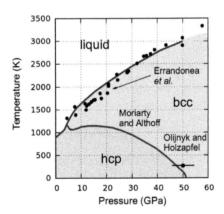

Fig. 8.8   The phase diagram of Mg from first principles. The left panel shows the calculations of the hcp–bcc phase boundary of Mg by Mehta *et al.* (2006). The use of different DFT exchange and correlation functionals (indicated by 'LDA' and 'GGA') can lead to different results for the phase boundaries. We can take these differences as an estimate of the error bar of the *ab initio* predictions. The right panel shows a calculation of the hcp–bcc–liquid phase diagram of Mg by Moriarty and Althoff (1995). The calculated phase boundaries (blue lines) are in very good agreement with the measurements (discs) by Errandonea *et al.* (2001) and Olijnyk and Holzapfel (1985).

introduced a sophisticated procedure whereby explicit analytical functions were fitted to the thermodynamic quantities calculated using DFT, and subsequently the phase boundaries were determined accurately using these analytical fits.

The right panel of Figure 8.8 shows a calculation of the entire phase diagram of Mg, including also a transition to the liquid, by Moriarty and Althoff (1995). In this case the authors used DFT calculations in order to parametrize an interatomic potential, that is an explicit function giving the force between any two atoms in terms of their distance. This mapping of the DFT total energy into a simplified expression (where the electrons no longer appear explicitly) is useful for studying very large systems. For example, Moriarty and Althoff (1995) were able to study directly the melting of Mg by using a model containing 1,024 atoms. The melting can be simulated by solving Newton's equation of motion for the atoms as function of time. By increasing the temperature in the simulations it is possible to explore the melting line, as shown in the figure. In the same figure we show a comparison with experimental measurements of the melting curve (Errandonea *et al.*, 2001). The measurements agree quite remarkably with the calculations, and it is worth noting that the calculations were performed several years *before* the measurements; i.e. this is truly a prediction from first principles.

After a long discussion of the structural, mechanical, vibrational and thermodynamic properties of materials, it is now time to move away from the nuclei. In the following chapters (9–11) we will 'forget' about the nuclei and focus on the energetics and dynamics of the electrons.

9
Band structures and photoelectron spectroscopy

In Chapters 4–8 we saw how to use DFT calculations in order to study the structure of materials, their deformations, their vibrations and phase diagrams. All our discussions referred to electrons in their *ground state*. In this chapter we want to discuss what DFT can tell us about the energetics and dynamics of electrons. This information is important in the study of the electrical properties of materials, optical properties (Chapter 10) and magnetic properties (Chapter 11).

For this discussion we will assume that $T = 0$ and that the nuclei can be considered as classical particles and remain in their equilibrium positions. While we know from Chapter 8 that this approximation is not adequate in the study of vibrational spectroscopy and phase diagrams, it turns out that it works fairly well in the study of electronic and optical properties. In Section 10.3.2 we will see one example where this approximation is inadequate, but for now we will be content to leave the nuclei completely immobile. At variance with the approach of Chapters 4–8 where our presentation started with molecules and ended with solids, in this chapter we specifically focus on solids. This choice is useful in order to connect with the notion of *band structures*, which should be already familiar from introductory textbooks on solid-state physics, such as Kittel (1976) or Ashcroft and Mermin (1976), or electrical engineering (Sze, 1981). As usual we will get started with a heuristic approach by leveraging on our intuition, and later on we will provide a more formal justification of the main concepts.

9.1 Kohn–Sham energies and wavefunctions

In order to concentrate on the electrons we assume that for a given material the coordinates of the nuclei are known from experiment, e.g. XRD, or have been calculated using the methods discussed in Chapter 4.

In Section 3.4 we have seen that, for each given nuclear configuration, DFT allows us to calculate the total energy of the electrons, E, and the electron density, $n(\mathbf{r})$, in the ground state (eqn 3.10 and eqns 3.27–3.32). Let us concentrate on the single-particle Kohn–Sham equations, which we reproduce here from Chapter 3:

$$-\frac{1}{2}\nabla^2 \phi_i(\mathbf{r}) + V_{\text{tot}}(\mathbf{r})\phi_i(\mathbf{r}) = \varepsilon_i \phi_i(\mathbf{r}), \tag{9.1}$$

$$V_{\text{tot}}(\mathbf{r}) = V_n(\mathbf{r}) + V_H(\mathbf{r}) + V_{xc}(\mathbf{r}). \tag{9.2}$$

We recall that in the first of these equations the $\phi_i(\mathbf{r})$ are the Kohn–Sham wavefunctions, the ε_i the corresponding eigenvalues and $V_{\text{tot}}(\mathbf{r})$ is the 'effective' potential experienced by the electrons. The second equation shows that V_{tot} includes the nuclear contribution, V_n (eqn 3.29), the Hartree term, V_H (eqn 2.47), and the exchange and correlation term, V_{xc} (eqn 3.31). The Kohn–Sham wavefunctions, $\phi_i(\mathbf{r})$, were introduced in Chapters 2 and 3 as a practical approximation for representing the many-body electronic wavefunction, $\Psi(\mathbf{r}_1, \ldots, \mathbf{r}_N)$ (see for example eqns 2.35 and 2.42).

Strictly speaking the Kohn–Sham wavefunctions are only a mathematical tool for building the charge density, $n(\mathbf{r})$, using eqn 3.32. In fact we should bear in mind that DFT is a theory of the total energy, the electron density and their relation in the ground state, and it does not tell us anything about the uses and meaning of Kohn–Sham energies and wavefunctions. However, we already have seen on one occasion that these wavefunctions and their eigenvalues can be very useful for interpreting complex experimental data. In fact in Section 5.5 we tentatively used Kohn–Sham wavefunctions in the Tersoff–Hamann theory (eqn 5.23) and discovered that they can reproduce measured STM maps very accurately. This and many other successes of Kohn–Sham states prompted researchers to ask whether the $\phi_i(\mathbf{r})$ and ε_i might carry some kind of physical meaning.

A solid answer to this question would require a detour in the world of quantum field theory (Mattuck, 1976), which is too advanced for this presentation. For now we simply anticipate that the answer is 'approximately yes': in many cases Kohn–Sham states and energies happen to be reasonably good starting points for describing things such as the electrical and optical properties of materials.

In the following we start with a heuristic approach which will allow us to discuss the nature of Kohn–Sham states and energies without being overwhelmed by the formalism. In Exercise 9.1 it is shown that the DFT electronic total energy in eqn 3.10 can be rewritten as:

$$E = \underbrace{\sum_i f_i \varepsilon_i}_{\text{band structure term}} - \underbrace{\left[E_H + \int d\mathbf{r}\, V_{xc}(\mathbf{r}) n(\mathbf{r}) - E_{xc} \right]}_{\text{double-counting terms}}, \tag{9.3}$$

where $f_i = 1$ for occupied Kohn–Sham states, and 0 for unoccupied states. The first term on the right is the sum of the energies of all the occupied Kohn–Sham states, and in the solid-state physics literature is referred to as the *band structure energy*. The second term contains some corrections necessary to avoid double counting of contributions in the total energy. For example, the band structure term already contains twice the Hartree energy (see Exercise 9.1), and therefore we must remove E_H. The same reasoning applies to the exchange and correlation terms in eqn 9.3.

Exercise 9.1 We want to derive eqn 9.3 using what we have learned in Chapter 3. ▶Combine eqn 3.10 and 3.32 to show that:

$$E = \sum_i f_i \int d\mathbf{r}\, \phi_i^*(\mathbf{r}) \left[-\frac{\nabla^2}{2} + V_n(\mathbf{r}) + \frac{1}{2} \int d\mathbf{r}' \frac{n(\mathbf{r}')}{|\mathbf{r} - \mathbf{r}'|} \right] \phi_i(\mathbf{r}) + E_{xc}[n],$$

where $f_i = 1$ if the state ϕ_i is occupied, and 0 otherwise. ☐By comparing this result with eqn 2.47 we can identify the integral inside the square brackets with the Hartree potential, $V_H(\mathbf{r})$. If we now add and subtract $V_H/2 + V_{xc}$ we can reconstruct the total potential, V_{tot}, of eqn 3.28 inside the brackets:

$$E = \sum_i f_i \int d\mathbf{r}\, \phi_i^*(\mathbf{r}) \left[-\frac{\nabla^2}{2} + V_{tot}(\mathbf{r}) \right] \phi_i(\mathbf{r}) - \int d\mathbf{r}\, n(\mathbf{r}) \left[\frac{1}{2} V_H(\mathbf{r}) + V_{xc}(\mathbf{r}) \right] + E_{xc}[n].$$

▶Remembering the definition of Kohn–Sham eigenvalues in eqn 3.27, make the final step in order to go from this last expression to eqn 9.3.

If we now take the partial derivatives of the total energy in eqn 9.3 with respect to the occupations, f_i, we find:

$$\frac{\partial E}{\partial f_i} = \varepsilon_i. \tag{9.4}$$

This result is useful for establishing a connection between the Kohn–Sham eigenvalues and the total energy of the system when electrons are added, removed, or excited to higher-energy states. Let us consider for example the situation where we start from a system of N electrons with energy E_N, and we add one electron in the (previously unoccupied) state ϕ_i. The resulting total energy can be denoted $E_{N+1,i}$. The change in total energy from this process can simply be obtained by integrating eqn 9.4:

$$E_{N+1,i} - E_N = \int_0^1 df_i \frac{\partial E}{\partial f_i} = \int_0^1 df_i\, \varepsilon_i.$$

Intuitively this integral corresponds to gradually adding very small fractions, df_i, of an electron until the total adds up to 1. The delicate aspect of this equation is that a change in the occupation, f_i, will induce a change in the eigenvalue, ε_i. In fact, as we fill up the 'electron sea', we also modify the electron density and hence the total potential, V_{tot}, in eqn 9.2. This in turn modifies the spectrum of Kohn–Sham eigenvalues. However, if we limit our discussion to solids or large nanostructures, the number of electrons is typically so large that the change in density will be very small. As a result, in a first approximation, we can consider the eigenvalue ε_i independent of the occupation f_i. With this approximation the equation above simplifies as:

$$E_{N+1,i} - E_N \simeq \varepsilon_i. \tag{9.5}$$

This result tells us that, whenever we add an electron to a solid, the total energy changes by an amount *approximately* corresponding to the Kohn–Sham eigenvalue of the new electron. It is easy to see that a similar result applies to the case where an electron is removed from the system. In other words we can *intuitively* imagine that each electron carries an energy corresponding to the eigenvalue of the Kohn–Sham state that it occupies.

The heuristic arguments proposed here form the basis for understanding the concept of energy levels and band structures in solids, and establish the link between density functional theory and the band theory presented in solid-state physics books, such as those by Kittel (1976) or Ashcroft and Mermin (1976).

A few caveats are now in order. In going from eqn 9.3 to eqn 9.4 we have tacitly taken for granted that the derivative of the double-counting term in eqn 9.3 vanishes. The formal proof that this is indeed the case can be found in the original work by Janak (1978). In addition, the connection between the Kohn–Sham energies and the total energy given by eqn 9.5 should only be seen as a pedagogical simplification, and should be treated with extreme caution.

The study of the relation between total energy, single-particle eigenvalues, and electron occupations is still an active research area. The interested reader will find illuminating presentations in the classic works of Slater and Johnson (1972) and Perdew and Zunger (1981).

9.2 Calculation of band structures using DFT

After having attached a meaning, albeit approximate, to Kohn–Sham eigenvalues, we now want to discuss what these eigenvalues and their eigenstates look like in a few prototypical solids. In this section and in Sections 9.3 and 9.4 we restrict the discussion to the case of crystals in order to connect with the standard notion of *band structures* encountered in solid-state physics courses.

In order to adapt eqns 3.27–3.32 to the case of crystalline solids (Appendix D) we make use of the *Bloch theorem* (Bloch, 1928). This states that the single-particle electronic wavefunctions in a crystal can be expressed as the product of a function periodic in the unit cell and a plane wave:

$$\phi_i(\mathbf{r}) \to \phi_{i\mathbf{k}}(\mathbf{r}) = e^{i\mathbf{k}\cdot\mathbf{r}} u_{i\mathbf{k}}(\mathbf{r}), \tag{9.6}$$

where

$$u_{i\mathbf{k}}(\mathbf{r}+\mathbf{T}) = u_{i\mathbf{k}}(\mathbf{r}), \text{ with } \mathbf{T} = n_1\mathbf{a}_1 + n_2\mathbf{a}_2 + n_3\mathbf{a}_3, \tag{9.7}$$

(n_1, n_2, n_3 are integers). Note that the 'i' in $\exp(i\mathbf{k}\cdot\mathbf{r})$ is the imaginary unit, while the 'i' in $\phi_{i\mathbf{k}}$ is the eigenstate index. For completeness the Bloch theorem is derived in Exercise 9.2. Our derivation is slightly different from those given by Kittel (1976) or Ashcroft and Mermin (1976), insofar as we start from the charge density, which is the central quantity in DFT.

Exercise 9.2 Here we derive the Bloch theorem in the case of a one-dimensional crystal with lattice parameter a. The derivation for the most general three-dimensional case proceeds exactly along the same lines but is slightly more tedious. We start by considering the wavefunctions, the Laplacian and the total potential in the Kohn–Sham equation (eqn 9.1) as dependent only on the one-dimensional variable x. The periodic repetition of the crystal structure implies that the electron density is also periodic: $n(x+a) = n(x)$. ▶Show that the periodicity of the electron density implies that also the total Kohn–Sham potential, V_{tot}, is periodic. □Using the fact that V_{tot} is periodic, and that the Laplace operator is insensitive to translations, show that if $\phi_i(x)$ is a solution of the Kohn–Sham equation:

$$-\frac{1}{2}\nabla^2\phi_i(x) + V_{tot}(x)\phi_i(x) = \varepsilon_i\phi_i(x),$$

then also $\phi_i(x+a)$ is a solution of the same equation. □The fact that $\phi_i(x)$ and $\phi_i(x+a)$ are solutions of the same linear and homogeneous differential equation implies that these two solutions can differ only by a multiplicative constant, say $\phi_i(x+a) = A_i\phi_i(x)$ (this is

strictly true only if the Kohn–Sham eigenvalues are not degenerate, i.e. they are all different).
▶Rewrite the condition of periodicity of the electron density in terms of the Kohn–Sham wavefunctions, and show that:

$$\sum_i f_i(1 - |A_i|^2)|\phi_i(x)|^2 = 0.$$

☐Since this condition must hold for every x in the crystal, the only possibility (apart from the uninteresting case $\phi_i = 0$) is that $|A_i|^2 = 1$ for every i, i.e. that every constant A_i belongs to the unit circle in the complex plane. Let us define for convenience $A_i = \exp(ik_ia)$, with k_i a real number. This allows us to rewrite the relation between $\phi_i(x + a)$ and $\phi_i(x)$ as:

$$\phi_i(x + a) = e^{ik_ia}\phi_i(x).$$

At this point in order to proceed it is convenient to expand the wavefunctions, ϕ_i, using a *continuous Fourier transform*. This is nothing but a Fourier series taken over a dense distribution of wavevectors from $-\infty$ to $+\infty$, and has exactly the same properties as the discrete Fourier transform:

$$\phi_i(x) = \int_{-\infty}^{+\infty} dq\, c_i(q)\, e^{iqx}. \tag{9.8}$$

Here the quantity $c_i(q)$ plays the role of Fourier coefficient corresponding to the wavevector q.
▶By combining the previous two equations show that the following identity holds:

$$\int_{-\infty}^{+\infty} dq\, c_i(q) \left(e^{iqa} - e^{ik_ia}\right) e^{iqx} = 0.$$

☐Since this identity must hold for every x, the only possibility is that $c_i(q)(e^{iqa} - e^{ik_ia}) = 0$ for every q. Therefore the coefficients $c_i(q)$ can be non-zero only when the term in parentheses vanishes, i.e. when $q = 2\pi n/a + k_i$, with n any integer. As a result the only non-vanishing contributions to the integral in eqn 9.8 are of the form:

$$\exp\left(i\frac{2\pi n}{a}x\right)\exp(ik_ix).$$

The complex exponential on the left is periodic, with period a, and the one on the right is common to every term in the integral in eqn 9.8. Therefore upon adding up all the contributions to that integral we find the product of a periodic function, which we call $u_i(x)$ for definiteness, and $\exp(ik_ix)$:

$$\phi_i(x) = u_i(x)\, e^{ik_ix}, \text{ with } u_i(x + a) = u_i(x).$$

This last result is precisely the Bloch theorem. It is worth noting that this derivation is not limited to the solutions of the Kohn–Sham equation, and applies generally to any periodic Hamiltonian.

If we make the replacement defined by eqn 9.6 inside the Kohn–Sham equation (eqn 9.1) and multiply both sides by $\exp(-i\mathbf{k}\cdot\mathbf{r})$ we obtain:

$$-e^{-i\mathbf{k}\cdot\mathbf{r}}\frac{1}{2}\nabla^2\left[e^{i\mathbf{k}\cdot\mathbf{r}}u_{i\mathbf{k}}(\mathbf{r})\right] + V_{\text{tot}}(\mathbf{r})u_{i\mathbf{k}}(\mathbf{r}) = \varepsilon_{i\mathbf{k}}u_{i\mathbf{k}}(\mathbf{r}).$$

An explicit evaluation of the second derivative of a product yields:

$$e^{-i\mathbf{k}\cdot\mathbf{r}}\nabla^2\left[e^{i\mathbf{k}\cdot\mathbf{r}}u_{i\mathbf{k}}(\mathbf{r})\right] = (\nabla + i\mathbf{k})^2u_{i\mathbf{k}}(\mathbf{r}),$$

which leads immediately to the result:

$$\left[-\frac{1}{2}(\nabla + i\mathbf{k})^2 + V_{\text{tot}}(\mathbf{r}) \right] u_{i\mathbf{k}}(\mathbf{r}) = \varepsilon_{i\mathbf{k}} u_{i\mathbf{k}}(\mathbf{r}). \tag{9.9}$$

In this 'crystal version' of the Kohn–Sham equations we see that the complex exponential $\exp(i\mathbf{k}\cdot\mathbf{r})$ has disappeared, and that the function to be determined is only the periodic part, $u_{i\mathbf{k}}$, of the Kohn–Sham state. A more elegant way of stating this result is that that the periodic part of the Kohn–Sham wavefunction is an eigenstate of a modified Hamiltonian, $\hat{H}_{\mathbf{k}}$:

$$\hat{H}_{\mathbf{k}} u_{i\mathbf{k}} = \varepsilon_{i\mathbf{k}} u_{i\mathbf{k}}, \qquad \hat{H}_{\mathbf{k}} = -\frac{1}{2}(\nabla + i\mathbf{k})^2 + V_{\text{tot}}. \tag{9.10}$$

Since a periodic model of a crystalline solid is by construction infinitely extended, it is convenient to normalize the functions $u_{i\mathbf{k}}(\mathbf{r})$ inside one unit cell (UC) of the crystal:

$$\int_{\text{UC}} |u_{i\mathbf{k}}(\mathbf{r})|^2 d\mathbf{r} = 1,$$

meaning that each wavefunction can accommodate one electron per unit cell.

From a practical point of view the result in eqn 9.10 is very important, since it tells us that in order to study electrons in crystals we need to solve Schrödinger equations only *inside one crystalline unit cell* and apply periodic boundary conditions. Once we have determined $u_{i\mathbf{k}}(\mathbf{r})$ inside one unit cell, we know from eqn 9.7 that the solution in any other crystal cell is simply a replica of this function. This observation affords us an enormous saving of computational resources. In fact the numerical solution of eqn 9.10 will require the description of $u_{i\mathbf{k}}$ on a discrete mesh of points spanning only one unit cell.

As an example, the unit cell of fcc Cu has a volume of 11.2 Å^3. If we were to describe the wavefunctions using a uniform mesh with a spacing of 0.1 Å between adjacent points, we would need only $11.2/0.1^3 = 11,200$ points in total. Storing the values of the wavefunction, $u_{i\mathbf{k}}$, at these points would require only about 175 kB of memory, and calculations on a standard laptop computer would take only a few minutes. Conversely, if we were to describe a large block of the crystal covering many unit cells, the calculations could become very time-consuming.

If we now compare eqn 9.6 with eqn 7.44, and eqn 9.10 with eqn 7.45, we see that there exists a strong analogy between electronic waves and vibration waves in crystals. As we might expect, this analogy stems from the common assumption that the crystal lattice is periodic. Building on such analogy, in Chapter 7 we saw that the diagonalization of the dynamical matrix for wavevectors $\mathbf{q}$ and $\mathbf{q}' = \mathbf{q} + \mathbf{G}$, with $\mathbf{G}$ a reciprocal vector, yields the same vibrational eigenmodes and eigenfrequencies (Section 7.4). That observation led us to consider only phonons for wavevectors $\mathbf{q}$ inside the first Brillouin zone, and discard all the duplicate solutions for $\mathbf{q}' = \mathbf{q} + \mathbf{G}$. In complete analogy with that result, also in the study of electrons in crystals it turns out that the solutions of eqn 9.10 for $\hat{H}_{\mathbf{k}+\mathbf{G}}$ are simply duplicates of the solutions for $\hat{H}_{\mathbf{k}}$ (see Exercise 9.3). This observation allows us to restrict the 'useful' range of wavevectors $\mathbf{k}$ (for which eqn 9.10 has to be solved) to the first Brillouin zone (Appendix D).

Exercise 9.3 In this exercise we want to show that the solutions of $\hat{H}_{\mathbf{k}+\mathbf{G}}$ and $\hat{H}_{\mathbf{k}}$ are simply duplicates if $\mathbf{G}$ is a reciprocal lattice vector. ▶Show that the eigenfunction $\phi_{i\mathbf{k}+\mathbf{G}}$ of $\hat{H}_{\mathbf{k}+\mathbf{G}}$ can be written as

$$\phi_{i\mathbf{k}+\mathbf{G}}(\mathbf{r}) = f(\mathbf{r})\,e^{i\mathbf{k}\cdot\mathbf{r}},$$

where $f(\mathbf{r}) = e^{i\mathbf{G}\cdot\mathbf{r}} u_{i\mathbf{k}+\mathbf{G}}(\mathbf{r})$ is a function with the periodicity of the crystal lattice. ▶By replacing $f(\mathbf{r})\exp(i\mathbf{k}\cdot\mathbf{r})$ inside the original Kohn–Sham equation (eqn 9.1) and repeating the steps which led to eqn 9.9 show that $f(\mathbf{r})$ must also be a solution of eqn 9.10. ☐Now we make the assumption that there can be only one eigenvalue for each wavevector $\mathbf{k}$ in eqn 9.10. This is not strictly necessary but affords us some simplification. Since $u_{i\mathbf{k}}(\mathbf{r})$ and $f(\mathbf{r})$ are both the only solution to eqn 9.10, the only possibility is that they differ at most by an arbitrary constant, $f(\mathbf{r}) = A\,u_{i\mathbf{k}}(\mathbf{r})$. We choose this constant to be $A = 1$. As a curiosity, this choice is referred to as the 'periodic gauge' (Resta, 2000). ▶Now combine the results of this exercise in order to show that $\phi_{i\mathbf{k}+\mathbf{G}} = \phi_{i\mathbf{k}}$.

Once we have determined the solutions of the Kohn–Sham equations in a crystal it is possible to move forward and construct the electron density $n(\mathbf{r})$. The density can be calculated as in eqn 3.32, provided we replace ϕ_i with $\phi_{i\mathbf{k}}$ as in eqn 9.6, and sum over both the index, i, and the wavevector, $\mathbf{k}$:

$$n(\mathbf{r}) = \sum_i \int_{BZ} \frac{d\mathbf{k}}{\Omega_{BZ}} f_{i\mathbf{k}} |u_{i\mathbf{k}}(\mathbf{r})|^2. \tag{9.11}$$

In order to avoid duplicate wavefunctions, only the wavevectors inside the first Brillouin zone are considered. The occupation numbers are $f_{i\mathbf{k}} = 1$ for occupied states and 0 otherwise. Clearly the electron density in eqn 9.11 inherits the periodicity of the wavefunctions, $u_{i\mathbf{k}}$; therefore, it is the same in every unit cell of the crystal. The knowledge of the electron density enables in turn the calculation of the total Kohn–Sham potential, V_{tot}, using eqns 3.28–3.31.

The generalization of eqn 9.3 to the case of crystalline solids is found easily:

$$E = \sum_i \int_{BZ} \frac{d\mathbf{k}}{\Omega_{BZ}} f_{i\mathbf{k}}\varepsilon_{i\mathbf{k}} - \left[E_H + \int d\mathbf{r}\, V_{xc}(\mathbf{r})n(\mathbf{r}) - E_{xc}\right], \tag{9.12}$$

where all quantities now indicate energies per unit cell. Since the terms inside the brackets are completely determined by the density, $n(\mathbf{r})$, the ground state of the system is obtained by choosing the occupation numbers so as to fill the electron sea, starting from the lowest-energy states. This filling procedure has to satisfy the constraint that the electron density adds up to the number of electrons per unit cell:

$$N = \int_{UC} n(\mathbf{r}) = \int_{BZ} \frac{d\mathbf{k}}{\Omega_{BZ}} \sum_i f_{i\mathbf{k}},$$

where use of eqn 9.11 was made.

The energy of the highest occupied eigenstate at $T = 0$ is referred to as the *Fermi level* and indicated as ϵ_F (see Sections 3.3.1 and 5.5). It is easy to verify that the total energy, E, of eqn 9.12 is minimized when every Kohn–Sham state below the Fermi level is occupied ($f_{i\mathbf{k}} = 1$ if $\varepsilon_{i\mathbf{k}} \le \epsilon_F$) and every state above is empty ($f_{i\mathbf{k}} = 0$ if $\varepsilon_{i\mathbf{k}} > \epsilon_F$). In fact, if this was not the case, it would be possible to lower the energy of the system by moving an electron from an occupied state above the Fermi level to an empty state below this level.

Practical DFT calculations for crystalline solids proceed essentially in the same way as described in Figure 3.3. The only notable differences are that: (i) the Kohn–Sham wavefunctions are obtained by solving eqn 9.9 instead of eqn 3.27, and (ii) the density is obtained via eqn 9.11 instead of eqn 3.32. The Brillouin zone integral in eqn 9.11 is evaluated numerically by considering a discrete mesh of wavevectors $\mathbf{k}$, spanning the Brillouin zone. Typical calculations use meshes including approximately $10\times10\times10$ wavevectors, although the exact size depends on the crystal.

9.2.1   The band structure of copper

As an example of the calculation of band structures using DFT we examine the Kohn–Sham eigenstates and eigenvalues of copper. At ambient conditions copper crystallizes in a fcc lattice with one atom per unit cell and lattice constant $a = 3.61$ Å. Copper is a ductile metal and is among the best electrical and thermal conductors (the electrical and thermal conductivities at room temperature are $5.9 \cdot 10^5$ ohm^{-1}cm^{-1} and 400 W m^{-1}K^{-1}, respectively). Owing to these properties and to its abundance in the Earth's crust, Cu is one of the most popular industrial metals and is practically ubiquitous in electrical wiring, as well as plumbing.

The electronic configuration of the Cu atom is $[\mathrm{Ar}]4s^13d^{10}$. As we saw in Section 5.1 for N, also in this case it is convenient to freeze the [Ar] core electrons in their ground-state configuration, and describe explicitly only the $4s^13d^{10}$ valence electrons. This choice corresponds to having 11 electrons in every crystalline unit cell.

Figure 9.1 shows the energy, $\varepsilon_{i\mathbf{k}}$, vs. wavevector, $\mathbf{k}$, dispersion relations of copper, calculated as described above. The actual calculation consists of two steps. First, the electron density is determined self-consistently, as illustrated in Figure 3.3. Once the ground-state density, $n(\mathbf{r})$, and the associated Kohn–Sham potential, $V_{\mathrm{tot}}(\mathbf{r})$, have been obtained, the Hamiltonian in eqn 9.10 is completely specified. At this point one sweeps the wavevector, $\mathbf{k}$, across the Brillouin zone, and for each wavevector eqn 9.10 is solved in order to determine the eigenvalues $\varepsilon_{i\mathbf{k}}$, $i = 1, 2, 3, \ldots$ This second step is typically referred to as a 'non-self-consistent calculation', meaning that the electron density and the total potential are now left unchanged. Simply put, this step corresponds to finding the eigenvalues and eigenstates of a Schrödinger equation where the potential, $V_{\mathrm{tot}}(\mathbf{r})$, is known. When the discrete eigenvalues $\varepsilon_{1\mathbf{k}}$, $\varepsilon_{2\mathbf{k}}$, ... calculated for each wavevector along a given path are joined together, a continuous dispersion plot emerges, as shown in Figure 9.1. The curves thus obtained are referred to as *bands*, and the entire plot goes under the name of *band structure*. It is worth noting the analogy between the electronic band structure in solids, such as that in Figure 9.1, and the phonon dispersions discussed in Chapter 8, such as those in Figure 8.3.

In Figure 9.1 the band structure of copper is compared with the angle-resolved photoelectron spectroscopy data reviewed by Courths and Hüfner (1984). The basic concepts of photoemission will be presented in Section 9.3, and for now we only make two simple observations. First, the fact that the band structure can be measured should tell us that the Kohn–Sham eigenvalues must carry (at least in a first approximation)

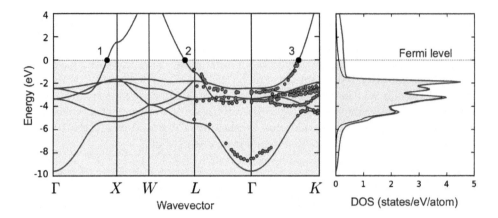

Fig. 9.1 Electronic band structure (left) and electronic density of states (right) of copper. The Kohn–Sham eigenvalues are plotted along typical high-symmetry paths of the Cu Brillouin zone, similarly to the phonon dispersions of diamond in Figure 8.3. The highlighted wavevectors are Γ: $\mathbf{q} = 0$, X: $\mathbf{q} = (2\pi/a)\mathbf{u}_y$, W: $\mathbf{q} = (\pi/a)(\mathbf{u}_x + 2\mathbf{u}_y)$, L: $\mathbf{q} = (\pi/a)(\mathbf{u}_x + \mathbf{u}_y + \mathbf{u}_z)$ and K: $\mathbf{q} = (3\pi/2a)(\mathbf{u}_x + \mathbf{u}_y)$. The shaded region in the left panel indicates the occupied bands, and the horizontal dashed line marks the Fermi level (the zero of the energy axis in this plot). If we concentrate on the $L\Gamma K$ segment we can identify some kind of parabolic dispersion from -10 eV to +4 eV (as in the free electron gas of eqn 3.14), interrupted by a 'spaghetti-like' set of bands between -6 eV and -2 eV. The DOS on the right shows some rather intense peaks in correspondence with this latter set of bands. An analysis of the wavefunction indicates that these bands are to be associated with combinations of Cu $3d^{10}$ atomic wavefunctions. The fraction of d-character in these wavefunctions is represented by the shaded yellow region in the DOS. The meaning of the three points labelled 1, 2, 3 at the Fermi level is explained in the caption to Figure 9.2. The red discs are from the experimental angle-resolved photoemission data collected in the review by Courths and Hüfner (1984, Figure 96).

some kind of physical reality, as discussed in Section 9.1. Second, the agreement between the DFT/LDA calculations in Figure 9.1 and experiment is rather satisfactory, with deviations below 0.4 eV in the energy range $-5 < E < 0$ eV, and below 1.2 eV for $-10 < E < -5$ eV. While considering these figures we should keep in mind that the computed band structures derive entirely from the first principles of quantum mechanics, the only parameter in the calculation being the atomic number of Cu. The reader can find a comprehensive comparison between the DFT/LDA band structure and photoelectron spectra of copper in a paper by Marini *et al.* (2001).

In analogy with the vibrational density of states introduced in Chapter 8 (eqn 8.29) we can define a density of electronic states:

$$\rho(E) = \sum_i \int_{\mathrm{BZ}} \frac{d\mathbf{k}}{\Omega_{\mathrm{BZ}}} \delta(E - \varepsilon_{i\mathbf{k}}). \tag{9.13}$$

The quantity $\rho(E)\,dE$ tells us how many Kohn–Sham states exist in the energy interval between E and $E + dE$, and is normalized in such a way that:

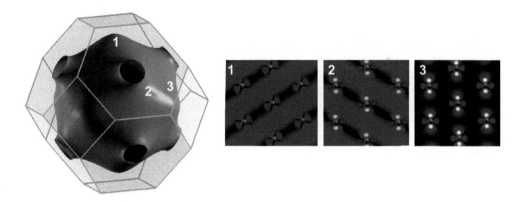

Fig. 9.2 Fermi surface (left) and some representative Kohn–Sham wavefunctions (right) of copper. The Fermi surface is the set of wavevectors **k** in the Brillouin zone (represented by a transparent truncated octahedron) such that $\varepsilon_{n\mathbf{k}} = \epsilon_F$. The labels 1, 2, 3 on this surface correspond to the wavevectors indicated in Figure 9.1, and the corresponding Kohn–Sham wavefunctions are shown in the three panels on the right. Each wavefunction plot is $|u_{n\mathbf{k}}(\mathbf{r})|^2$, with **r** in a (111) plane cutting through the Cu atoms. In all panels we see that the wavefunctions are a mixture of Cu d electronic states located on the Cu atoms and additional electrons in between the atoms. Even if we only concentrate on the shape of the wavefunctions around the Cu atoms we can see significant differences between states across the Fermi surface. For example, in panel 1 we can recognize d_{xy} atomic orbitals, while in panels 2 and 3 we see d_{z^2} orbitals.

$$\int_{-\infty}^{\epsilon_F} \rho(E)\,dE = N.$$

The definition of electronic DOS provided by eqn 9.13 corresponds to the generalization to the case of crystalline solids of eqn 5.20 encountered in the theory of STM. The Kohn–Sham density of states of Cu is shown in the right panel of Figure 9.1.

In Figure 9.1 we see that the Fermi level marks the separation between the occupied Kohn–Sham bands (below) and the unoccupied bands (above). In this example we can identify three bands crossing the Fermi level. In general, for a crystalline solid in three dimensions the locus of all the wavevectors, **k**, such that $\varepsilon_{i\mathbf{k}} = \epsilon_F$, defines a two-dimensional surface, $\mathcal{S}_F = \{\mathbf{k} \text{ such that } \varepsilon_{i\mathbf{k}} = \epsilon_F\}$. Such a surface is called the *Fermi surface*, and plays a very important role in the study of the electrical conductivity of metals. In fact, the concept of the Fermi surface is so crucial to the theory of metals, that in the textbook by Kittel (1976) metals are defined as 'solids with a Fermi surface'. A review of the connection between the theory of conduction in metals and the Fermi surface can be found, among others, in Chapter 15 of the book by Ashcroft and Mermin (1976).

The Fermi surface of Cu calculated using DFT/LDA is shown in Figure 9.2. This surface is reminiscent of a sphere contained inside the first Brillouin zone, bulging out towards the (111) directions until it pierces the hexagonal faces, leading to characteristic 'necks'. All these features have been confirmed by de Haas–van Alphen measurements (Shoenberg, 1962; Roaf, 1962), and also by angle-resolved photoemission spectroscopy (Aebi *et al.*, 1994; Stampfl *et al.*, 1995).

In Figure 9.2 we also show a few selected Kohn–Sham wavefunctions with energies on the Fermi surface. From these plots it should be clear that Kohn–Sham wavefunctions are relatively complicated objects, even in such a simple example of a crystalline solid with one atom per unit cell. In practice it would be impossible to capture all the fine features shown in Figure 9.2 with paper and pencil, and numerical methods are an absolute necessity.

Fermi surface and wavefunction plots like those in Figure 9.2 have become routine nowadays, and can be obtained in a matter of minutes on a standard laptop computer. The processing of DFT calculations in order to obtain such appealing images is typically performed using highly optimized scientific rendering software, such as the program 'XCrysDen' (Kokalj, 2003), which was used indeed for generating Figure 9.2.

Exercise 9.4 In Figure 9.1 we made the observation that, apart from the spaghetti-like set of bands between -6 eV and -2 eV, the dispersions of fcc Cu resemble those of a free electron gas. In order to quantify this similarity we want to directly compare the dispersions of a free electron gas with those reported in Figure 9.1. ►After setting $V_{\text{tot}} = -9.6$ eV in eqn 9.10, show that the free electron eigenvalues are given by $\varepsilon_{\mathbf{k}} = 0.29 \, a^2 |\mathbf{k}|^2 - 9.6$ eV. ►Plot the eigenvalues, $\varepsilon_{\mathbf{k}}$, along the reciprocal space paths $\Gamma \to X$ and $\Gamma \to K$, and superimpose your curves onto the band structure in Figure 9.1. ►Determine the wavevectors at which your free electron bands cross the Fermi level, and compare your results with the crossings labelled 1 and 3 in Figure 9.1.

9.3   Basics of angle-resolved photoelectron spectroscopy

In many cases the band structure of crystalline solids can be probed directly using an experimental technique named 'angle-resolved photoelectron spectroscopy', in short ARPES. We already have seen an example of ARPES band structure in Figure 9.1 (red discs). Probably the most distinctive advantage of this technique is that it can measure at once the energies and wavevectors of many electrons inside a material, thereby making it possible to map out entire band structures in a matter of minutes. Given the increasing importance of photoelectron spectroscopy in condensed matter physics and nanoscience, we here give a brief account of the basic principles underlying this experimental technique.

Photoelectron spectroscopy (PES) is based on the photoelectric effect (Hertz, 1887; Einstein, 1905*a*), whereby it is possible to extract electrons from a material by illuminating the sample with light of appropriate wavelength. The systematic use of PES as a tool for studying the electronic structure of materials, from molecules to solids, started in the mid-1950s, and a review of the historical developments in the early days can be found in Siegbahn (1982).

The key difference between PES and ARPES is that in the latter the intensity of electrons extracted by illumination is recorded as a function of the emission angle. As a mnemonic expedient we can think of PES as a tool for measuring the electronic DOS (e.g. the right panel of Figure 9.1), and ARPES as a tool for determining the electron band structures (e.g. the left panel of the same figure).

Angle-resolved photoemission spectroscopy experienced a significant growth in the mid-1990s, following the development of high-resolution hemispherical electron analysers and the increased availability of synchrotron light sources. During this period the main focus has been on high-temperature superconducting materials such as the copper oxides, but more recently the interest has been moving towards the so-called 'Dirac materials', i.e. graphene and topological insulators. These aspects are captured in the reviews by Damascelli *et al.* (2003) and by Lu *et al.* (2012), respectively.

A complete theory of photoemission spectroscopy requires the use of advanced quantum field theory methods (Bardyszewski and Hedin, 1985), and is beyond the scope of this presentation. Here we provide a much simplified description, with the goal of grasping the main idea behind ARPES, without any pretence of rigour.

Figure 9.3 shows a simplified scheme of the working principle of an ARPES experiment. When light of appropriate wavelength (to be defined below) illuminates the sample, electrons will absorb some energy from the electromagnetic field. A certain fraction of the electrons will acquire enough energy to escape the potential barrier which contains them inside the sample. These electrons are referred to as 'photoelectrons'. By analysing the trajectory of the photoelectrons inside an electric field it is possible to reconstruct the energy and wavevector of the electrons prior to extraction.

In the setup illustrated in the left panel of Figure 9.3, the trajectory of the photoelectrons inside the analyser is bent by an electric field. This field is obtained by establishing an electrostatic potential difference between two metallic hemispheres. The length of the trajectory depends on the kinetic energy of the photoelectrons, and can be recorded by the impact point of the photoelectron at the detector. In a similar way the take-off angle of the photoelectrons can be analysed by recording the impact positions in the direction perpendicular to the plane of the trajectory. The typical diameter of the largest hemisphere is around 40 cm, and the detector is a small electron multiplier about 2 cm wide, coupled to a phosphorescent screen and a CCD camera.

The quantitative relation between the kinetic energy of the photoelectrons and the length of their trajectory can be determined as follows. Let us denote by d the separation between the two hemispheres, and by V_s the applied potential. We make the approximation that at any point in between the hemispheres the situation corresponds to a parallel-plate capacitor. In this approximation the electric field is oriented along the radial direction and has magnitude V_s/d. A photoelectron entering the analyser with velocity v and kinetic energy:

$$E_{\text{kin}} = \frac{1}{2}m_e v^2,$$

experiences a centripetal force and follows a circular trajectory. We denote the radius of this trajectory R. By equating the magnitude of the centripetal force and that of the electrostatic force we obtain:

$$e\frac{V_s}{d} = m_e \frac{v^2}{R}.$$

The combination of the previous two results yields:

$$E_{\text{kin}} = \frac{eV_s}{2d} R. \tag{9.14}$$

This relation enables the determination of the kinetic energy of the photoelectron by measuring the radius of its trajectory. In reality the electrostatic potential in between the hemispheres depends on the radius, and hence eqn 9.14 is only approximate. The rigorous theory of the hemispherical capacitor (Purcell, 1938) yields a relation which reduces to our approximate result in the limit $d \ll R$.

At this point we need to establish the relation between the kinetic energy of the extracted photoelectron and the energy of the same electron prior to extraction. For

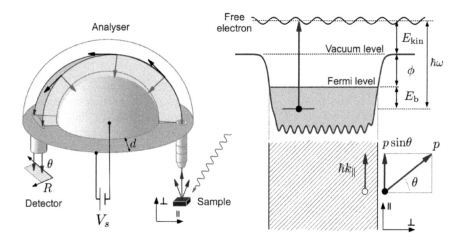

Fig. 9.3 Schematic of an ARPES experiment, from the extraction of a photoelectron to the detector. The left panel shows a simplified scheme of the hemispherical analyser. This is a capacitor consisting of two concentric hemispheres held at the electrostatic potential difference V_s. The difference between the radius of the outer and the inner hemisphere is d. When light with energy $\hbar\omega \geq \phi$ illuminates the sample, photoelectrons are emitted and a fraction of these enters the analyser. In the crudest approximation the photoelectrons follow circular trajectories inside the analyser, whose radius, R, depends on the initial kinetic energy. This radius is measured by recording the coordinate of the impact point at the detector along the axis denoted R in the figure. The take-off angle with respect to the surface normal is measured by recording the coordinate of the impact along the axis denoted θ in the figure. The red arrows along the trajectory represent the centripetal electrostatic field which bends the electron trajectory. In the top right corner is a scheme of the energy levels. Any electron with energy above the vacuum level is free to travel outside of the sample, and is schematically represented by a plane wave. The bottom right corner shows how the component of the momentum parallel to the sample surface is conserved during the extraction of the photoelectron. The perpendicular component is not conserved since the potential barrier (top panel) establishes a retention force which reduces the electron momentum along this direction. The first analyser of this kind was introduced by Beamson *et al.* (1990).

this purpose it is necessary to introduce the notion of *photons*. Simply stated, photons are the quanta of the electromagnetic field, in much the same way that the phonons of Chapter 8 are the quanta of vibrations. In complete analogy with the case of vibrations, an electromagnetic field of frequency ω can be thought of as consisting of energy quanta, $\hbar\omega$, called photons. An introduction to the quantum theory of light can be found, among others, in Chapter 22 of Merzbacher (1998). In the following we consider for simplicity only the case of one-photon processes, similar to what was done in Chapter 8 for Raman and neutron spectroscopy.

The energetics of photon absorption by the sample can be analysed as follows. We start from a system comprising one photon with energy $\hbar\omega$ and N electrons with total energy E_N, and we end up with a system comprising $N-1$ electrons with total energy E_{N-1} inside the sample, one free electron with energy E_{kin}, and no photons. If this system is reasonably isolated from external disturbances, we can state the conservation of energy as follows:

$$\hbar\omega + E_N = E_{N-1} + E_{\text{kin}}. \tag{9.15}$$

The difference in the total energies, $E_N - E_{N-1}$, can be expressed in terms of the single-particle energy, ε, of the electron prior to extraction using eqn 9.5:

$$E_N = E_{N-1} + \varepsilon.$$

Here ε is taken with respect to the *vacuum level*, which is defined as the energy of an electron at rest very far away from the sample (see Figure 9.3). It is usually convenient to split the energy eigenvalue, ε, as:

$$-\varepsilon = E_b + \phi.$$

The first term, E_b, is the energy of the electron measured from the Fermi level, and is called the *binding energy*. The second term, ϕ, is the minimum amount of energy necessary to promote an electron into vacuum, and is referred to as the *work function* (Kittel, 1976). We already encountered the work function in Section 5.5. The rationale for this splitting will become clear shortly.

If we now combine the last three equations we find:

$$\hbar\omega = E_b + \phi + E_{\text{kin}}. \tag{9.16}$$

By definition we must have $E_{\text{kin}} \geq 0$ and $E_b \geq 0$; therefore a necessary condition for electron extraction is:

$$\hbar\omega \geq \phi,$$

i.e. only electromagnetic fields with frequency $\omega \geq \phi/\hbar$ can extract electrons. This observation is at the core of the theory of the photoelectric effect by Einstein (1905a), and provides a simple and effective procedure for measuring the work function.

At this point we can combine eqns 9.14 and 9.16 to obtain our final result:

$$E_b = \hbar\omega - \phi - \frac{eV_s}{2d}R, \tag{9.17}$$

relating the binding energy, E_b, of the electron with the location, R, of the impact point at the detector.

The procedure discussed so far gives E_b, which is a proxy for the electron eigenvalue, $\varepsilon_{i\mathbf{k}}$, in a band structure plot. However, the wavevector $\mathbf{k}$ of the electron is still to be determined. With reference to Figure 9.3, the component of the electron momentum parallel to the surface of the sample, $p\sin\theta$, is conserved when the electron is extracted. In fact there are no forces acting on the photoelectron along the surface plane. Assuming that the electron inside the sample behaves like an almost-free electron, $\phi_{i\mathbf{k}}(\mathbf{r}) = \exp(i\mathbf{k}\cdot\mathbf{r})$, its momentum prior to extraction will be $\hbar\mathbf{k}$. If we denote the component of $\mathbf{k}$ parallel to the surface by $k_\parallel$, then the conservation of momentum gives:

$$\hbar k_\parallel = p\sin\theta = \sqrt{2m_e E_{\text{kin}}}\,\sin\theta. \tag{9.18}$$

By recording the position of the photoelectron at the detector it is possible to measure the take-off angle, θ, and, by virtue of the last equation, the wavevector, $k_\parallel$.

Taken together, eqns 9.17 and 9.18 show that a two-dimensional map of the impact points (R,θ) of the photoelectrons provides a direct representation of the E_b vs. $k_\parallel$ dispersion relations, i.e. the band structure.

At the end of this brief survey we point out that, while only the momentum *parallel* to the surface is strictly conserved, procedures for reconstructing also the perpendicular component of the momentum have been developed, and complete three-dimensional mapping of the band structure, $\varepsilon_{i\mathbf{k}}$ vs. $\mathbf{k}$, is possible nowadays (Himpsel, 1983). Current ARPES experiments are very accurate and in some cases can reach an energy resolution better than 1 meV (Liu *et al.*, 2008).

The band structure of copper given by the red discs in Figure 9.1 is precisely an example of ARPES measurement.

Exercise 9.5 An ARPES experiment uses photons of energy 60 eV from a synchrotron light source in order to probe a copper film exposing the (100) surface. The photoelectrons are collected into a hemispherical analyser where the two hemispheres have radii 15 cm and 20 cm, respectively, and are held at an electrostatic potential difference $V_s = 30$ V. ▶Find the work function of the Cu(100) surface by consulting for example Kittel (1976). ▶Using eqn 9.17 show that this experiment can probe the band structure of copper in the energy range $\epsilon_F - \Delta\varepsilon < \varepsilon_{i\mathbf{k}} < \epsilon_F$, where ϵ_F is the Fermi level and $\Delta\varepsilon = 10.3$ eV. ▶Now we allow the voltage between the hemispheres to vary. Calculate the maximum voltage, V_s, for which electrons are collected at the detector, as well as the voltage which allows us to probe the widest energy range in the band structure. ☐In this experiment the detector is able to resolve electrons landing at a distance of 100 μm. ▶Using eqn 9.17 show that the energy resolution of the analyser is 30 meV.

As a second example, Figure 9.4 shows the ARPES spectrum of diamond (Yokoya *et al.*, 2006), as well as a comparison with the Kohn–Sham band structure obtained from DFT/LDA calculations. From this figure we see that the DFT band structure provides a reasonable description of the *occupied* electronic states of diamond. The most obvious discrepancy which can be seen in Figure 9.4 is that the DFT bands are somewhat 'compressed' along the energy axis with respect to the measured bands. If we look at the band width, i.e. the energy difference between the top and bottom of the dispersions, we find 21.7 eV in the case of DFT/LDA calculations, and about 24 eV for experiment. This allows us to estimate the 'compression' to be of the order of $(24 - 21.7)/24 \simeq 10\%$. This level of agreement between DFT and experiment is

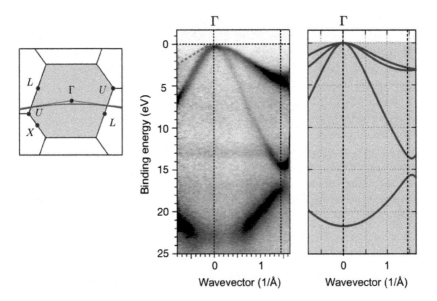

Fig. 9.4 Comparison between the ARPES spectrum of diamond and DFT/LDA band structure. The calculations (right panel) were performed as described in Section 9.2, using the same computational setup as for Exercise 7.11. The path of wavevectors, **k**, in reciprocal space is shown by a red line (theory) or a blue line (experiment) in the two-dimensional cut through the Brillouin zone on the left. The experimental spectrum in the central panel is from Yokoya *et al.* (2006). The experiment was performed on boron-doped diamond; however, the presence of B does not alter significantly the band structure of pristine diamond, therefore the spectrum should be representative of pure diamond. In the ARPES spectrum we can see that some bands are less visible than others. This is because the probability of extracting electrons with a given energy and wavevector depends on their coupling with the incident electromagnetic field. We will discuss these aspects in more detail in Chapter 10. Note also that the measured spectrum exhibits a faint signal in the top left corner, indicated by a dashed line. Comparison with the calculated band structure indicates that this signal is to be associated with the two highest-energy bands to the left of the Γ point (Yokoya *et al.*, 2006). Experimental spectrum courtesy of T. Yokoya, adapted by permission from the Institute of Physics Publishing: *Science and Technology of Advanced Materials* (Yokoya *et al.*, 2006), National Institute for Materials Science, copyright 2006.

similar to what was already observed in the case of copper (Figure 9.1). The main difference is that in the case of Cu the bands were 'stretched' by about 10%.

Another interesting difference between experiment and calculations is that the measured bands are somewhat blurred along the energy axis. This broadening is not arising from the instrument resolution (which for the experiment in Figure 9.4 is 250 meV), but results instead from the following process. When an electron is extracted from the sample, the system tends to go back to its ground state by filling the vacant electronic state. While this is happening the band structure changes slightly, since one electron is now missing. As a consequence, the photoelectrons will exhibit a distribution of energies peaked around the original band. A proper discussion of the concept of spectral broadening requires the use of quantum field theory techniques, which are beyond the scope of this presentation. The interested reader will find a rigorous analysis in the classic text by Hedin and Lundqvist (1969).

At the end of this section on PES we stress one point which has not come up yet explicitly. From Figures 9.1, 9.3, and 9.4 it should be clear that PES only probes quantum states which are occupied by electrons. The natural question is therefore whether it is possible to probe unoccupied electronic states in a similar way. The answer is yes, there exists a complementary technique, referred to as *inverse photoelectron spectroscopy* or IPES, which probes unoccupied electronic states. The technique consists of directing electrons onto a sample, and recording the photons which are radiated upon electron capture. We will not describe this technique, but simply refer the reader to the review by Himpsel (1990).

9.4 Metals, insulators and semiconductors

In the previous section we have seen the band structures of two crystalline solids, copper and diamond. We are certainly familiar with the notion that copper is a metal while diamond is an insulator. But how do we exactly distinguish metals from insulators by only looking at their band structures? On page 161 it was mentioned that a metal is a solid with a Fermi surface. In the case of diamond (Figure 9.4) the Fermi level at $T=0$ lies right at the top of the occupied bands; therefore in this case the Fermi surface degenerates into one point and the density of electronic states at the Fermi level vanishes. In principle this observation could be enough to distinguish metals from insulators; however, a point-like Fermi surface is not a very useful concept. A much more useful property of the electronic structure of insulators is that the unoccupied bands are separated from the occupied bands by an energy gap or *band gap*. This can be seen in Figure 9.5, where the DFT/LDA band structures of diamond and silicon are shown.

The existence of a band gap bears directly on the ability of materials to conduct electricity. A proper discussion of this aspect would require the study of the quantum theory of electron transport (Ziman, 1960), or at least the introductory exposition given in Chapter 13 of Ashcroft and Mermin (1976). Here we only make a heuristic observation in order to stress the key difference between metals and insulators.

Within a simplified single-particle description, we can think of electrical conduction in the following terms. In absence of an external bias voltage, a material will be in a steady-state condition, whereby each electron occupies a state $\phi_{i\mathbf{k}}$ in the band structure. The average velocity of the electrons in this sample is obtained by adding up the quantum-mechanical expectation values of the velocities of all the occupied electronic states,

$$\mathbf{v}_{\text{ave}} = \frac{1}{N} \sum_i \int \frac{d\mathbf{k}}{\Omega_{\text{BZ}}} f_{i\mathbf{k}} \langle \phi_{i\mathbf{k}} | \hat{\mathbf{v}} | \phi_{i\mathbf{k}} \rangle, \tag{9.19}$$

where $\hat{\mathbf{v}}$ is the electron velocity operator (Ashcroft and Mermin, 1976). In absence of bias it can be shown that the average velocity is zero, as we intuitively expect. In

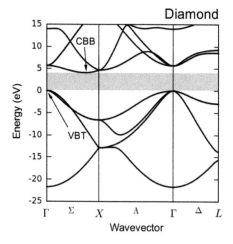
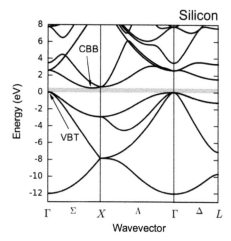

Fig. 9.5 The band structures of diamond and silicon calculated using DFT/LDA. The calculations were performed as described in Section 9.2, using the same computational setups as for Exercise 7.11 and Figure 5.2, respectively. The path of the wavevectors **k** in reciprocal space is chosen to be the same as in Figure 8.3. Diamond and silicon share the same crystal structure (apart from the lattice constant), hence the same Brillouin zone as in Figure 8.3. The grey regions indicate the band gap, and the arrows point to the 'top of the valence band' (VBT) and to the 'bottom of the conduction band' (CBB). The VBT is at Γ in both cases, and the CBB lies along the Σ line of the Brillouin zone. In diamond the CBB is found at the wavevector $\mathbf{k}_{\mathrm{CBB}} = 0.74(2\pi/a_{\mathrm{C}})\mathbf{u}_x$. In silicon we have $\mathbf{k}_{\mathrm{CBB}} = 0.84(2\pi/a_{\mathrm{Si}})\mathbf{u}_x$ (here a_{C} and a_{Si} are the lattice constants of diamond and silicon). In both cases there are six equivalent minima in the first Brillouin zone, whose wavevectors are obtained by replacing $\mathbf{u}_x$ in the previous expressions by $\pm\mathbf{u}_x$, $\pm\mathbf{u}_y$, or $\pm\mathbf{u}_z$ (Kittel, 1976). Note the different energy scales in the two panels.

order to drive a current we need to have $|\mathbf{v}_{\mathrm{ave}}| > 0$. This can be achieved by promoting some electrons into unoccupied electronic states, e.g. by having some occupations, $f_{i\mathbf{k}}$, change from 1 to 0 and some others from 0 to 1 inside eqn 9.19. These electronic 'transitions' can be induced by means of an external bias.

Now, in a metal there is no energy separation between occupied and unoccupied states, and therefore a very small bias is enough to promote electronic transitions and drive a current. As an example, a copper wire of diameter 1 mm and length 1 m needs only a bias of $\sim$20 mV in order to carry a current of 1 A.

The picture is very different in the case of insulators, where the occupied and the unoccupied quantum states are separated by the band gap. In order to drive a current in this case, it is necessary to promote electronic transitions across the band gap, and the bias required for this may be very large. For example, in a diamond specimen of 1 mm, essentially no current is observed until the bias reaches 1 MV, at which point dielectric breakdown occurs.

In practice the existence of a band gap is so central to the properties of insulators that, in analogy with the case of metals and Fermi surfaces, we could define an *insulator as a solid with a band gap.* Band gaps can be measured by combining PES and IPES spectra, or by using optical absorption experiments as we will see in Chapter 10. The measured band gaps of diamond and silicon are 5.4 eV (Clark *et al.*, 1964) and 1.2 eV (Macfarlane *et al.*, 1958), respectively. When we compare these values with those calculated in Figure 9.5, 4.1 eV and 0.5 eV, we realize that our calculations underestimate quite substantially the experimental band gaps. The underestimation of band gaps in DFT/LDA is a general trend, and we will come back to this point in Section 9.5.

The classification of metals as solids with a Fermi surface, and insulators as solids with a band gap, leaves aside the important category of *semiconductors*, of which silicon is the archetypal example. The scheme in Figure 9.6 may help clarify the relation between these families. The schematic band structure shown in this figure consists of two main bands separated by an energy gap. The leftmost panel is representative of a metal. Here the Fermi level is well inside one band. As a consequence the shape and size of the Fermi surface and the electron transport properties are not very sensitive to small changes in the Fermi level.

The central panel in Figure 9.6 is representative of an insulator. In this case there is no Fermi surface and consequently the DOS at the Fermi level vanishes. In order to modify the dynamics of the electrons it is necessary to overcome the energy band gap. This is very difficult to achieve using, say, a standard electricity supply of 220 V and 50 Hz.

The rightmost panel in Figure 9.6 shows a situation intermediate between an insulator and a metal. If we imagine adding electrons to the insulator in the middle panel, they will occupy the next available states in the upper band. This time the Fermi level is very near the edge of the band but not exactly at the bottom. For all practical purposes

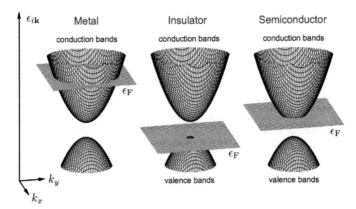

Fig. 9.6 Simplified scheme for classifying solids into metals (left), insulators (middle) and semiconductors (right), based on the placement of the Fermi level. In this example the wavevector is in two dimensions for visualization purposes, and the Fermi surface is given by the intersection between the parabolic bands (black) and the plane of constant energy at ϵ_F (blue). In the case of insulators and semiconductors it is usual to distinguish between valence bands (below the band gap) and conduction bands (above).

this situation corresponds to having a metal with a very small Fermi surface. In the simplest case of parabolic bands, and measuring the Fermi level from the bottom of the band, eqn 3.15 tells us that the density of electrons in this band is proportional to $\epsilon_F^{3/2}$. As a consequence, the transport properties will be very sensitive to the exact location of the Fermi level, and this can be used to tune the material from an insulator to a metal. This system is called a *semiconductor*, and the two sets of bands in the example are referred to as the *valence band* and the *conduction band*.

In the case of silicon the filling of the conduction band is typically achieved via phosphorus doping. In fact P has the electronic configuration $[Ne]3s^2p^3$, and hence it carries one more electron than Si.

In analogy with the definitions of metals and insulators given above, we could say that semiconductors are solids with tunable Fermi surfaces. However, in practice it is more common to think of a semiconductor as a *solid with tunable carrier concentration* (in semiconductor physics the term 'carrier' refers to the fact that electrons 'carry' electric charge).

A few words of caution are now in order. The classification of solids in metals, insulators and semiconductors discussed here is somewhat oversimplified. In reality the boundaries between these families are not always clear cut. One extreme example where these three categories do overlap is the case of graphene. Graphene is one atomic layer of carbon atoms arranged in a honeycomb lattice (Figure D.1). Graphene was obtained by Novoselov *et al.* (2005) using micromechanical cleavage of graphite, and since then it has become the subject of most intensive research programmes worldwide (Geim, 2011). The band structure of graphene (Figure D.2) exhibits a vanishing band gap, suggesting that this material should be considered a metal. At the same time the Fermi surface is only one point: therefore graphene is also an insulator. In addition the carrier concentration can easily be tuned, hence this material is a semiconductor. This conundrum shows that the definitions given above are merely a matter of language, and serve the sole purpose of organizing concepts.

Another delicate aspect is that the middle panel of Figure 9.6 refers to what we call 'band insulators'. There are at least two other important categories of insulators which do not fit in this simplified picture, namely 'Mott insulators' and 'topological insulators'. The discussion of these topics would require a significant detour; therefore we limit ourselves to pointing the reader to the papers by Anisimov *et al.* (1991) and Hasan and Kane (2010), respectively.

Exercise 9.6 In this exercise we want to understand how the concept of *effective mass* arises in the study of semiconductors (Kittel, 1976). Let us consider phosphorus-doped silicon. As explained above, in P-doped Si the valence band is entirely filled, and the Fermi level lies slightly above the bottom of the conduction band (CBB). The electrical properties of P-doped Si are mostly determined by the band structure near the CBB; therefore in many applications it is useful to find simple parametrizations of this band structure around the CBB. If we change the axes in Figure 9.5 in such a way to have $\mathbf{k} = 0$ and $\varepsilon = 0$ at the CBB, a Taylor expansion of the Kohn–Sham energy for small $\mathbf{k}$ gives:

$$\varepsilon = \frac{1}{2}\sum_{\alpha,\beta}\frac{\partial^2\varepsilon}{\partial k_\alpha \partial k_\beta}k_\alpha k_\beta + \mathcal{O}(|\mathbf{k}|^3), \tag{9.20}$$

where α and β run over the Cartesian coordinates as usual, and the term linear in $\mathbf{k}$ vanishes since we are expanding around a minimum. This expression can be made very similar to the dispersions of a free electron gas (see eqn 3.14) if we choose a reference frame where the matrix of partial derivatives is diagonal. Let us call $S_{\alpha\beta}$ the orthogonal matrix which diagonalizes $\partial^2\varepsilon/\partial k_\alpha \partial k_\beta$, and $K_\alpha = \sum_\beta S_{\alpha\beta} k_\beta$ the transformed coordinates. In the new reference frame the above equation takes the form:

$$\varepsilon = \frac{1}{2}\left(\frac{\partial^2\varepsilon}{\partial K_1^2}K_1^2 + \frac{\partial^2\varepsilon}{\partial K_2^2}K_2^2 + \frac{\partial^2\varepsilon}{\partial K_3^2}K_3^2\right) + \mathcal{O}(|\mathbf{K}|^3).$$

If we now define:

$$\frac{1}{m_\alpha^*} = \frac{1}{\hbar^2}\frac{\partial^2\varepsilon}{\partial K_\alpha^2} \quad \text{for } \alpha = 1, 2, 3, \tag{9.21}$$

then the previous equation becomes similar to a three-dimensional free electron gas:

$$\varepsilon = \frac{\hbar^2 K_1^2}{2m_1^*} + \frac{\hbar^2 K_2^2}{2m_2^*} + \frac{\hbar^2 K_3^2}{2m_3^*}. \tag{9.22}$$

In fact in the case of a free electron gas the dispersion relation is $\varepsilon = \hbar^2|\mathbf{K}|^2/2m_e$ (see eqn 3.14), which corresponds to eqn 9.22 if we set $m_1^* = m_2^* = m_3^* = m_e$. In general the 'effective' masses, m_α^*, will not be the same as in the free electron gas, but will be determined by the curvature of the bands near the CBB. ▶We want to calculate the effective masses of silicon in the conduction band by evaluating numerically the mixed partial derivatives in eqn 9.20. Following the same procedure as in Exercise 7.6, show that for small q

$$\frac{\partial^2\varepsilon}{\partial k_x^2} = \frac{1}{q^2}[\varepsilon(q,0,0) - 2\varepsilon(0,0,0) + \varepsilon(-q,0,0)] + \mathcal{O}(q^2), \tag{9.23}$$

$$\frac{\partial^2\varepsilon}{\partial k_x \partial k_y} = \frac{1}{4q^2}[\varepsilon(q,q,0) + \varepsilon(-q,-q,0) - \varepsilon(q,-q,0) - \varepsilon(-q,q,0)] + \mathcal{O}(q^2), \tag{9.24}$$

where $\varepsilon(q,0,0)$ stands for $\varepsilon(\mathbf{k} = q\mathbf{u}_x)$ and so on. ☐All the other partial derivatives can be obtained in the same way. By considering all the possible derivatives we see that our calculation requires the evaluation of ε at 19 wavevectors: 12 wavevectors of coordinates $(\pm q, \pm q, 0)$, $(\pm q, 0, \pm q)$, $(0, \pm q, \pm q)$, 6 with coordinates $(\pm q, 0, 0)$, $(0, \pm q, 0)$, $(0, 0 \pm q)$, as well as the origin $(0, 0, 0)$. The following table reports the Kohn–Sham energy eigenvalues evaluated at these points around the CBB of Si, with $q = 0.01 \cdot 2\pi/a$. All the values are in meV:

$(0, 0, 0)$	0.000				
$(+q, 0, 0)$	0.539	$(0, +q, 0)$	2.722	$(0, 0, +q)$	2.722
$(-q, 0, 0)$	0.539	$(0, -q, 0)$	2.722	$(0, 0, -q)$	2.722
$(+q, +q, 0)$	2.308	$(+q, 0, +q)$	2.308	$(0, +q, +q)$	5.408
$(+q, -q, 0)$	2.308	$(+q, 0, -q)$	2.308	$(0, +q, -q)$	5.408
$(-q, +q, 0)$	4.213	$(-q, 0, +q)$	4.213	$(0, -q, +q)$	5.408
$(-q, -q, 0)$	4.213	$(-q, 0, -q)$	4.213	$(0, -q, -q)$	5.408

▶Using eqns 9.23 and 9.24 calculate the matrix of the partial second derivatives and show that the result is as follows:

$$\left(\frac{2\pi}{a}\right)^2\frac{\partial^2\varepsilon}{\partial k_\alpha \partial k_\beta} = \begin{pmatrix} 10.78 & 0 & 0 \\ 0 & 54.44 & 0 \\ 0 & 0 & 54.44 \end{pmatrix}\text{eV}.$$

In this case the matrix of partial derivatives is already diagonal; therefore the rotation of the reference frame is unnecessary. ▶Using the last result and eqn 9.21, show that the bottom of the conduction band of silicon can be described using two effective masses, $m_l = 0.96\, m_e$ and $m_t = 0.19\, m_e$. The former is referred to as the 'longitudinal effective mass', the latter is the 'transverse effective mass'. ☐These results indicate that, near the bottom of the conduction band, silicon behaves almost as a free electron gas. The key differences with respect to the free electron gas are that the masses are slightly smaller, and that they also depend on the direction of propagation. ▶Using eqn 9.22 and your calculated effective masses, plot the parabolic dispersion relations of Si for $\mathbf{k} = q\mathbf{u}_x$ with $0 < q < 2\pi/a$, and superimpose your plot onto the complete band structure in Figure 9.5. ▶In a sample of P-doped Si at $T = 0$ the Fermi level is $\epsilon_F = \varepsilon_{CBB} + \varepsilon_0$ with $\varepsilon_0 > 0$. Using eqn 9.22 and your calculated effective masses, show that the Fermi surface of this system consists of a prolate spheroid centred at the wavevector of the CBB, with the major axis oriented along the Σ line of the Brillouin zone. ▶Show that the size of the major axis of this spheroid in reciprocal space is

$$\Delta q = 2\frac{m_l}{m_e}\sqrt{\frac{2m_e\varepsilon_0}{\hbar^2}}.$$

▶Show that, for $\varepsilon_0 = 10$ meV, which corresponds approximately to a P concentration of $5 \cdot 10^{18}$ cm^{-3} at $T = 0$, the size of the major axis is $\Delta q = 0.08 \cdot (2\pi/a)$. ☐Since in the Brillouin zone of Si there are six equivalent Σ lines, in reality the Fermi surface consists of six such spheroids (Kittel, 1976).

In Exercise 9.6 we calculated the effective conduction masses of silicon using the Kohn–Sham energy eigenvalues, and obtained $m_l = 0.96\, m_e$ and $m_t = 0.19\, m_e$. Using cyclotron resonance experiments, Dresselhaus *et al.* (1955) determined the effective masses $m_l/m_e = 0.97 \pm 0.02$ and $m_t/m_e = 0.19 \pm 0.01$. Therefore our calculations are in good agreement with experiment.

In general, the agreement between effective masses calculated using DFT and experiment is not as good. For example, Oshikiri *et al.* (2002) and Kim *et al.* (2010) report deviations from experiment of up to 100% in some cases. Research efforts are currently underway to develop more accurate DFT approaches for the calculation of effective mass parameters.

9.5   The band gap problem

In Section 9.4 we observed that the band gaps of diamond and silicon obtained from the Kohn–Sham band structure (Figure 9.5) underestimate significantly the experimental values. This is actually a very general trend, as we can see from the comparison reported in Figure 9.7. As a rule of thumb the Kohn–Sham band gaps are found to underestimate the measured gaps by about 40%.

As we will see in Chapter 10, the band gap is a critical quantity when it comes to the dielectric and optical properties of materials; therefore the inability to reliably predict band gaps represents a rather serious shortcoming. The underestimation of band gaps is possibly the most notorious difficulty in DFT calculations, to the point that it is usually referred to as 'the band gap problem'. It is not surprising then that since the early days of DFT many strategies have been devised to circumvent this difficulty, from *ad hoc* corrections to advanced quantum field theory methods.

Here we want to shed some light on what actually is the 'problem', and how it can be fixed. This topic is rather advanced, and therefore we will keep the discussion to a qualitative level, pointing the reader to the specialized literature where appropriate.

In Section 9.4 the band gap was introduced as the smallest energy separation between unoccupied and occupied Kohn–Sham eigenvalues:

$$E_g^{KS} = \varepsilon_{CBB} - \varepsilon_{VBT}, \tag{9.25}$$

where the superscript 'KS' is to remind us that this quantity is obtained by calculating the Kohn–Sham band structure. We can rewrite this equation by expressing ε_{CBB} and ε_{VBT} in terms of total energies, by means of eqn 9.5:

$$E_{N+1} \simeq E_N + \varepsilon_{CBB}, \tag{9.26}$$
$$E_{N-1} \simeq E_N - \varepsilon_{VBT}, \tag{9.27}$$

where E_N and $E_{N\pm1}$ are the ground-state total energies of the neutral system and that with one electron added or removed, respectively. Note the '$\simeq$' sign, which comes from the fact that eqn 9.5 is not an exact relation. By combining eqns 9.25–9.27 we obtain:

$$E_g^{KS} \simeq \underbrace{(E_{N-1} - E_N)}_{\substack{\text{work function or} \\ \text{ionization potential}}} - \underbrace{(E_N - E_{N+1})}_{\text{electron affinity}}. \tag{9.28}$$

In this expression the first term on the right-hand side is precisely the work function, ϕ (see Section 9.3); therefore it can be measured directly in PES experiments. In the case of atoms and molecules the same quantity is referred to as the 'ionization potential'. The second term is called the 'electron affinity', and can similarly be measured using IPES experiments.

Now we can see the origin of the band gap problem: the quantity which is measured by experiment is:

$$E_g^{qp} = (E_{N-1} - E_N) - (E_N - E_{N+1}), \tag{9.29}$$

and is referred to as the 'quasiparticle gap' or 'electrical gap'. By comparing this definition with eqn 9.28 we clearly see that the Kohn–Sham gap provides *only an approximation* to the quasiparticle gap, $E_g^{KS} \simeq E_g^{qp}$. However, in general we cannot expect E_g^{KS} to accurately predict the experimental gaps.

From this analysis it may appear that the 'band gap problem' is only a matter of semantics: in order to connect with experiment we need to calculate the quasiparticle gap of eqn 9.29 instead of the Kohn–Sham gap of eqn 9.25. In reality things are more complicated than this, as we explain in the following.

If we consider a small molecule, the calculation of the DFT total energies in the neutral state (E_N), in the cation (E_{N-1}) and in the anion (E_{N+1}) is certainly possible; therefore we can obtain E_g^{qp} directly from differences in *total energy*, without resorting to Kohn–Sham eigenvalues. This procedure goes under the name of the 'ΔSCF (self-consistent field) method', and is reasonably successful for small molecules (Rostgaard *et al.*, 2010).

The difficulty arises when we want to use the same procedure in the case of solids. The key difference between a small molecule and a solid is that the change in the electron density upon addition or removal of one electron is exceedingly small. For example, in a 1 mm^3 specimen of silicon there are about $2 \cdot 10^{20}$ valence electrons. Therefore,

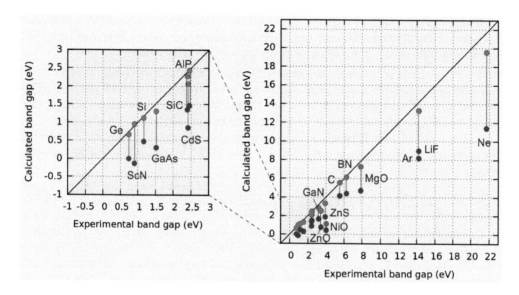

Fig. 9.7 Comparison between measured band gaps, DFT/LDA calculations and *GW* calculations (see eqn 9.32) for solids. The blue discs are DFT/LDA Kohn–Sham band gaps and the red discs are from *GW* calculations. The points belonging to the same solid are connected by a vertical line. The black line at 45° indicates where the data-points should be if they were in perfect agreement with experiment. The smaller panel (on the left) is an expanded view of the lower range of band gaps. All the data are from Table I of Tran and Blaha (2009).

by adding or removing one electron, the density will change by a negligible amount, $\Delta n \sim 10^{-20} n$. By replacing the DFT total energy of eqn 9.3 inside eqn 9.29, and taking the limit $\Delta n \to 0$, it can be shown that (Perdew *et al.*, 1982; Perdew and Levy, 1983):

$$\lim_{\Delta n \to 0} E_g^{\text{qp}} = E_g^{\text{KS}} + \Delta_{xc}, \tag{9.30}$$

where the extra energy, Δ_{xc}, is given by:

$$\Delta_{xc} = \lim_{\Delta n \to 0} V_{xc}[n + \Delta n] - V_{xc}[n - \Delta n]. \tag{9.31}$$

This result indicates that the quasiparticle band gap and the Kohn–Sham band gap of a solid differ by a constant, Δ_{xc}. Now, if we could use the 'exact' exchange and correlation functional E_{xc}, we would obtain a *discontinuous jump* of V_{xc} in eqn 9.31, i.e. $\Delta_{xc} \neq 0$ (Sham and Schlüter, 1983).

The trouble is that we do not know the exact functional. When we use instead the standard LDA approximation, by construction V_{xc} is a continuous function of the density (see for example eqn 3.23) and therefore there is no discontinuity: $\Delta_{xc} = 0$. This shows that DFT/LDA calculations *cannot* yield the correct quasiparticle gap.

In practice the band gap problem of DFT is actually a problem of the Kohn–Sham formulation of DFT, and in particular of the approximations made for the exchange and correlation potential, V_{xc}. It is now established that the LDA and similar approximations for the exchange and correlation energy are inadequate when we need to calculate the energy for adding or removing electrons in solids.

In order to get out of this impasse, among the many possible improvements, one which has emerged as reliable and accurate is the so-called 'quasiparticle *GW* method'. This method is based on quantum field theory techniques, and, interestingly enough, it is as old as DFT itself, dating back to the work of Hedin (1965).

From the point of view of practical calculations, the *GW* method consists of replacing the exchange and correlation potential, V_{xc}, in eqn 9.1 by a function called the 'self-energy', denoted by Σ:

$$V_{xc}(\mathbf{r})\phi_i(\mathbf{r}) \to \int d\mathbf{r}' \Sigma(\mathbf{r}, \mathbf{r}', \varepsilon_i)\phi_i(\mathbf{r}'). \tag{9.32}$$

In this expression the self-energy depends both on $\mathbf{r}, \mathbf{r}'$ and on the energy, ε_i; therefore the numerical procedure becomes significantly more involved. The name *GW* arises from the fact that Σ is obtained as the product of two quantities called the Green's function (G) and the screened Coulomb interaction (W). The precise meaning of these quantities can be found in the review by Hedin and Lundqvist (1969). At variance with V_{xc}, Σ captures the finite discontinuity Δ_{xc} in eqn 9.30, and therefore yields correctly the quasiparticle band gap.

Figure 9.7 shows that *GW* calculations of band gaps correct most of the error found in DFT/LDA, and restore the predictive power of the method. The price to pay for this improvement is that such calculations are considerably more time-consuming than in DFT/LDA, and can be performed only on relatively small systems (e.g. up to 100 atoms).

The interested reader will find useful insights into the *GW* method in the seminal works by Strinati *et al.* (1982), Hybertsen and Louie (1986) and Godby *et al.* (1986). In-depth discussions of the formalism, its advantages and its limitations can be found in the reviews by Hedin and Lundqvist (1969), Aryasetiawan and Gunnarsson (1998) and Onida *et al.* (2002).

10
Dielectric function and optical spectra

In this chapter we discuss the use of DFT calculations for studying the optical properties of materials. A precise understanding of these properties and how they arise from the underlying atomistic structure is important in many applications, ranging from optoelectronics to photovoltaics. For example, in the design of solar cells one recurring question is what fraction of the incident sunlight can be absorbed by the optically active material, and how this material can be modified in order to absorb more light. Similarly, in the area of solid-state lighting it is important to understand how to optimize materials for transforming the energy of electrons into light. More fundamentally, optical properties form an essential component of our experience. For example, we can think of the colour of objects, or even the fact that certain materials are transparent to light while others are opaque.

Nowadays techniques based on DFT are employed routinely to gain insight into the optical properties of materials. Given the complexity of the theory and of the computational schemes, it is not easy to achieve quantitative agreement with experiment, and it is common to complement DFT calculations with semi-empirical corrections. Such a 'mixed first-principles/semi-empirical' approach can be very powerful for investigating the trends of dielectric and optical properties, and for formulating hypotheses regarding the underlying atomistic mechanisms.

The goal of this chapter is to show how DFT can be used to calculate optical properties, the level of accuracy that can be expected and the main limitations. As usual we will start from a heuristic perspective, and we will follow this up with the general formalism and a few examples of DFT calculations.

10.1 The dielectric function of a model solid

In this section we introduce the concept of *dielectric function* by considering the simplest possible system, a cubic lattice of H atoms. The only reason for such a choice is that this system is simple enough for us to see very clearly how the notion of dielectric function emerges from the first principles of quantum mechanics. In practice we will first derive the 'electric polarizability' of a hydrogen atom, and then we will use the Clausius–Mossotti relation in order to obtain the dielectric function of hypothetical solid H. As a curiosity, the possibility of realizing solid atomic hydrogen at very high pressures was pointed out long ago by Wigner and Huntington (1935), and only recently there have been hints of the possible existence of such a system

(Eremets and Troyan, 2011). The study of solid hydrogen is not only of interest for testing fundamental theories, but may also be relevant to the study of the core of giant planets such as Jupiter and Saturn.

This first section can be regarded as a long exercise in basic quantum mechanics, and serves the purpose of introducing a number of important concepts. The reader already familiar with the quantum theory of atomic polarizability might want to skip this section altogether.

10.1.1   Electron dynamics in a radiation field

Let us start by considering an isolated H atom, with the proton held immobile at the centre of our reference frame. For times $t < 0$ we have the atom in its ground state, with the electron occupying the $1s$ orbital, ϕ_{1s}, of energy E_{1s}. We want to study what happens if, at $t = 0$, we switch on an external radiation field. We take this external field to be a uniform electric field directed along the x axis:

$$\mathbf{E}(t) = \mathcal{E}f(t)\,\mathbf{u}_x, \quad \text{with } f(t) = \begin{cases} 0 & \text{if } t < 0, \\ \cos(\omega t) & \text{if } t \geq 0. \end{cases} \tag{10.1}$$

This equation describes monochromatic electromagnetic radiation of frequency ω, as could be obtained, for example, using a laser. In the following, in order to avoid ambiguity, we will not make use of Hartree atomic units. From classical electrostatics we know that the potential energy of an electron located at point $\mathbf{r}$ in this field is:

$$V(\mathbf{r}, t) = e\,\mathbf{E}(t)\cdot\mathbf{r}. \tag{10.2}$$

In order to study the time evolution of the electron wavefunction, $\psi(\mathbf{r}, t)$, we need to solve the *time-dependent Schrödinger equation* (Merzbacher, 1998):

$$i\hbar\frac{\partial}{\partial t}\psi(\mathbf{r}, t) = [\hat{H} + V(\mathbf{r}, t)]\psi(\mathbf{r}, t). \tag{10.3}$$

In this equation the Hamiltonian, $\hat{H}$, is that for the hydrogen atom in the absence of external fields:

$$\hat{H} = -\frac{\hbar^2}{2m_e}\nabla^2 - \frac{e^2}{4\pi\epsilon_0|\mathbf{r}|}, \tag{10.4}$$

and the additional potential, V, is from eqn 10.2. The starting point of the time evolution is given by our initial condition $\psi(\mathbf{r}, t < 0) = \phi_{1s}(\mathbf{r})$.

For simplicity we now consider that the Hamiltonian, $\hat{H}$, only admits the quantum states $1s$, $2s$, $2p_x$, $2p_y$, and $2p_z$. This is similar to what was done in Exercise 4.5 for the van der Waals interaction. In principle we could include many more states, such as $3s$, $3p$, $3d$, and so on; however, this is not really needed for the present discussion. It can immediately be verified by direct substitution that the function $\phi_{1s}(\mathbf{r})\exp(-iE_{1s}t/\hbar)$, the function $\phi_{2s}(\mathbf{r})\exp(-iE_{2s}t/\hbar)$, and so on, are all solutions of eqn 10.3 when $V=0$. This observation suggests that a solution of eqn 10.3 for $t > 0$ may be expressed as a linear combination of such functions:

$$\psi(\mathbf{r}, t) = c_{1s}(t)\phi_{1s}(\mathbf{r})e^{-\frac{i}{\hbar}E_{1s}t} + \cdots + c_{2p_z}(t)\phi_{2p_z}(\mathbf{r})e^{-\frac{i}{\hbar}E_{2p_z}t}. \tag{10.5}$$

In general, it is always possible to write $\psi(\mathbf{r}, t)$ as a linear combination of atomic orbitals, since the eigenstates of $\hat{H}$ form a complete Hilbert space. Note that the coefficients $c_{1s}(t), \ldots, c_{2p_z}(t)$ depend on the time variable. Since the wavefunction ψ corresponds to one electron, we can replace eqn 10.5 inside the normalization condition $\int \psi^*(\mathbf{r}, t)\psi(\mathbf{r}, t)d\mathbf{r} = 1$ to obtain:

$$|c_{1s}(t)|^2 + |c_{2s}(t)|^2 + |c_{2p_x}(t)|^2 + |c_{2p_y}(t)|^2 + |c_{2p_z}(t)|^2 = 1. \qquad (10.6)$$

This condition is useful to form an intuitive idea of the shape of $\psi(\mathbf{r}, t)$. For example, if we have $|c_{1s}(t)|^2 = 0.9$ and $|c_{2p_x}(t)|^2 = 0.1$, then we can tell that all the other coefficients are zero and the wavefunction is a 90%/10% mixture of $1s$ and $2p_x$ orbitals. The shape and energy of hydrogen orbitals are well known (see for example Exercise 10.1, below); therefore the only unknowns in eqn 10.5 are the time-dependent coefficients. We can determine these coefficients as follows. We first replace eqn 10.5 inside eqn 10.3. Then we multiply both sides by $\phi_{1s}(\mathbf{r})$ and we integrate over the space variable. By considering that each orbital is an eigenstate of the Hamiltonian, $\hat{H}$, and that different orbitals are orthogonal, for $t \geq 0$ we find:

$$i\hbar\frac{dc_{1s}(t)}{dt} = e\mathcal{E}\, x_{1s,1s}\, \cos(\omega t)\, e^{\frac{i}{\hbar}(E_{1s}-E_{1s})t}c_{1s}(t)$$

$$+ e\mathcal{E}\, x_{1s,2s}\, \cos(\omega t)\, e^{\frac{i}{\hbar}(E_{1s}-E_{2s})t}c_{2s}(t)$$

$$+ e\mathcal{E}\, x_{1s,2p_x}\cos(\omega t)\, e^{\frac{i}{\hbar}(E_{1s}-E_{2p_x})t}c_{2p_x}(t)$$

$$+ e\mathcal{E}\, x_{1s,2p_y}\cos(\omega t)\, e^{\frac{i}{\hbar}(E_{1s}-E_{2p_y})t}c_{2p_y}(t)$$

$$+ e\mathcal{E}\, x_{1s,2p_z}\cos(\omega t)\, e^{\frac{i}{\hbar}(E_{1s}-E_{2p_z})t}c_{2p_z}(t), \qquad (10.7)$$

where we have introduced the 'matrix elements' $x_{1s,1s}, \ldots, x_{1s,2p_z}$ of the position operator x. For example, the matrix element $x_{1s,2p_x}$ is defined as:

$$x_{1s,2p_x} = \int d\mathbf{r}\, \phi_{1s}(\mathbf{r})\, x\, \phi_{2p_x}(\mathbf{r}), \qquad (10.8)$$

and similarly for all the others. We note that this matrix element already made its debut in eqn 4.33.

Equation 10.7 has been obtained by multiplying both sides of eqn 10.3 by ϕ_{1s} and then integrating over $\mathbf{r}$. Clearly we could repeat exactly the same operation for $\phi_{2s}, \ldots, \phi_{2p_z}$ and obtain similar equations. This procedure would lead us to a linear system of differential equations like eqn 10.7, one for every coefficient $c_{1s}, \ldots, c_{2p_z}$.

The advantage of the procedure leading to eqn 10.7 is that in practical applications many matrix elements are identically zero, and hence only a few terms survive on the right-hand side of the equation. As an example, it should be easy to see that:

$$x_{1s,1s} = \int d\mathbf{r}\, \phi_{1s}(\mathbf{r})\, x\, \phi_{1s}(\mathbf{r}) = 0,$$

because $|\phi_{1s}(\mathbf{r})|^2$ is an even function, while x is an odd function. By inspecting the matrix elements one by one we can quickly verify that all of them are zero, with the exception of $x_{1s,2p_x}$ and $x_{2s,2p_x}$.

Exercise 10.1 The $1s$, $2s$ and $2p_x$ wavefunctions of the H atom are (Merzbacher, 1998):

$$\phi_{1s}(\mathbf{r}) = \frac{a_0^{-\frac{3}{2}}}{\sqrt{\pi}} \exp\left(-\frac{r}{a_0}\right),$$

$$\phi_{2s}(\mathbf{r}) = \frac{a_0^{-\frac{3}{2}}}{\sqrt{\pi}} \frac{1}{4\sqrt{2}} \left(2 - \frac{r}{a_0}\right) \exp\left(-\frac{r}{2a_0}\right),$$

$$\phi_{2p_x}(\mathbf{r}) = \frac{a_0^{-\frac{3}{2}}}{\sqrt{\pi}} \frac{1}{4\sqrt{2}} \frac{x}{a_0} \exp\left(-\frac{r}{2a_0}\right),$$

with $r = |\mathbf{r}|$ and a_0 the Bohr radius. By using spherical coordinates (r, θ, φ) with the polar axis along $\mathbf{u}_x$, and remembering that the volume element is $d\mathbf{r} = r^2 \sin\theta \, dr \, d\theta \, d\varphi$, show that the matrix elements $x_{1s,2p_x}$ and $x_{2s,2p_x}$ of the position operator are:

$$x_{1s,2p_x} = \frac{2^8}{3^5 \sqrt{2}} a_0 \simeq 0.4 \text{ Å},$$

$$x_{2s,2p_x} = -3 a_0 \simeq -1.6 \text{ Å}.$$

It can be useful to consider the following integrals:

$$\int_0^\infty u^4 e^{-u} du = 24, \quad \int_0^\infty u^5 e^{-u} du = 120, \quad \int_0^\pi \cos^2(u)\sin(u)du = \frac{2}{3}.$$

In Exercise 10.1 we found that $x_{1s,2p_x}$ and $x_{2s,2p_x}$ are both of the order of the Bohr radius, as we might have expected. Since the only orbitals with non-zero matrix elements between them are the $1s$, $2s$ and $2p_x$, we are left with three equations to solve.

In order to simplify our discussion we make the approximation that the $2s$ orbital is not very important and can be ignored. This approximation does not hold in general. However, it can be shown that the time required by the electron to move between the $2s$ and $2p_x$ states is much longer than the time required to go between $1s$ and $2p_x$. In the following we want to focus on the latter process; therefore we can safely set $c_{2s}(t) \simeq 0$. Using this simplification there remain only two equations, one for c_{1s} and the other for c_{2p_x}:

$$i\hbar \frac{dc_{1s}(t)}{dt} = e\mathcal{E} \, x_{1s,2p_x} \cos(\omega t) e^{i(E_{1s}-E_{2p_x})t/\hbar} c_{2p_x}(t), \tag{10.9}$$

$$i\hbar \frac{dc_{2p_x}(t)}{dt} = e\mathcal{E} \, x_{1s,2p_x} \cos(\omega t) e^{i(E_{2p_x}-E_{1s})t/\hbar} c_{1s}(t). \tag{10.10}$$

These equations must be solved with the initial conditions $c_{1s}(0) = 1$ and $c_{2p_x}(0) = 0$. Before proceeding it is helpful to introduce the frequency units:

$$\omega_0 = \frac{e\mathcal{E} \, x_{1s,2p_x}}{\hbar}, \tag{10.11}$$

$$\omega_{12} = \frac{E_{2p_x} - E_{1s}}{\hbar}. \tag{10.12}$$

The first frequency gives a measure of the strength of the electric field. Typical electric fields used in electronics are of the order of 1 V/μm; therefore the energy $\hbar\omega_0$ must be of the order of 0.1 meV. This value corresponds to a timescale of 100 fs. The frequency ω_{12} in eqn 10.12 is associated with the energy difference between the hydrogen $1s$ and $2p$ states, $\hbar\omega_{12} = 10.2$ eV, and corresponds to a period of oscillation of about 0.1 fs. We now focus on a situation where $c_{2p_x}(t)$ remains 'small' at all times, in such a way that the right-hand side of eqn 10.9 is practically zero and $c_{1s}(t) \simeq 1$. If we replace the value 1 for $c_{1s}(t)$ in eqn 10.10 we obtain:

$$\frac{dc_{2p_x}}{dt} = -i\omega_0 \cos(\omega t)\, e^{i\omega_{12}t}. \tag{10.13}$$

This equation can be solved without difficulty by writing the cosine in terms of complex exponentials. After a few tedious manipulations we find:

$$c_{2p_x} = -i\omega_0 \left\{ \frac{e^{i(\omega_{12}+\omega)t/2}\sin[(\omega_{12}+\omega)t/2]}{\omega_{12}+\omega} + \frac{e^{i(\omega_{12}-\omega)t/2}\sin[(\omega_{12}-\omega)t/2]}{\omega_{12}-\omega} \right\}. \tag{10.14}$$

In this expression the modulus of the trigonometric functions is always ≤ 1. Therefore the magnitude of c_{2p_x} is bound as follows:

$$|c_{2p_x}(t)| \leq \frac{2\omega_0}{|\omega_{12}-\omega|}.$$

In the simplest case of $\omega = 0$, corresponding to a static electric field, this upper bound gives:

$$|c_{2p_x}(t)| \leq 2\frac{\hbar\omega_0}{\hbar\omega_{12}} \sim 2\frac{0.1\text{ meV}}{10.2\text{ eV}} \sim 10^{-5}. \tag{10.15}$$

This reasoning shows that indeed $c_{2p_x}(t)$ remains very small and our procedure is correct. In reality this criterion breaks down when $\omega \sim \omega_{12}$, in which case the coefficient grows steadily with time. Physically this situation corresponds to a 'transition' of the electron from the $1s$ state to the $2p_x$ state. In order to describe properly the case $\omega \sim \omega_{12}$ we need to make a small adjustment to the theory, as will be explained in Section 10.1.2.

By collecting our results thus far, we end up with the following approximate solution for eqn 10.3, corresponding to the initial condition where the H atom is in its ground state:

$$\psi(\mathbf{r}, t) = e^{-\frac{i}{\hbar}E_{1s}t}\left[\phi_{1s}(\mathbf{r}) + e^{-i\omega_{12}t}c_{2p_x}(t)\phi_{2p_x}(\mathbf{r}) \right], \tag{10.16}$$

with c_{2p_x} given by eqn 10.14. Figure 10.1 shows the time evolution of the electron density, $|\psi(\mathbf{r}, t)|^2$, calculated from this equation as the oscillating external field completes one cycle. In the figure we see that the electronic charge undergoes oscillations along the direction of the applied electric field, as we would expect from a classical model of a charge connected to a fixed point by a spring.

The result obtained in this section has been discussed in a number of textbooks under various forms (e.g. Ziman, 1972; Merzbacher, 1998; Novotny and Hecht, 2006; Sakurai and Napolitano, 2011). The main difference between our discussion and other presentations is that we purposely avoided the use of two important theoretical tools of quantum mechanics, namely 'time-dependent perturbation theory' and the 'interaction

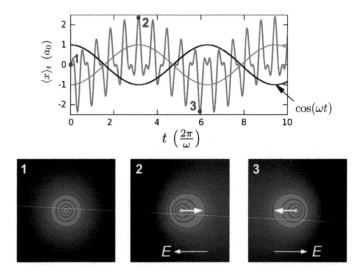

Fig. 10.1   Time evolution of the electron in the H atom subject to an oscillating electric field. The black line in the top panel shows the time dependence of the electric field, $\mathbf{E} = \mathcal{E}\cos(\omega t)\mathbf{u}_x$. The thick red line in the same panel is the instantaneous expectation value of the electron position (see eqn 10.18). This curve results from the combination of two oscillatory motions, one of frequency ω represented by the thin red line, and the other of frequency ω_{12} superimposed on the former. This second component is not very important in practice as it is typically damped (see Section 10.1.2). Had we considered also the $2s$ orbital in eqn 10.10, we would see an additional high-frequency component in this plot; however, the fundamental harmonic would stay unchanged.

The three bottom panels show snapshots of the electron density, $|\psi(\mathbf{r}, t)|^2$, in a cut through the xy plane, for the three times labelled '1, 2, 3' in the top panel. We can see that in panels 2 and 3 the electron charge is displaced opposite to the electric field. This displacement generates an electric dipole moment which counters the external field, and is indicated by white arrows centred on the proton. In this example we used $\hbar\omega = 0.1 \cdot \hbar\omega_{12} \sim 1$ eV, and $\hbar\omega_0 = 0.8 \cdot \hbar\omega_{12} \sim 8$ eV. The external field chosen for this figure is extremely strong and unrealistic, and should be considered only as an expedient for visualizing the displacement of the electron charge. Under typical electric fields the change of the electron charge density would not be discernible to the eye.

picture'. These latter techniques are mathematically very elegant; however, they make it harder to grasp the key mechanisms and develop some intuition. The reader interested in a more formal approach will find a good starting point in Chapter 5 of Sakurai and Napolitano (2011).

10.1.2   Electron dynamics and emission of radiation

The knowledge of the time evolution of the electron wavefunction, $\psi(\mathbf{r}, t)$, allows us to track the average position of the electron at all times. The instantaneous expectation value of the electron position is given by:

$$\langle x \rangle_t = \int x |\psi(\mathbf{r}, t)|^2 \, d\mathbf{r}. \tag{10.17}$$

This quantity can be expressed in terms of the coefficient $c_{2p_x}(t)$ by replacing the wavefunction of eqn 10.16 inside eqn 10.17:

$$\langle x \rangle_t = 2 \, x_{1s,2p_x} \operatorname{Re}\left[c_{2p_x}(t) e^{-i\omega_{12}t} \right], \tag{10.18}$$

where 'Re' indicates the real part of a complex number. If we substitute for $c_{2p_x}(t)$ from eqn 10.14 we obtain:

$$\langle x \rangle_t = -x_{1s,2p_x} \, \omega_0 \left(\frac{1}{\omega_{12} + \omega} + \frac{1}{\omega_{12} - \omega} \right) [\cos(\omega t) - \cos(\omega_{12} t)]. \tag{10.19}$$

This result indicates that the electron undergoes oscillations around the proton with the same frequency, ω, as the driving field, $\mathbf{E}(t)$. The evolution of the electron position, $\langle x \rangle_t$, as a function of time is shown by the thick red line in Figure 10.1. The additional term, $\cos(\omega_{12} t)$, in eqn 10.19 is not very important, and disappears in a more accurate description of the time evolution (see Exercise 10.2, below). This extra term relates to the fact that in our model the electric field jumps discontinuously from 0 to E at $t = 0$, as can be seen from eqn 10.1.

From eqn 10.19 we see that the amplitude of the electron displacement during each cycle is:

$$x_{1s,2p_x} \frac{\omega_0 \omega_{12}}{\omega_{12}^2 - \omega^2}.$$

As a consequence, the electron reaches regions further away from the nucleus as the laser frequency, ω, approaches ω_{12}. This situation corresponds to the laser 'resonating' with a natural frequency of the atom. Since for $\omega = \omega_{12}$ the amplitude of oscillation is infinite, it is tempting to conclude that, at resonance, the electron could actually escape from the proton. It turns out that this conclusion does not correspond to what is observed in experiments; therefore our theory must be somewhat incomplete.

From classical electrodynamics it is known that an electric charge in accelerated motion loses energy by emitting electromagnetic waves. In particular, an electron whose position fluctuates with time according to the law $x(t) = x_0 \cos(\omega t)$ radiates energy with a power output, P, (energy per unit time) given by Larmor's formula (Jackson, 1998):

$$P = \frac{2}{3} \frac{e^2}{4\pi\epsilon_0} \frac{x_0^2 \omega^4}{c^3}. \tag{10.20}$$

This phenomenon is common in our daily experience. In fact this is precisely the physical principle underlying the operation of antennas in mobile phones and wireless devices. The emission of radiation by the electron in the H atom can be included phenomenologically in our description by adding a *damping* term in eqn 10.13:

$$\frac{dc_{2p_x}}{dt} = -i\omega_0 \cos(\omega t) \, e^{i\omega_{12}t} \underbrace{-\gamma \, c_{2p_x}(t)}_{\text{damping term}}. \tag{10.21}$$

The role of the new term is to ensure that, when the electric field is turned off by setting $\omega_0 = 0$, the energy lost by emission of radiation is no longer compensated for by the electric field, and the electron goes back to the ground state:

$$\text{if } \omega_0 = 0, \text{ then } \frac{dc_{2p_x}}{dt} = -\gamma\, c_{2p_x}; \text{ therefore } c_{2p_x} = \text{const} \cdot e^{-\gamma t}.$$

The constant γ is called the *rate of decay*. In absence of a driving field the damping term in eqn 10.21 suppresses completely the coefficient c_{2p_x} within a time of the order of $1/\gamma$.

A rigorous discussion of the emission of radiation from excited atomic states would require us to study the quantum theory of the electromagnetic field by Dirac (1927), and is beyond the scope of this presentation. For a very accessible yet rigorous introduction to this subject the reader is referred to the review by Fermi (1932). In particular, in paragraphs 7 and 8 of that review it is shown clearly how an excited atom can emit radiation (even in the absence of a driving field) and what the characteristics are of the emitted field.

The solution to eqn 10.21 can be found by using the method of variation of parameters, and is outlined in Exercise 10.2. Using this solution, the expectation value of the electron position becomes:

$$\langle x \rangle_t = \text{Re}\left[-x_{1s,2p_x}\omega_0 \left(\frac{1}{\omega_{12} + \omega + i\gamma} + \frac{1}{\omega_{12} - \omega - i\gamma} \right) e^{-i\omega t}\right]. \tag{10.22}$$

Clearly in this case the resonance condition, $\omega = \omega_{12}$, does not lead to a divergence of the oscillation amplitude. Apart from this aspect, eqn 10.22 is similar to our previous result in eqn 10.19.

Exercise 10.2 We want to solve eqn 10.21 by using the method of variation of parameters. This method is based on the observation that the homogeneous differential equation associated with eqn 10.21 admits the solution $A\exp(-\gamma t)$, with A a constant. The idea is to find a solution to the complete equation by letting A depend on time, i.e. to look for a solution of the form $A(t)\exp(-\gamma t)$. ▶By replacing the trial solution $A(t)\exp(-\gamma t)$ in eqn 10.21 show that $A(t)$ has to satisfy the equation:

$$\frac{dA}{dt} = -i\omega_0 \cos(\omega t)\, e^{i(\omega_{12}-i\gamma)t} \text{ with the initial condition } A(0) = 0.$$

▶After expressing the cosine in terms of complex exponentials, solve for $A(t)$ and show that the time evolution of the coefficient c_{2p_x} is given by:

$$c_{2p_x}(t) = -\frac{\omega_0}{2}\left[\frac{e^{i\omega t}}{\omega_{12} + \omega - i\gamma} + \frac{e^{-i\omega t}}{\omega_{12} - \omega - i\gamma}\right]e^{i\omega_{12}t} + \frac{\omega_0}{2}\left[\frac{1}{\omega_{12} + \omega - i\gamma} + \frac{1}{\omega_{12} - \omega - i\gamma}\right]e^{-\gamma t}.$$

☐From this last result it should be immediately possible to obtain eqn 10.22 by replacing c_{2p_x} inside eqn 10.18, and considering times $t \gg 1/\gamma$ so that the term on the right can safely be ignored. We note that this latter term gives the $\cos(\omega_{12}t)$ oscillation in eqn 10.19 when $\gamma = 0$.

10.1.3 Electric dipole moment and polarizability

The oscillations of the electronic charge induced by the electric field (Figure 10.1) yield a fluctuating electric dipole moment, $\mathbf{p}(t)$, in the H atom. From classical electrostatics we know that such a dipole corresponds to the instantaneous expectation value of the electron position, $\langle x \rangle_t$, times the electron charge (Jackson, 1998):

$$\mathbf{p}(t) = -e\langle x \rangle_t \mathbf{u}_x. \tag{10.23}$$

By using our result in eqn 10.22 and the definitions in eqns 10.1, 10.11 and 10.12, we are now in a position to calculate the induced dipole of the H atom in an electric field:

$$\mathbf{p}(t) = \mathrm{Re}\left[\alpha(\omega)\mathcal{E}e^{-i\omega t}\right]\mathbf{u}_x, \tag{10.24}$$

with

$$\alpha(\omega) = \frac{e^2 x_{1s,2p_x}^2}{\hbar}\left(\frac{1}{\omega_{12} + \omega + i\gamma} + \frac{1}{\omega_{12} - \omega - i\gamma}\right). \tag{10.25}$$

The quantity $\alpha(\omega)$ is called the *polarizability*, and contains most of the information that we need to understand the dielectric and optical properties of atoms and molecules. Figure 10.2 shows the polarizability of the H atom vs. the frequency, ω, of the driving field, calculated from eqn 10.25. In the figure we clearly see a 'resonance' in correspondence of the energy $\hbar\omega_{12} = E_{2p_x} - E_{1s} = 10.2$ eV, both in the real and in the imaginary part of the complex polarizability.

Since it is known from experiment that the decay time, $1/\gamma$, is of the order of 1 ns (Chupp *et al.*, 1968), and therefore $\gamma/\omega_{12} \sim 10^{-7}$, it is common to neglect γ^2 with respect to ω_{12}^2, and to rewrite the polarizability as follows:

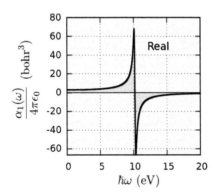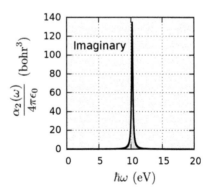

Fig. 10.2 The electric polarizability of the H atom vs. excitation energy, $\hbar\omega$, as obtained from eqn 10.25. The real part, $\alpha_1(\omega)$, and the imaginary part, $\alpha_2(\omega)$, both exhibit structure in correspondence with the energy difference, $E_{2p_x} - E_{1s}$, between the $2p_x$ and the $1s$ electronic eigenstates of the Hamiltonian in eqn 10.4. In this plot we used an exaggerated decay rate, $\gamma = 0.02\,\omega_{12}$, for visualization purposes. Using a realistic value of γ the imaginary part of the polarizability would look like an extremely sharp peak. In a complete calculation we would obtain many more 'resonances', as shown by eqn 10.29 below.

$$\alpha(\omega) = \alpha_1(\omega) + i\alpha_2(\omega),$$

with

$$\alpha_1(\omega) = f_{12} \frac{e^2}{m_e} \frac{1}{\omega_{12}^2 - \omega^2}, \tag{10.26}$$

$$\alpha_2(\omega) = f_{12} \frac{e^2}{m_e} \frac{\pi}{2\omega} \delta(\omega - \omega_{12}). \tag{10.27}$$

In the last equation we made use of the definition of the Dirac delta in terms of the Lorentzian function from eqn 5.10. The dimensionless quantity f_{12} is defined as:

$$f_{12} = \frac{2m_e}{\hbar} \omega_{12} x_{1s,2p_x}^2, \tag{10.28}$$

and is called the *oscillator strength* of the transition $1s \rightarrow 2p_x$. A simple classical model of atomic polarizability, whereby the electron and the proton are connected by a classical spring, would give precisely eqn 10.26 with $f_{12} = 1$ (Jackson, 1998). Therefore we can interpret f_{12} as a 'correction factor' which takes into account the quantum nature of the electron.

We can now compare our result with experiment. The measured polarizability of atomic hydrogen when the frequency of the driving field is $\hbar\omega = 2.1$ eV is $4.61 \pm 0.07\, a_0^3$ (Marlow and Bershader, 1964). After replacing the values of the physical constants and using the result of Exercise 10.1 for $x_{1s,2p_x}$, eqn 10.26 gives:

$$f_{12} = 0.416, \qquad \frac{\alpha}{4\pi\epsilon_0} = 3.09\, a_0^3 \quad \text{for} \quad \hbar\omega = 2.1 \text{ eV}.$$

Clearly, our calculation underestimates the measured value by about 50%. The origin of this discrepancy lies in our initial choice of taking the electron wavefunction to be a linear combination of only the orbitals $1s, \ldots, 2p_z$ (eqn 10.5). Had we included more orbitals we would have obtained a sum of terms similar to the one in eqn 10.25:

$$\alpha(\omega) = \sum_n \frac{e^2 x_{0,n}^2}{\hbar} \left(\frac{1}{\omega_{0,n} + \omega + i\gamma} + \frac{1}{\omega_{0,n} - \omega - i\gamma} \right). \tag{10.29}$$

Here, n runs over all the quantum states of the H atom with wavefunction $\phi_n(\mathbf{r})$ and energy E_n; the matrix elements of the position are $x_{0,n} = \langle \phi_{1s}|x|\phi_n \rangle$ and the resonance frequencies are $\omega_{0,n} = (E_n - E_{1s})/\hbar$. In our simplified calculation leading to eqn 10.25 we have only one term of the above sum, corresponding to the $2p_x$ state.

The calculation of the sum in eqn 10.29, which involves an infinite number of terms, can be performed using a method by Dalgarno and Lewis (1955). The result of the complete calculation, $4.66\, a_0^3$, is in good agreement with the measured value.

From eqn 10.29 we see that the resonance $\omega_{0,n}$ contributes to the sum only if the corresponding matrix element, $x_{0,n}$, is non-zero. In this case we speak of a *dipole-allowed* transition. When $x_{0,n} = 0$ we have instead a *dipole-forbidden* transition, which does not contribute to the polarizability.

10.1.4   Power dissipation

The knowledge of the time evolution of the electron wavefunction under an external electric field allows us to calculate the power dissipated by the laser source. For this purpose we evaluate the instantaneous work, $W(t)$, done by the field on the electron:

$$W(t) = F(t)\frac{d}{dt}\langle x \rangle_t, \tag{10.30}$$

with the force $F(t) = -e\mathcal{E}\cos(\omega t)$. By replacing in this expression the instantaneous position, $\langle x \rangle_t$, from eqns 10.23 and 10.24 we obtain:

$$W(t) = \mathcal{E}^2\omega\cos^2(\omega t)\,\alpha_2(\omega) - \frac{1}{2}\omega\mathcal{E}^2\sin(2\omega t)\,\alpha_1(\omega). \tag{10.31}$$

If we average the work over one laser cycle, the second term vanishes, and the $\cos^2(\omega t)$ yields $1/2$. Therefore the average power dissipated by the driving field is:

$$\overline{W} = \frac{\mathcal{E}^2}{2}\omega\,\alpha_2(\omega). \tag{10.32}$$

This result shows that the quantity $\omega\,\alpha_2(\omega)$ describes the amount of energy lost by the radiation field and absorbed by the atom.

It is useful to restate this result in the standard language of electrodynamics. The energy carried by the oscillating field per unit area and per unit time is given by the average magnitude of the Poynting vector, $\overline{S}$ (Jackson, 1998):

$$\overline{S} = \epsilon_0 c \frac{\mathcal{E}^2}{2},$$

where c is the speed of light. By defining the *absorption cross-section*:

$$\sigma(\omega) = \frac{\omega\,\alpha_2(\omega)}{\epsilon_0 c},$$

we obtain:

$$\overline{W} = \overline{S}\,\sigma(\omega). \tag{10.33}$$

According to this relation we can interpret the absorption cross-section, $\sigma(\omega)$, as the 'effective area' of the atom for the absorption of energy: the power $\overline{W}$ absorbed by the atom is equal to the incoming power per unit area, $\overline{S}$, times the effective area, $\sigma(\omega)$, of the atom. The absorption cross-section is the key quantity in the analysis of the *optical absorption spectra* of atoms and molecules.

It is interesting to observe that the absorption cross-section is non-zero only when $\alpha_2 \neq 0$, and this can happen only when the excited electronic state $2p_x$ has a finite lifetime, i.e. $\gamma > 0$. Therefore, in this simple example, the source of power dissipation is precisely the emission of radiation by the atom.

10.1.5 Dielectric function of the model solid

After having analysed the response of an isolated H atom to an oscillating electric field, we can now move on and consider a hypothetical solid of hydrogen atoms. In the case of solids the analogue of the atomic polarizability, $\alpha(\omega)$, is the 'dielectric function', $\epsilon(\omega)$. This quantity describes the relation between the electric displacement field, $\mathbf{D}$, and the electric field, $\mathbf{E}$, in materials: $\mathbf{D} = \epsilon_0 \, \epsilon \, \mathbf{E}$. Here it is assumed that the reader is familiar with the laws of electrodynamics in dielectric media; otherwise, a comprehensive review can be found in the textbook by Jackson (1998).

Since $\mathbf{D}$ and $\mathbf{E}$ are three-dimensional vectors we expect the dielectric function to be a tensor, as in the case of stress and strain in Chapter 6. However, if we limit our discussion to solid possessing cubic symmetry, the picture simplifies drastically and ϵ reduces to a scalar function.

In order to calculate the dielectric function of solid H we use a classical argument, and add up the polarizabilities of each H atom using the Clausius–Mossotti relation (Ashcroft and Mermin, 1976):

$$\frac{\epsilon - 1}{\epsilon + 2} = \frac{4\pi}{3} \frac{N}{V} \frac{\alpha}{4\pi\epsilon_0}.$$

As usual, N and V are the number of electrons and the crystal volume, respectively. By combining this relation with eqns 10.26 and 10.27 we obtain, after a few manipulations:

$$\epsilon(\omega) = \epsilon_1(\omega) + i\epsilon_2(\omega),$$

$$\epsilon_1(\omega) = 1 + \tilde{f}_{12} \frac{e^2}{m_e} \frac{1}{\tilde{\omega}_{12}^2 - \omega^2}, \tag{10.34}$$

$$\epsilon_2(\omega) = \tilde{f}_{12} \frac{e^2}{m_e} \frac{\pi}{2\omega} \delta(\omega - \tilde{\omega}_{12}), \tag{10.35}$$

with the definitions:

$$\tilde{f}_{12} = \frac{1}{\epsilon_0} \frac{N}{V} f_{12} \quad \text{and} \quad \tilde{\omega}_{12} = \frac{\omega_{12}}{\sqrt{1 + [\epsilon(0) - 1]/3}}. \tag{10.36}$$

From these relations we clearly see that the dielectric function of our model solid has essentially the same dependence on frequency as the atomic polarizability. A close comparison between eqns 10.34 and 10.35 and their counterpart eqns 10.26 and 10.26 indicates that the only differences are: (i) the presence of an extra '1' in the real part of the dielectric function, (ii) the slightly different definition of the oscillator strength, which for the solid takes into account the number of atoms per unit volume, and (iii) a small redshift of the resonant frequency. The last effect is due to the electric field generated on a given atom by all the other atomic dipoles in the crystal.

Figure 10.3 shows the dielectric function of hypothetical bcc hydrogen calculated using eqns 10.34 and 10.35. As we expected, the curves are qualitatively similar to those for the polarizability in Figure 10.2, with one notable exception: at very high frequency, $\epsilon_1(\omega)$ tends to unity. Intuitively this corresponds to the situation where the driving field oscillates so rapidly that the electrons do not have enough time to follow it. In this case the permittivity is the same as that of free space.

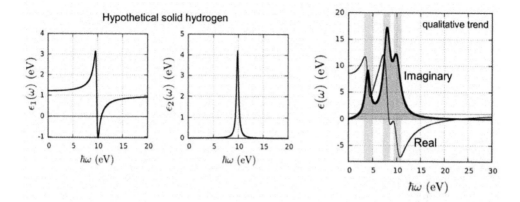

Fig. 10.3 The two panels on the left show the real and the imaginary part of the dielectric function of a hypothetical solid, a bcc lattice of H atoms. In this example the lattice parameter is $a = 3$ Å, and the curves are obtained from eqns 10.34 and 10.35. DFT calculations on the dielectric screening in this hypothetical material can be found in the work by Kioupakis *et al.* (2008). The right panel shows how a more realistic calculation of the dielectric function of a solid would look. In this example we included three main transition energies, marked by the vertical grey bars, and used rather broad Lorentzian lineshapes. The results of DFT calculations of $\epsilon(\omega)$ for real solids will be presented in Section 10.3.

10.2 General properties of the dielectric function

In this section we generalize what we have learned in the previous section, and discuss the key equations for calculating the dielectric function of solids from first principles. In order to keep the notation light we restrict our discussion to solids with cubic symmetry. This will allow us to deal only with a scalar dielectric function, $\epsilon(\omega)$.

The most direct approach for generalizing eqns 10.34 and 10.35 to any solid is to think of solids as 'giant molecules'. For this purpose we need to consider the time-dependent Schrödinger equation for all the electrons in the system, in the presence of the oscillating field of eqn 10.1. This is an immediate generalization of eqn 2.28:

$$i\hbar \frac{\partial}{\partial t}\Psi = \left[-\sum_i \frac{\hbar^2 \nabla_i^2}{2m_e} + \sum_i eV_n(\mathbf{r}_i) + \frac{1}{2}\sum_{i \neq j} \frac{e^2}{4\pi\epsilon_0 |\mathbf{r}_i - \mathbf{r}_j|} \right] \Psi + \sum_i e\mathcal{E}f(t)x_i \, \Psi.$$

(10.37)

The term in the square brackets is the standard Hamiltonian for the system without an electric field, discussed in Chapter 2 (unlike in eqn 2.28 here we are not using Hartree atomic units). The term containing the electric field, on the right, is the energy of all the electrons arising from the presence of the driving field, and is the analogue of the one appearing in eqn 10.3. In the present discussion we are ignoring for simplicity the effect of the driving field on the nuclei. We will come back to the nuclei in Section 10.3.2.

A formal solution of eqn 10.37 can be obtained, in complete analogy with the case of the H atom, by writing the time-dependent many-body wavefunction as a linear combination of the solutions of the equation with $\mathcal{E} = 0$. In the following we denote these solutions $\Psi_n(\mathbf{r}_1, \ldots, \mathbf{r}_N)e^{-iE_nt/\hbar}$, with the total energies $E_0 < E_1 < E_2$ and so on. By definition, the state Ψ_0 represents the ground state of the system. Our trial solution is therefore:

$$\Psi(\mathbf{r}_1, \ldots, \mathbf{r}_N, t) = \sum_n c_n(t)e^{-iE_nt/\hbar}\Psi_n(\mathbf{r}_1, \ldots, \mathbf{r}_N),$$

where the time-dependent coefficients, $c_n(t)$, are the counterpart of those in eqn 10.5. These coefficients can be determined in the same way as we did for the H atom, under the assumption that the wavefunction remains close to the ground state, i.e. $c_0(t) \simeq 1$ and $c_n(t) \ll 1$ for $n = 1, 2, \ldots$ The resulting equations for the coefficients have the same form as in eqn 10.13 and so we will not bother rewriting them here.

Once we have determined the time-dependent coefficients $c_n(t)$, we can calculate the average position of all the electrons, $\langle x_1 + \cdots + x_N \rangle_t/N$, under the influence of the driving field. This allows us to obtain the total volumetric dipole moment, in analogy with eqn 10.23:

$$\mathbf{P} = -\frac{e}{V}\langle x_1 + \cdots + x_N \rangle_t \mathbf{u}_x. \tag{10.38}$$

This quantity is referred to as the *macroscopic (electronic) polarization*, and corresponds to the sum of the dipole moments of all the electrons. By repeating the same steps which led to eqn 10.24 in the case of the H atom, we obtain the time evolution of the polarization:

$$\mathbf{P} = \mathrm{Re}\left[\epsilon_0(\epsilon - 1)\mathcal{E}e^{-i\omega t}\right]\mathbf{u}_x, \tag{10.39}$$

where we have introduced the dielectric function $\epsilon = \epsilon_1 + i\epsilon_2$:

$$\epsilon_1(\omega) = 1 + \frac{1}{\epsilon_0}\frac{N}{V}\sum_n f_{0,n}\frac{e^2}{m_e}\frac{1}{\omega_{0,n}^2 - \omega^2}, \tag{10.40}$$

$$\epsilon_2(\omega) = \frac{1}{\epsilon_0}\frac{N}{V}\sum_{c,v} f_{0,n}\frac{e^2}{m_e}\frac{\pi}{2\omega}\delta(\omega - \omega_{0,n}). \tag{10.41}$$

In this case the resonance frequencies and the oscillator strengths are constructed in analogy with eqn 10.12, eqns 10.28 and 10.36:

$$\omega_{0,n} = \frac{E_n - E_0}{\hbar}, \tag{10.42}$$

$$f_{0,n} = \frac{2m_e}{\hbar}\omega_{0,n}\frac{1}{N}|\langle\Psi_0|\,x_1 + \cdots + x_N\,|\Psi_n\rangle|^2. \tag{10.43}$$

It is now useful to pay attention to the physical units in eqn 10.40. We already know from Section 10.1 that the oscillator strengths, $f_{0,n}$, are dimensionless; therefore the quantity $Ne^2/(V\varepsilon_0 m_e)$ must have the dimensions of the square of a frequency. The quantity:

$$\omega_p = \sqrt{\frac{e^2}{\varepsilon_0 m_e}\frac{N}{V}}, \tag{10.44}$$

is called the *plasma frequency* of the solid (Ashcroft and Mermin, 1976), and can be seen as a proxy for the average density of electrons. Using this definition eqns 10.40 and 10.41 take the following compact form:

$$\epsilon_1(\omega) = 1 + \sum_n f_{0,n} \frac{\omega_p^2}{\omega_{0,n}^2 - \omega^2}, \tag{10.45}$$

$$\epsilon_2(\omega) = \frac{\pi}{2} \sum_n f_{0,n} \frac{\omega_p^2}{\omega} \delta(\omega - \omega_{0,n}). \tag{10.46}$$

The qualitative shape of the real and imaginary parts of the dielectric function of a solid are illustrated by the right-hand panel of Figure 10.3. Generally speaking, the presence of many non-zero oscillator strengths gives rise to an essentially continuous spectrum.

Before moving on to employ DFT in calculations of the dielectric function, it is convenient to state a useful rule that $\epsilon(\omega)$ has to satisfy. In Exercise 10.3 it is shown that the oscillator strengths, $f_{0,n}$, add up to 1:

$$\sum_n f_{0,n} = 1. \tag{10.47}$$

This result is known as the 'Thomas–Reiche–Kuhn sum rule' or simply the 'f-sum rule' (Merzbacher, 1998). By combining the f-sum rule with eqn 10.46 we find immediately:

$$\int_0^\infty \omega\, \epsilon_2(\omega)\, d\omega = \frac{\pi}{2}\omega_p^2. \tag{10.48}$$

This means that the integral of $\omega\, \epsilon_2(\omega)$ is completely defined by the electron density via ω_p (see eqn 10.44). This relation is useful when we want to check whether a calculation of $\epsilon(\omega)$ using some approximations for the many-body wavefunctions and energies is sensible, and how far we are from the 'exact' result.

Exercise 10.3 In this exercise we want to derive the f-sum rule expressed by eqn 10.47. For this purpose we define the shorthand notation $\hat{X} = x_1 + x_2 + \cdots + x_N$ to signify the 'many-body position operator'. ▶Using the definition of the oscillator strength in eqn 10.43 and the Dirac notation introduced in Chapter 2, show that:

$$\sum_n f_{0,n} = \frac{2m_e}{\hbar^2} \sum_n (E_n - E_0)\langle\Psi_0|\hat{X}|\Psi_n\rangle\langle\Psi_n|\hat{X}|\Psi_0\rangle.$$

☐The stationary eigenstates, Ψ_n, are solutions of the time-independent Schrödinger equation given by eqn 2.30: $\hat{H}|\Psi_n\rangle = E_n|\Psi_n\rangle$. ▶Using this observation show that:

$$(E_n - E_0)\langle\Psi_0|\hat{X}|\Psi_n\rangle = \langle\Psi_0|\hat{X}\hat{H} - \hat{H}\hat{X}|\Psi_n\rangle,$$

$$(E_n - E_0)\langle\Psi_n|\hat{X}|\Psi_0\rangle = \langle\Psi_n|\hat{H}\hat{X} - \hat{X}\hat{H}|\Psi_0\rangle.$$

▶Now combine the previous three equations and use the completeness of the eigenstates, $\sum_n |\Psi_n\rangle\langle\Psi_n| = 1$, in order to show that:

$$\sum_n f_{0,n} = -\frac{m_e}{\hbar^2}\langle\Psi_0|\hat{X}(\hat{X}\hat{H} - \hat{H}\hat{X}) - (\hat{X}\hat{H} - \hat{H}\hat{X})\hat{X}|\Psi_0\rangle. \tag{10.49}$$

☐At this point it is necessary to evaluate explicitly the operators inside the bracket. At first glance this seems a bit scary, but the task is a simple one. For example, the operator $\hat{X}\hat{H}$ is simply $x_1 \hat{H} + \cdots + x_N \hat{H}$. Let us focus on the first term in the sum, $x_1 \hat{H}$. The variable x_1 can certainly be interchanged with the potential terms in the Hamiltonian $\hat{H}$, and also with the kinetic energy terms associated with the variables $x_2, \ldots, x_N$. However, we cannot exchange ∇_1^2 and x_1, as the differential operator ∇_1 acts on the position variable x_1. Therefore we have:

$$x_1 \hat{H} - \hat{H} x_1 = -\frac{\hbar^2}{2m_e} \left(x_1 \nabla_1^2 - \nabla_1^2 x_1 \right) = \frac{\hbar^2}{m_e} \nabla_1. \tag{10.50}$$

At this point it should be easy to see that $\hat{X}\hat{H} - \hat{H}\hat{X} = \hbar^2(\nabla_1 + \cdots + \nabla_N)/m_e$. If we replace this expression inside eqn 10.49, and we evaluate similarly all the terms like $x_1 \nabla_1 - \nabla_1 x_1$, we obtain the f-sum rule given by eqn 10.47.

It is worth noting that the evaluation of the terms inside the bracket in eqn 10.49 can be performed almost instantaneously by making use of standard quantum-mechanical commutators; see for instance the presentation by Merzbacher (1998). In this exercise we refrained from using the commutators in order to keep the discussion self-contained.

10.2.1 Power dissipation and optical absorption coefficient

Now that we have a formal recipe for calculating the dielectric function of a solid, we can repeat the same procedure followed for deriving the power dissipation for the H atom in Section 10.1.4. The generalization of eqn 10.30 and the ensuing steps lead promptly to the following result:

$$\overline{W} = V \frac{\epsilon_0 \mathcal{E}^2}{2} \omega \, \epsilon_2(\omega), \tag{10.51}$$

where $\overline{W}$ is the average power transferred from the driving field to the solid.

From classical electrodynamics we can recognize in the term $\epsilon_0 \mathcal{E}^2/2$ the time-averaged energy density of the external electromagnetic field (see for example Section 7.1 of Jackson, 1998). Therefore the product $V \times \epsilon_0 \mathcal{E}^2/2$ is the total time-averaged energy of the external field throughout the solid. If we now make a heuristic leap to a quantum description of the electromagnetic field, we can say that this energy corresponds to N_p photons of energy $\hbar\omega$, and that the power dissipation corresponds to a loss of photons by the external field:

$$V \frac{\epsilon_0 \mathcal{E}^2}{2} = N_\mathrm{p} \hbar\omega, \qquad \overline{W} = -\frac{d}{dt}(N_\mathrm{p} \hbar\omega).$$

Using this heuristic association we can rewrite eqn 10.51 as follows:

$$\frac{dN_\mathrm{p}}{dt} = -\omega \, \epsilon_2(\omega) N_\mathrm{p}. \tag{10.52}$$

This expression has an appealing intuitive interpretation: $\omega \, \epsilon_2(\omega)$ represents the fraction of the total number of photons which is absorbed by the solid per unit time.

In the study of the optical properties of materials it is very common to discuss the absorption of radiation in terms of a quantity related to $\omega \, \epsilon_2(\omega)$ and called the *absorption coefficient*. This quantity is given by:

$$\kappa(\omega) = \frac{\omega \, \epsilon_2(\omega)}{c \, n(\omega)}, \tag{10.53}$$

where n is the real part of the refractive index, $\tilde{n} = n + \kappa$, and is related to the dielectric function as follows:

$$n^2 = \frac{\sqrt{\epsilon_1^2 + \epsilon_2^2} + \epsilon_1}{2}, \quad \text{or equivalently} \quad \epsilon = \tilde{n}^2.$$

The absorption coefficient describes the spatial attenuation of the intensity of an electromagnetic wave travelling through the sample. In fact, if we consider a plane electromagnetic wave, $\mathcal{E}(x,t) = \mathcal{E}_0 \exp[i(qx - \omega t)]$, and we take into account the dispersion relation $\omega = (c/\tilde{n}) q$ resulting from Maxwell's wave equations, we obtain the decaying intensity $|\mathcal{E}(x,t)|^2 = \mathcal{E}_0^2 \exp(-\kappa x)$.

The absorption coefficient, κ, in eqn 10.53 has the dimensions of the inverse of a length. The quantity $1/\kappa$, called the *optical penetration depth*, provides information on the thickness of a film that will absorb most of the incident light. A detailed discussion of the propagation of light waves in materials and the notion of absorption coefficient can be found in Chapter 9 of Landau and Lifshitz (1960).

Since a direct measurement of the absorption coefficient would require in many cases extremely thin samples, in practice the dielectric function of solids is typically determined by measuring the reflection of light from a surface, using a technique named *spectroscopic ellipsometry*. In this technique a beam of linearly polarized light is directed towards the sample, and the reflected beam is analysed. The name of the technique derives from the fact that, at oblique incidence, the reflected beam carries elliptical (as opposed to linear) polarization. From a measurement of the fraction of light reflected from the sample, it is possible to obtain the dielectric function $\epsilon(\omega)$ of the material through the standard Fresnel's equations of geometrical optics. Light sources for spectroscopic ellipsometry span the whole range from commercial halogen lamps to synchrotron radiation. The interested reader will find a brief description of spectroscopic ellipsometry in Chapter 6 of the textbook by Yu and Cardona (2010), and a comprehensive presentation in the book by Fujiwara (2007).

10.3 Using DFT to calculate dielectric functions

A close look at eqns 10.37 and 10.42 reveals that, while the theory presented in Section 10.2 appears general and elegant, we had to introduce the time-dependent many-body wavefunction $\Psi(\mathbf{r}_1, \ldots, \mathbf{r}_N, t)$ and the *excited electronic states* Ψ_1, Ψ_2 and so on. This is somewhat in contrast to the fact that most of this book is devoted to the properties of the electronic ground state, $\Psi_0(\mathbf{r}_1, \ldots, \mathbf{r}_N)$.

In fact, the notion of electronic excitations is something of a taboo in the world of DFT, because the Hohenberg–Kohn theorem (Section 3.1.1) crucially relies on the premise that we are considering only the ground state, Ψ_0. This is the point where the use of DFT as an 'educated guess' for describing material properties (as opposed to a rigorous tool) begins. In practice, while the study of electronic excitations is outside of the remit of DFT, this theory can still be used as a tool for approximate calculations, or as a starting point for more sophisticated theories of excited electrons.

The simplest possible way to use DFT in order to calculate the dielectric function in eqns 10.45 and 10.46 is to replace the many-body wavefunctions, $\Psi_n(\mathbf{r}_1, \ldots, \mathbf{r}_N)$, by Slater determinants of Kohn–Sham states (see eqn 2.42). Most purists of DFT will see this as a barbaric approximation. However, more sophisticated calculations are much more time-consuming (see Section 10.3.2); therefore, in practice, this is the first point of call whenever we are interested in calculating optical properties of materials. Besides, as we will see below in some examples, the results are *qualitatively* reasonable, at least as long as we are not interested in investigating the fine features of the optical spectra.

In Exercise 10.4 it is shown that, within this approximation to the many-body wavefunctions, the dielectric function acquires the following form:

$$\epsilon_1(\omega) = 1 + \sum_{c,v} f_{cv} \frac{\omega_p^2}{\omega_{cv}^2 - \omega^2}, \tag{10.54}$$

$$\epsilon_2(\omega) = \frac{\pi}{2} \sum_{c,v} f_{cv} \frac{\omega_p^2}{\omega} \delta(\omega - \omega_{cv}), \tag{10.55}$$

where the sums run over the occupied states, v, and all the unoccupied states, c, up to infinity (the indices stand for 'valence' and 'conduction', respectively). For the sake of definiteness we can think of a semiconductor such as silicon, with a band gap separating the valence states from the conduction states. However, these equations are completely general, and are easily modified to deal with metals for example. The plasma frequency, ω_p, in eqns 10.54 and 10.55 maintains the same meaning as in the previous section, while the resonance frequencies and the oscillator strengths are modified as follows:

$$\omega_{cv} = \frac{\varepsilon_c - \varepsilon_v}{\hbar}, \tag{10.56}$$

$$f_{cv} = \frac{1}{N} \frac{2m_e}{\hbar} \omega_{cv} |\langle \phi_c| x |\phi_v \rangle|^2. \tag{10.57}$$

In the first equation the energies ε_v and ε_c are the usual Kohn–Sham eigenvalues discussed in Chapters 3 and 9. The oscillator strengths in the second equation are expressed by using the associated Kohn–Sham single-particle wavefunctions, $\phi_v(\mathbf{r})$ and $\phi_c(\mathbf{r})$. It is interesting to observe that, after all the effort made in Section 10.2 to develop a many-body theory, we are essentially ending up with equations very similar to our initial example for the H atom (e.g. see eqn 10.28).

Exercise 10.4 In this exercise we want to adapt eqn 10.45 to the case where we approximate the many-body wavefunctions, Ψ_n, as Slater determinants of Kohn–Sham states. For simplicity we consider a system with only two electrons and four possible Kohn–Sham states, and later deduce the general rule by extrapolation. The Kohn–Sham states will be denoted by $\phi_1(\mathbf{r})$, $\phi_2(\mathbf{r})$, $\phi_3(\mathbf{r})$, and $\phi_4(\mathbf{r})$, with the eigenvalues $\varepsilon_1 < \varepsilon_2 < \varepsilon_3 < \varepsilon_4$. Let us use the following notation for the Slater determinants constructed out of these states:

$$|\Psi_0\rangle = |1100\rangle = \frac{1}{\sqrt{2}} [\phi_1(\mathbf{r}_1)\phi_2(\mathbf{r}_2) - \phi_1(\mathbf{r}_2)\phi_2(\mathbf{r}_1)],$$

and similarly for the other possibilities. We will calculate the total energy of this state by simply adding up the Kohn–Sham eigenvalues, $E_0 = \varepsilon_1 + \varepsilon_2$ (from eqn 9.12 we know that this is not correct; however, here it does not matter much since we will have to deal only with differences of total energies). Clearly, $|\Psi_0\rangle$ is the ground state of the system since the electrons occupy the states with the lowest-energy eigenvalues. The number of all possible Slater determinants that we can construct corresponds to the number of ways in which we can arrange two marbles in four boxes, i.e. $4!/[2!(4-2)!] = 6$.

▶Write the Slater determinants and the total energies corresponding to the following states: $|1010\rangle$, $|1001\rangle$, $|0110\rangle$, $|0101\rangle$, and $|0011\rangle$.

▶Now show, by performing an explicit evaluation, that:

$$\langle 1100|x_1 + x_2|1010\rangle = \langle \phi_2|x|\phi_3\rangle, \qquad \omega_{1100,1010} = \frac{\varepsilon_3 - \varepsilon_2}{\hbar}.$$

Note that the integral on the left is performed over the space variables $\mathbf{r}_1$ and $\mathbf{r}_2$, while the one on the right is performed over $\mathbf{r}$. For this exercise you will need to explicitly write the Slater determinants in terms of Kohn–Sham states. This will result in eight separate terms in each case, six of which are zero due to the orthogonality of the Kohn–Sham states. ☐Similar results can be obtained for $\langle 1100|x_1 + x_2|1001\rangle$, $\langle 1100|x_1 + x_2|0110\rangle$, and $\langle 1100|x_1 + x_2|0101\rangle$.

▶Using the same procedure as above, show that:

$$\langle 1100|x_1 + x_2|0011\rangle = 0.$$

☐In this exercise the valence states are ϕ_1 and ϕ_2, since they are occupied in the ground state. Similarly the conduction states are ϕ_3 and ϕ_4. Therefore the results above can be described in terms of single-particle transitions as follows: all the transitions whereby one electron is promoted from $v = 1\,\text{or}\,2$ to $c = 3\,\text{or}\,4$ are allowed (unless $\langle \phi_c|\,x\,|\phi_v\rangle = 0$), while the transitions where two electrons are promoted yield a zero matrix element.

We can extrapolate this rule to the general case of N electrons as follows: we need to consider transitions between the ground state and all the Slater determinants obtained from the ground state by removing one electron from a valence state, v, and placing it in a conduction state, c. If we denote these determinants by $|\Psi_{cv}\rangle$, replace inside eqn 10.45, and use the results above for the position matrix elements, we obtain immediately eqns 10.54–10.57.

As a curiosity, the rules for the dipole matrix elements between Slater determinants that we 'discovered' in this exercise form part of a family of relations which are widely used in quantum chemistry, and are referred to as 'Slater–Condon rules' (Slater, 1929; Condon, 1930).

Even if we are making an important approximation by replacing the exact many-body wavefunctions with singly excited Slater determinants, we have the good news that the f-sum rule is preserved. In fact, some simple albeit rather tedious algebra shows that the oscillator strengths in eqn 10.57 also obey the sum rule:

$$\sum_{cv} f_{cv} = 1.$$

This f-sum rule implies that the integral of $\omega\epsilon_2(\omega)$ is correctly normalized as in eqn 10.48, even when we describe electronic excitations by means of Kohn–Sham wavefunctions.

10.3.1 The case of extended solids

We started Section 10.2 by proposing that we could formally describe solids as giant molecules. That point of view is perfectly legitimate when we are discussing the basic formalism; however, it needs to be slightly modified if we want to perform actual calculations.

As we have seen in Section 9.2, it is often very convenient to describe solids as an infinite number of periodic replicas of one unit cell. The advantage of this approach is that one can determine the wavefunctions for the entire solid by calculating solutions to the Kohn–Sham equations inside one unit cell and using periodic boundary conditions (eqn 9.10). Within this extended model of a solid, if we attempt an evaluation of the oscillator strengths in eqn 10.57, the first difficulty we encounter is that the integral is not well defined. In fact, the position operator, x, is not bound, i.e. it becomes infinite at $\pm\infty$. In order to avoid this difficulty it is useful to replace the position operator using the standard quantum-mechanical commutator:

$$[x, \hat{H}] = x\hat{H} - \hat{H}x = \frac{i\hbar}{m_e}p_x. \tag{10.58}$$

This is just a more elegant way of rewriting eqn 10.50, with the difference that here $\hat{H}$ represents the Kohn–Sham Hamiltonian (eqn 3.27). The usefulness of this substitution can be seen by considering the matrix elements of the momentum, $p_x = -i\hbar\partial/\partial x$ (see for instance Merzbacher, 1998, page 465):

$$\langle\phi_c|p_x|\phi_v\rangle = \frac{m_e}{i\hbar}\langle\phi_c|[x, \hat{H}]|\phi_v\rangle = \frac{m_e}{i\hbar}\langle\phi_c|x\hat{H} - \hat{H}x|\phi_v\rangle = \frac{m_e}{i\hbar}(\varepsilon_v - \varepsilon_c)\langle\phi_c|x|\phi_v\rangle.$$

Using this result we can rewrite the oscillator strengths of eqn 10.57 in such a way that the position does not appear *explicitly*:

$$f_{cv} = \frac{1}{N}\frac{2}{\hbar m_e}\frac{1}{\omega_{cv}}|\langle\phi_c|p_x|\phi_v\rangle|^2. \tag{10.59}$$

This alternative recipe for calculating the optical oscillator strengths solves the problem of the position operator becoming divergent at infinity.

It is important to observe that eqn 10.59 is a direct consequence of the fact that the position, x, commutes with the Kohn–Sham potential, $V_{tot}(\mathbf{r})$, of eqn 3.28. In practical implementations the use of *non-local pseudopotentials* (Appendix E) introduces an additional term, which requires a careful evaluation. This extra term was first analysed by Starace (1971) and more recently by Ismail-Beigi *et al.* (2001) and Pickard and Mauri (2003); its effects on the matrix elements were quantified within DFT/LDA by Read and Needs (1991). This additional term also leads to a modification of the f-sum rule (eqn 10.47), as shown by Starace.

Another aspect which is specific to DFT calculations for extended solids is that we want to be able to perform all calculations within one unit cell only. In this case the oscillator strengths can conveniently be rewritten by exploiting the Bloch theorem expressed by eqn 9.6. In Exercise 10.5 it is shown that the momentum matrix element between two Bloch states, $\psi_{v,\mathbf{k}_v}$ and $\psi_{c,\mathbf{k}_c}$, simplifies as follows:

$$\langle\psi_{c,\mathbf{k}_c}|p_x|\psi_{v,\mathbf{k}_v}\rangle = \delta(\mathbf{k}_v, \mathbf{k}_c)\langle u_{c,\mathbf{k}_c}|p_x|u_{v,\mathbf{k}_v}\rangle, \tag{10.60}$$

where $u_{v,\mathbf{k}_v}$ and $u_{c,\mathbf{k}_c}$ are the Bloch-periodic components of the wavefunctions, as in eqn 9.6. The most important aspect of this identity is that whenever the two states do not have the same wavevector, the matrix element vanishes.

Exercise 10.5 In this exercise we want to derive eqn 10.60. We start by writing $\psi_v(\mathbf{r})$ and $\psi_c(\mathbf{r})$ in the form of Bloch wavefunctions (eqn 9.6): $\psi_{v,\mathbf{k}_v} = \exp(i\mathbf{k}_v \cdot \mathbf{r})u_{v,\mathbf{k}_v}(\mathbf{r})$ and $\psi_{v,\mathbf{k}_c} = \exp(i\mathbf{k}_c \cdot \mathbf{r})u_{c,\mathbf{k}_c}(\mathbf{r})$, with $u_{v,\mathbf{k}_v}$ and $u_{c,\mathbf{k}_c}$ having the periodicity of the crystal lattice.
▶Show that:

$$\langle \psi_{c,\mathbf{k}_c}|p_x|\psi_{v,\mathbf{k}_v}\rangle = \int d\mathbf{r}\, e^{i(\mathbf{k}_v-\mathbf{k}_c)\cdot\mathbf{r}}(u^*_{c,\mathbf{k}_c}p_x u_{v,\mathbf{k}_v} + \hbar k_x u^*_{c,\mathbf{k}_c}u_{v,\mathbf{k}_v}),$$

where the integral extends over the entire crystal. ☐Let us denote the term within the brackets $f(\mathbf{r})$. This function also has the periodicity of the lattice, i.e. $f(\mathbf{r}+\mathbf{R}) = f(\mathbf{r})$, where $\mathbf{R}$ is a vector of the direct lattice. If we split the integral into the sum of the integrals over each unit cell we find:

$$\langle \psi_{c,\mathbf{k}_c}|p_x|\psi_{v,\mathbf{k}_v}\rangle = \sum_{\mathbf{R}}\int_{\mathrm{UC}} d\mathbf{r}\, e^{i(\mathbf{k}_v-\mathbf{k}_c)\cdot(\mathbf{r}-\mathbf{R})} f(\mathbf{r}-\mathbf{R}) = \sum_{\mathbf{R}} e^{-i(\mathbf{k}_v-\mathbf{k}_c)\cdot\mathbf{R}}\int_{\mathrm{UC}} d\mathbf{r}\, e^{i(\mathbf{k}_v-\mathbf{k}_c)\cdot\mathbf{r}} f(\mathbf{r}).$$

Now we observe that the term $\sum_{\mathbf{R}}\exp[i(\mathbf{k}_v - \mathbf{k}_c)\cdot\mathbf{R}]$ differs from zero only if $\mathbf{k}_v = \mathbf{k}_c + \mathbf{G}$, with $\mathbf{G}$ a reciprocal lattice vector (Appendix D). This can be verified rapidly in one dimension by rewriting the sum as $\sum_{m=-\infty}^{+\infty} \beta^m$ with $\beta = \exp[i(k_{v,x} - k_{c,x})a]$ and a the lattice constant. This sum vanishes unless $\beta = 1$, in which case the result is equal to the number of unit cells, N_{UC}, in the crystal. With a mild abuse of notation we can therefore write (assuming that $\mathbf{k}_v$ and $\mathbf{k}_c$ both belong to the first Brillouin zone):

$$\sum_{\mathbf{R}} e^{i(\mathbf{k}_v-\mathbf{k}_c)\cdot\mathbf{r}} = N_{\mathrm{UC}}\,\delta(\mathbf{k}_v, \mathbf{k}_c),$$

where the delta function is in the Kronecker sense (i.e. the Bloch vectors need to be considered as belonging to a discrete grid). ▶Using this last result and the orthogonality of $u_{v,\mathbf{k}_v}$ and $u_{c,\mathbf{k}_c}$ show that:

$$\langle \psi_{c,\mathbf{k}_c}|p_x|\psi_{v,\mathbf{k}_v}\rangle = N_{\mathrm{UC}}\,\delta(\mathbf{k}_v, \mathbf{k}_c)\langle u_{c,\mathbf{k}_c}|p_x|u_{v,\mathbf{k}_v}\rangle$$

where the bracket on the right is defined as the integral over one unit cell:

$$\langle u_{c,\mathbf{k}_c}|p_x|u_{v,\mathbf{k}_v}\rangle = \int_{\mathrm{UC}} d\mathbf{r}\, u^*_{c,\mathbf{k}_c}\, p_x\, u_{v,\mathbf{k}_v}.$$

At this point it is convenient to re-normalize the periodic components of the wavefunctions, $u_{v,\mathbf{k}_v}$ and $u_{c,\mathbf{k}_c}$, in such a way that they correspond to one electron per unit cell. This amounts to multiplying both by $N_{\mathrm{UC}}^{1/2}$, so that the previous result becomes:

$$\langle \psi_{c,\mathbf{k}_c}|p_x|\psi_{v,\mathbf{k}_v}\rangle = \delta(\mathbf{k}_v, \mathbf{k}_c)\langle u_{c,\mathbf{k}_c}|p_x|u_{v,\mathbf{k}_v}\rangle.$$

By combining eqns 10.44, 10.55, 10.59 and 10.60 and writing all the Kohn–Sham states ψ_v and ψ_c in Bloch notation, we can eventually produce a complete expression for the (imaginary part of) the dielectric function in extended solids (Ehrenreich and Cohen, 1959):

$$\epsilon_2(\omega) = \frac{\pi e^2}{\varepsilon_0 m_e^2 \Omega}\frac{1}{\omega^2}\sum_{cv}\int \frac{d\mathbf{k}}{\Omega_{\mathrm{BZ}}}|\langle u_{c,\mathbf{k}}|p_x|u_{v,\mathbf{k}}\rangle|^2 \delta(\varepsilon_{c\mathbf{k}} - \varepsilon_{v\mathbf{k}} - \hbar\omega). \qquad (10.61)$$

The derivation of this result is a bit tedious but straightforward. The real part can be obtained from this expression by comparison with eqns 10.54 and 10.55. From this equation we can see that the only ingredients which are necessary for calculating the optical properties of solids are the Kohn–Sham energies and wavefunctions. This is an example of why it is important to be able to calculate band structures, as discussed in Chapter 9. It is also worth keeping in mind that eqn 10.61 does not rely on any empirical assumptions about the materials under consideration, and everything has been obtained directly from the first principles of quantum mechanics.

As a word of caution it is worth anticipating that, in reality, eqn 10.61 is truly the simplest possible expression for the dielectric function, and for accurate calculations more complicated forms are needed, as we will discuss in Section 10.3.2.

After this rather lengthy formal introduction it is time to analyse the underlying physics, in the same way as we did for the H atom at the beginning of this chapter. By looking at eqn 10.61 we can immediately draw the following conclusions about the absorption of light, as given by $\omega \epsilon_2(\omega)$:

o Only 'transitions' from occupied Kohn–Sham states to empty Kohn–Sham states can contribute to the absorption of light.

o In these transitions the wavevector of the initial state, $u_{v\mathbf{k}}$, must be equal to the wavevector of the final state, $u_{c\mathbf{k}}$. We can reformulate this observation by stating that the 'crystal momentum', $\hbar \mathbf{k}$, is conserved during the process of light absorption.

o Only transitions where the difference between the energy of the empty state, $\varepsilon_{c\mathbf{k}}$, and that of the occupied state, $\varepsilon_{v\mathbf{k}}$, equals the energy of one photon, $\hbar\omega$, can contribute to the optical absorption by the material.

o The energy onset of the optical absorption corresponds to the smallest energy difference, $\varepsilon_{c\mathbf{k}} - \varepsilon_{v\mathbf{k}}$, between conduction and valence states with the same wavevector, i.e. to the *direct band gap* (unless the associated matrix element vanishes due to the symmetry of the wavefunctions).

As a mnemonic expedient it is useful to summarize these observations by saying that the absorption of light by materials conserves both the energy and the momentum of the combined system, 'solid + light'. Within this picture we can imagine that, upon absorption of one photon, one electron is promoted from an occupied single-particle state to an unoccupied state. This process is referred to as *direct absorption* (Kittel, 1976; Ashcroft and Mermin, 1976; Yu and Cardona, 2010). Apart from this mnemonic expedient we should always keep in mind that a solid is a many-electron system and the displacement of charge by external fields involves non-trivial quantum many-body effects.

Figure 10.4 shows a comparison between the measured absorption coefficient of silicon and the coefficient calculated using eqns 10.61 and 10.53. The top left panel shows that DFT/LDA yields an absorption coefficient which underestimates the experimental onset. This effect is a direct consequence of the band gap problem of DFT (Section 9.5).

In principle we could fix the band gap problem and the onset by performing a more sophisticated '*GW* calculation', as described in Section 9.5. However, for simplicity, we here use a semi-empirical approach and modify the band gap *ad hoc* in order to shift the onset towards higher energies. This procedure, which is commonly referred to as a 'scissor correction', trivially consists of increasing the energy of all the conduction

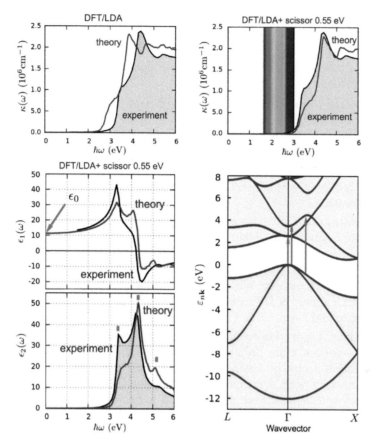

Fig. 10.4 Comparison between the optical properties of silicon from spectroscopic ellipsometry and DFT/LDA calculations based on eqn 10.61. The measurements (black lines) are taken from Table II of Aspnes and Studna (1983). The blue lines are the DFT/LDA calculations. The two panels at the top show the absorption coefficient, κ, as a function of photon energy, $\hbar\omega$. The panel on the left shows plain DFT/LDA calculations. Since DFT/LDA underestimates band gaps, in the right panel we show the absorption coefficient re-calculated after applying an *empirical* 'scissor correction' of +0.55 eV to all conduction bands. We also show the spectrum of visible light (approximately 1.7 to 3.1 eV) for comparison. The two panels at the bottom left show the real part, $\epsilon_1(\omega)$, and the imaginary part, $\epsilon_2(\omega)$, of the dielectric function of silicon. The calculation is also based on eqn 10.61 and includes the scissor correction of +0.55 eV. The band structure shown in the bottom right illustrates how certain 'transitions' (marked by red arrows) can be associated with specific peaks in $\epsilon_2(\omega)$ (marked by small red bars). The arrow in the plot of $\epsilon_1(\omega)$ indicates the value of the *static dielectric constant of silicon*, $\epsilon(0) = 11.3$. The calculations were performed using the same setup as in Figure 5.2, with the only difference that the Brillouin zone integral was evaluated using 10,000 **k**-vectors. In this case the use of a very fine grid is important for capturing the sharp features in $\epsilon_2(\omega)$. The calculations reported in this figure required about 10 hours on a laptop computer.

bands by a fixed amount. The shift is chosen in such a way as to match the experimental spectrum. In the case of Figure 10.4 we have chosen a scissor correction of 0.55 eV, leading to the theoretical absorption coefficient in the top right panel. By comparing calculations and experiment in Figure 10.4 we can make the following simple observations:

○ The calculated shape of the absorption coefficient is in reasonable agreement with experiment. In fact, we can recognize in both cases a shoulder at the onset 3.4 eV, a main peak at 4.2 eV and a small peak around 5 eV.

○ The (scissor-corrected) band gap of silicon is 1.1 eV; however, the onset of absorption is at 3.4 eV. This is not a surprise since, as anticipated on page 198, what matters here is the direct band gap, as opposed to the minimum band gap (see Figure 9.5).

○ The calculated *dielectric constant* of silicon, i.e. the value of the dielectric function for a static electric field, is $\epsilon(0) = 11.3$. This compares reasonably well with the measured dielectric constant, $\epsilon_0 = 11.8$ (Chandler-Horowitz and Amirtharaj, 2005). However, we stress that the calculations reported in Figure 9.5 use an *empirical* correction to the band gap. Without such a correction the comparison would worsen slightly. For a survey of DFT calculations of the dielectric constant of Si the reader is referred to Dal Corso *et al.* (1994).

○ The calculated absorption coefficient, $\kappa(\omega)$, underestimates the shoulder at 3.4 eV. This effect is even more pronounced when we consider the plot of $\epsilon_2(\omega)$, where the calculation is clearly missing the peak at 3.4 eV. In order to satisfy the f-sum rule, the missing peak is compensated by a larger value of ϵ_2 around 5 eV. This discrepancy will be discussed in more detail in Section 10.3.2.

The trends discussed for silicon are rather general, and a similar degree of accuracy can be expected for other materials. The key point to remember here is that the calculated optical properties of solids are only as good as the underlying band structure. Therefore, if, for a given class of materials, DFT does not provide reasonably accurate band structures, we must treat the calculated optical properties with caution.

Exercise 10.6 We want to develop an intuitive understanding of the shape of the dielectric function in Figure 10.4. For this purpose we make a drastic simplification and use the following 'single-oscillator' model, inspired by eqn 10.25, and eqns 10.54 and 10.55:

$$\epsilon(\omega) = 1 + f_0 \frac{\omega_p^2}{2\omega_0} \left(\frac{1}{\omega_0 + \omega + i\gamma} + \frac{1}{\omega_0 - \omega - i\gamma} \right). \tag{10.62}$$

In this expression we have one known parameter, the plasma frequency of silicon, ω_p, and three unknown parameters which need to be determined: f_0, ω_0 and γ. ▶Show that this expression reduces correctly to eqns 10.54 and 10.55 if in those equations we consider only one oscillator with resonance $\omega_{cv} = \omega_0$ and strength $f_{cv} = f_0$, and take the limit $\gamma \to 0$. ▶Show that, in the limit $\gamma \to 0$, the model dielectric function above satisfies the f-sum rule in eqn 10.48 when we set $f_0 = 1$. ▶Calculate the plasma frequency of silicon by considering only the eight valence electrons per unit cell and the experimental lattice parameter $a = 5.43$ Å. As a sanity check, the answer should be $\hbar\omega_p = 16.61$ eV. ☐We can determine the resonance frequency, ω_0, in eqn 10.62 by using the calculated value of $\epsilon(0)$ in Figure 10.4. ▶Show that for $\gamma \to 0$ we have:

$$\epsilon(0) = 1 + \frac{\omega_{\mathrm{p}}^2}{\omega_0^2}.$$

▶Calculate ω_0 using this expression and the value of $\epsilon(0)$ given in Figure 10.4, and show that the result is $\hbar\omega_0 = 5.17$ eV. ▶Now collect the results of this exercise, set $\hbar\gamma = 0.1$ eV, and plot the model dielectric function on top of those of Figure 10.4. Show that the simplified model in eqn 10.62 provides a good description of the DFT result at low energy ($\hbar\omega < 2$ eV), while the main peak in $\epsilon_2(\omega)$ is shifted towards higher energies by about 0.8 eV. □Despite its simplicity the model dielectric function of eqn 10.62 can be very useful for calculations where the precise details of the dielectric functions are not critical. This model dielectric function belongs to a family of approximations which are generally referred to as 'generalized plasmon-pole models' (Johnson, 1974; Hybertsen and Louie, 1986; Engel and Farid, 1993).

10.3.2   Electron–hole interaction and phonon-assisted absorption

In this section we briefly touch upon two slightly more advanced aspects which need to be considered in the calculation of the optical properties of materials.

Electron–hole interaction.   We start from the plot of $\epsilon_2(\omega)$ in Figure 10.4. As discussed in the previous section, the calculated curve misses the experimental peak at 3.4 eV. The origin of this discrepancy is now well understood, and relates to a missing piece of physics in our description (Hanke and Sham, 1980; Albrecht *et al.*, 1998). In our intuitive understanding of the optical absorption process, we can imagine an electron promoted from the valence band to the conduction band upon absorption of one photon. In this situation the excited electron will leave behind a 'hole' or lack of electron charge. After taking into account the positive charge of the nuclei, we see that the excited system will differ from the ground state by the presence of two puddles of charge, one positive and one negative. Already within the context of classical electrostatics we can understand that these two puddles will experience an attractive electrostatic force. This electrostatic interaction can modify the energetics of the absorption process, and hence distort the plot of $\epsilon_2(\omega)$.

A complete discussion of the role of *electron–hole interactions* in optical absorption is too advanced for this presentation, and the interested reader is referred to the comprehensive review by Onida *et al.* (2002). Here we limit ourselves to pointing out that our theory was essentially 'exact' up until eqns 10.45 and 10.46. The interaction between electron and holes was implicitly dropped when we decided to represent the many-body wavefunctions, $\Psi_n(\mathbf{r}_1, \ldots, \mathbf{r}_N)$, using Slater determinants of Kohn–Sham states, and their total energies, E_n, by the sum of Kohn–Sham energies (Exercise 10.4). After taking into account the electron–hole interaction, the agreement between calculated and measured optical absorption spectra improves considerably. As an example, Figure 10.5 shows a first-principles calculation of the dielectric function of silicon by Marini (2008) where the electron–hole interaction is included. From this figure it is clear that the missing peak at 3.4 eV in Figure 10.4 has been restored.

Phonon-assisted absorption.   Another aspect which is not captured by our simplified theory of Section 10.3.1 is the role of phonons in the optical properties of crystalline solids.

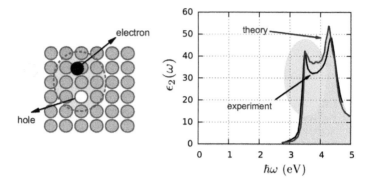

Fig. 10.5 The left panel shows a schematic representation of an excited electron in a solid and the hole left behind. The dashed red line indicates the attractive Coulomb interaction experienced by the electron and the hole. The right panel shows a comparison between the measured imaginary part of the dielectric function of silicon, and the one calculated including the Coulomb interaction between electron and hole. Both theoretical and experimental data are from Figure 1 of Marini (2008). The calculations in the figure were performed by Marini using a computational method which goes beyond standard DFT, and is generally referred to as the *Bethe–Salpeter approach* (Sham and Rice, 1966; Strinati, 1984). Further details on this method can be found in the original works on the subject (Sham and Rice, 1966; Onida *et al.*, 1995; Rohlfing and Louie, 1998; Benedict *et al.*, 1998). The Bethe–Salpeter calculation correctly captures the experimental peak at 3.4 eV in the dielectric function of silicon, unlike the DFT/LDA calculations in Figure 10.4.

Let us consider once again Figure 10.4. The top right panel of this figure shows that the visible portion of the electromagnetic spectrum extends up to photon energies around $\hbar\omega = 3.1$ eV. At the same time, the onset of optical absorption in crystalline silicon corresponds to the minimum direct gap, i.e. 3.1 eV (after the scissor correction used in Figure 10.4). Since, according to our calculation, silicon cannot absorb electromagnetic waves with $\hbar\omega < 3.1$ eV, we are led to conclude that silicon crystals should be transparent to visible light. Obviously this conclusion is in contrast to the empirical observation that silicon has a metallic greyish colour. Moreover, were silicon unable to absorb visible light, then there would be absolutely no point in making solar cells based on this material.

Some useful insight into this conundrum can be gained by plotting the measured absorption coefficient on a logarithmic scale. This is shown in Figure 10.6, where we see that for photon energies between ~1 eV and ~3 eV the measured absorption coefficient is very small but non-zero. If we consider again the band structure of silicon shown in Figure 9.5, we realize that this range corresponds to energies above the minimum band gap of silicon (1.1 eV after scissor correction) and below the direct band gap (3.1 eV). Therefore direct measurements seem to indicate that, in contrast to the rules stated on page 198, silicon can indeed absorb photons with energies *below* the direct band gap.

This phenomenon corresponds to the well-known *phonon-assisted optical absorption*, which is discussed in every textbook on semiconductors (e.g. Yu and Cardona, 2010).

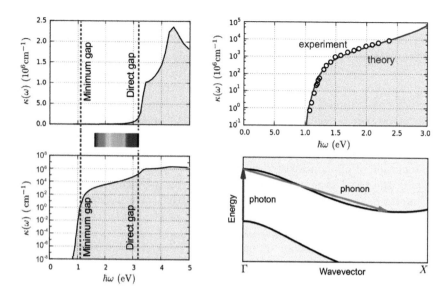

Fig. 10.6 The panels on the left show the measured absorption coefficient of silicon (Green and Keevers, 1995) on a linear scale (top) and a logarithmic scale (bottom). Here we see that the absorption coefficient between 1.1 eV and 3.1 eV is very small but non-zero. The colours of the visible spectrum are also given for comparison. The panel on the top right shows the comparison between the measured absorption coefficient (black circles) and DFT calculations including phonon-assisted processes (blue line). The data are taken from Figure 3 of Noffsinger *et al.* (2012). The panel on the bottom right shows a possible phonon-assisted optical transition: an electron first makes a photon-induced transition whereby the wavevector is conserved (blue), then a transition assisted by a phonon whereby the wavevector can change (red). In this case the absorption onset corresponds to the *minimum (indirect) band gap*. Intuitively we can expect a photon+phonon process to be less likely than a photon-only process. Accordingly phonon-assisted optical absorption is very weak.

Without going into the details of the theory of phonon-assisted absorption, we here point out that there is a small probability that photons are absorbed during transitions where the initial and final electronic states have different momenta, $\psi_{v\mathbf{k}_v} \to \psi_{c\mathbf{k}_c}$ with $\mathbf{k}_v \neq \mathbf{k}_c$. This phenomenon can take place if a *phonon* is also involved in the process. Once phonon-assisted processes are included in the theory, the calculated absorption coefficient exhibits very good agreement with experiment. This is shown in the top right panel of Figure 10.6, where the calculations by Noffsinger *et al.* (2012) are compared with the measured absorption coefficient of silicon in the visible spectrum. In this case the inclusion of phonon-assisted processes restores the agreement between theory and experiment, and the calculations no longer 'predict' that silicon is transparent to visible light. The theory of phonon-assisted absorption is beyond the scope of this presentation; therefore we stress only the following points:

o   The theory developed so far misses phonon-assisted absorption simply because in Section 10.2 we completely ignored the nuclei. In fact we assumed that the nuclei are held immobile, and we worried only about the many-body electronic wavefunction, $\Psi(\mathbf{r}_1, \ldots, \mathbf{r}_N)$. Had we started from the complete wavefunction of the electrons and the nuclei, $\Psi(\mathbf{r}_1, \ldots, \mathbf{r}_N, \mathbf{R}_1, \ldots, \mathbf{R}_M)$, we would have found an extra term in eqn 10.61 arising from phonons.

○ After including phonons, the shape of the dielectric function as given in eqn 10.61 does not change much. The only novelties are that in the Dirac delta functions we find also phonon energies, and in the matrix elements we find terms describing how the Kohn–Sham potential changes when we displace the nuclei. The complete expression can be found, among others, in Yu and Cardona (2010).

○ The inclusion of phonon-assisted absorption is rather complicated, since it requires not only the band structure as in Chapter 9, but also the phonon dispersions as in Chapter 8. A recent example of a complete DFT calculation including phonon-assisted absorption can be found in Noffsinger *et al.* (2012).

Exercise 10.7 As a consequence of optical absorption, the intensity of a plane and monochromatic electromagnetic wave inside a material decreases as $I(x) = I_0 \exp[-\kappa(\omega)x]$, where x indicates the distance from the surface (see page 193). ▶Using the data in Figure 10.6 calculate how thick a silicon film needs to be in order to absorb 99% of the incoming radiation, when the photon energy is (i) $\hbar\omega = 0.9$ eV, (ii) $\hbar\omega = 1.1$ eV and (iii) $\hbar\omega = 4.3$ eV. ▶Compare your results with the optimum thickness of a silicon wafer in solar cells, which is 100 μm (Tiedje *et al.*, 1984).

10.4 Advanced concepts in the theory of the dielectric function

At the end of this chapter it is useful to give a broader perspective on the concept of dielectric function. So far we limited ourselves to considering the *response* of a system to a uniform electric field oscillating in time. A more general approach would involve studying the response to a non-uniform electric field. For example, we can consider an external potential, $v_{\text{ext}}(\mathbf{r}, t)$, which depends both on the space variable, $\mathbf{r}$, and the time, t. When the electrons and the nuclei experience this potential, they rearrange in order to yield a new electrostatic potential, $v_{\text{scr}}(\mathbf{r}, t)$, where the subscript stands for 'screened'. The function, ϵ^{-1}, which links the external and the screened potential as follows:

$$v_{\text{scr}}(\mathbf{r}, t) = \int d\mathbf{r}' \int dt' \, \epsilon^{-1}(\mathbf{r}, \mathbf{r}', t - t') v_{\text{ext}}(\mathbf{r}', t'), \qquad (10.63)$$

is called the *inverse dielectric matrix*. If we ignore the space and time variables, eqn 10.63 clearly resembles the standard definition of the dielectric constant in electrostatics. The dependence of the inverse dielectric matrix on the time coordinates is simply a statement of the fact that electrons and nuclei have non-zero inertial mass; therefore it will take some time for them to adjust to the external potential. The dependence on the space variables, on the other hand, expresses the fact that at the atomic scale the electron density is not homogeneous. As a result, a localized perturbation will displace varying amounts of electron density from different regions and directions.

In order to study the interaction of materials with electromagnetic waves it is most often convenient to replace the external and screened potentials in eqn 10.63 by their

Fourier components $v_{scr}(\mathbf{q},\omega)\exp[i(\mathbf{q}\cdot\mathbf{r}-\omega t)]$ and $v_{ext}(\mathbf{q}',\omega)\exp[i(\mathbf{q}'\cdot\mathbf{r}'-\omega t')]$. This replacement leads naturally to the introduction of the inverse dielectric matrix in a 'wavevector-frequency representation':

$$\epsilon^{-1}(\mathbf{r},\mathbf{r}',t-t') = \int d\mathbf{q}\int d\mathbf{q}'\int d\omega\, e^{i(\mathbf{q}\cdot\mathbf{r}-\omega t)}\, \epsilon^{-1}(\mathbf{q},\mathbf{q}',\omega)\, e^{-i(\mathbf{q}'\cdot\mathbf{r}'-\omega t')}.$$

Within this more formal notation the 'dielectric function' discussed throughout this chapter is the reciprocal of $\epsilon^{-1}(\mathbf{q}=0,\mathbf{q}'=0,\omega)$, and describes the screening of perturbations of very long wavelength (i.e. uniform fields). Along the same lines, the 'dielectric constant' corresponds to the static limit of the latter quantity, i.e. it describes the screening of constant and uniform perturbations. The usual macroscopic dielectric tensor used in crystal optics can be obtained from the inverse dielectric matrix by using the relation between electric fields and potentials, e.g. $\mathbf{E}_{ext}(\mathbf{r},t) = -\nabla v_{ext}(\mathbf{r},t)$.

The study of ϵ^{-1} in its full complexity would require the use of theoretical tools like field quantization, and as such goes beyond the scope of our presentation. The interested reader will find an introductory exposition of this topic in Chapter 16 of the textbook by Ashcroft and Mermin (1976), and a more advanced discussion in Chapter 5.7 of Ziman (1975). These presentations are based on the simplified model of a homogeneous electron gas (Section 3.3). The theory for realistic models of solids can be found in the original works by Ehrenreich and Cohen (1959), Adler (1962) and Wiser (1963). In particular, the expression derived in eqn 10.61 corresponds to the formulation of Ehrenreich and Cohen. On the other hand, the result based on the many-body electronic wavefunctions in eqn 10.40 is completely general, and is a useful starting point for orienting oneself in the maze of the optical properties of materials.

An alternative option for describing the optical properties of materials is to use the so-called *time-dependent density functional theory*, or in short 'TD-DFT'. Time-dependent DFT is based on the Runge–Gross theorem (Runge and Gross, 1984), which generalizes the Hoenberg–Kohn theorem of Section 3.1.1 to the case of time-dependent external potentials. The key practical aspect of TD-DFT is that the many-body time-dependent Schrödinger equation (eqn 10.37) is replaced by a set of time-dependent Kohn–Sham equations, thereby making the calculations very affordable. TD-DFT has proven most successful in the case of molecules and clusters, and research is currently ongoing to improve its predictive power in the case of extended solids. The reader interested in TD-DFT will find an overview of the state-of-the-art in the reviews by Marques and Gross (2004) and Onida *et al.* (2002).

As a closing remark, throughout this chapter we focused on the dielectric function and absorption spectra in the visible portion of the electromagnetic spectrum. As one might have guessed, it is possible to study the dielectric function across a much broader spectral range. Typically, different portions of the spectrum are best addressed by making specific approximations. For example in the *mid infrared* and *far infrared* regions, i.e. for $\hbar\omega = 1-400$ meV, we find optical transitions between different vibrational quantum states of materials. These contributions do not appear in eqn 10.61 because in Section 10.2 we decided to hold the nuclei immobile in their equilibrium positions. The qualitative shape of the dielectric function in the mid and far infrared is similar to what we already have seen in Section 10.3. Full details on the theory can be found in Brüesch (1986), and an example of a DFT calculation of $\epsilon(\omega)$ in the infrared for amorphous SiO_2 can be found in Pasquarello and Car (1997).

For completeness we also point out that the theory of dielectric screening in insulators has undergone a number of innovations since the early 1990s. In particular, the problems connected with the fact that the position operator, x, diverges at infinity have been solved by introducing a 'Berry-phase' formulation of the position operator (King-Smith and Vanderbilt, 1993; Souza *et al.*, 2002; Umari and Pasquarello, 2002). This research area is commonly referred to as the 'modern theory of polarization', and a comprehensive review of the subject is given by Resta (1994).

11

Density functional theory and magnetic materials

After having discussed the interaction of materials with electric fields in Chapter 10, we now discuss the uses of DFT for studying the magnetic properties of materials. Magnetism is a very fascinating phenomenon, which has been stimulating human curiosity since the discovery of magnetite, Fe_3O_4, during the Bronze Age. Nowadays magnetic materials form the basis of much of current technology, from electronics to automotive applications.

The most obvious examples of magnetic materials are thin film coatings for data storage, such as those found in hard discs and credit cards. In these films the magnetization state of small structural domains is used to encode digital information. Other examples of technological uses of small magnets are the actuators for loudspeakers and headsets. At a larger scale we also find magnetic materials employed for induction engines. For example, the motors of some electric and hybrid vehicles are based on high-energy-density permanent magnets such as $Nd_2Fe_{14}B$ (Croat *et al.*, 1984; Herbst *et al.*, 1984).

Another area of materials science where magnetism plays a key role is the research and development of multiferroics and magnetoelectrics. These classes of materials broadly include those systems where ferroelectricity and ferromagnetism coexist, or systems where the dielectric and magnetic properties are tightly interconnected. The simultaneous control of electric polarization and magnetization in the same material would make it possible to realize many interesting devices, e.g. multi-state memories or spin valves. The classic example of a thin film multiferroic is $BiFeO_3$ (Wang *et al.*, 2003), and recent reviews of this emerging area are given by Eerenstein *et al.* (2006) and Ramesh and Spaldin (2007).

In this chapter we will concentrate on the calculation of the spontaneous magnetization of materials using DFT. The reader interested in an introduction to the theory of magnetism in solids is referred to the standard presentations by Ashcroft and Mermin (1976) or by Kittel (1976).

11.1   The Dirac equation and the concept of spin

The simplest way to start a discussion of magnetism is to remind ourselves that an electric current in a circular loop acts like a magnet. In fact, not only does this current generate a magnetic field, but it also experiences a torque when placed in a magnetic field. To see this, let us forget about quantum mechanics for a moment. If we consider an electron travelling along a circular orbit of radius a contained in the xy plane, with the velocity v winding around the z axis according to the right-hand rule (the thumb pointing towards positive values of z), classical magnetostatics tells us that the average magnetic dipole moment of this current loop is (Jackson, 1998):

$$\mathbf{m} = -\frac{e}{2m_e}\mathbf{L}, \tag{11.1}$$

with $\mathbf{L} = am_e v\,\mathbf{u}_z$ being the orbital angular momentum. This magnetic dipole generates a magnetic induction field, $\mathbf{B}(\mathbf{r})$, in complete analogy with eqn 8.1 seen for the electric dipole:

$$\mathbf{B}_{\text{dip}}(\mathbf{r}) = \frac{1}{c^2}\frac{3(\mathbf{m}\cdot\hat{\mathbf{r}})\,\hat{\mathbf{r}} - \mathbf{m}}{4\pi\epsilon_0\, r^3},$$

where c is the speed of light in vacuum, as usual. At the same time, if this current loop is immersed in a uniform external induction field, $\mathbf{B}$, it acquires a potential energy:

$$E_{\text{B}} = -\mathbf{m}\cdot\mathbf{B}. \tag{11.2}$$

This energy is minimized when the magnetic dipole is aligned parallel to the external field, precisely like a compass.

The fact that electrons in motion generate as well as experience magnetic fields carries implications for the electronic structure of materials, e.g. the energy of the ground state. The most important implication is that electrons have an intrinsic property called *spin*. The keyword 'spin' has already appeared in this book, starting with page 20 of Chapter 2, and most likely the reader is already familiar with it from elementary quantum mechanics. Nevertheless, since the concept of spin is central to the theory of magnetism, in this section we keep heuristics at a minimum and outline briefly the first-principles theory of spin developed by Dirac (1928).

For simplicity let us consider only one electron in an electromagnetic field. From standard electricity and magnetism it is known that the electric field, $\mathbf{E}(\mathbf{r},t)$, and the magnetic induction, $\mathbf{B}(\mathbf{r},t)$, can be expressed in terms of a vector, $\mathbf{A}(\mathbf{r},t)$, and a scalar function, $\varphi(\mathbf{r},t)$, as follows:

$$\mathbf{B}(\mathbf{r},t) = \nabla \times \mathbf{A}(\mathbf{r},t), \qquad \mathbf{E}(\mathbf{r},t) = -\nabla\varphi(\mathbf{r},t) - \frac{\partial\mathbf{A}(\mathbf{r},t)}{\partial t}, \tag{11.3}$$

where '$\times$' denotes the curl or cross-product. Formulating the theory of electricity and magnetism in terms of $\mathbf{E}$ and $\mathbf{B}$ or in terms of $\mathbf{A}$ and φ is perfectly equivalent, and the use of one representation or another is dictated by convenience. A thorough explanation of these concepts can be found in Chapter 6 of Jackson (1998). The reason for introducing eqn 11.3 in this chapter is that the classical Hamiltonian of an electron in an electromagnetic field can be written as:

$$\hat{H} = \frac{1}{2m_e}(\mathbf{p} + e\mathbf{A})^2 - e\varphi. \tag{11.4}$$

The validity of this prescription can be checked by using Hamilton's equations of Exercise 4.3. In fact, by inserting $\hat{H}$ inside eqns 4.15 and 4.16, and using eqn 11.3 one can find, after lengthy manipulations:

$$m_e \frac{d^2\mathbf{r}}{dt^2} = -e\left(\mathbf{E} + \frac{d\mathbf{r}}{dt} \times \mathbf{B}\right).$$

This is nothing but Newton's equation, here for the electron, with the Lorentz force on the right-hand side. The quantum-mechanical generalization of eqn 11.4 is straightforward and follows the steps discussed in Section 2.2. In the case of one electron in an electromagnetic field we have:

$$i\hbar \frac{\partial}{\partial t}\psi(\mathbf{r},t) = \left[\frac{1}{2m_e}(\mathbf{p} + e\mathbf{A})^2 - e\varphi\right]\psi(\mathbf{r},t),\tag{11.5}$$

where this time $\mathbf{p} = -i\hbar\nabla$ is the quantum-mechanical momentum operator, and $\psi(\mathbf{r},t)$ is the usual electron wavefunction. A quick comparison with Section 10.1.1 shows that we already encountered a simplified version of this equation. Indeed, if we set $\mathbf{A} = 0$ and $\varphi = -e^2/4\pi\epsilon_0|\mathbf{r}| + e\mathcal{E}f(t)\,\mathbf{r}$, then we recover immediately eqn 10.3.

Equation 11.5 is the most general form of the time-dependent Schrödinger equation for one electron in an electromagnetic field. Despite its enormous importance in quantum mechanics, this equation has one shortcoming: the electromagnetic field generated by the moving electron does not satisfy Einstein's principle of special relativity (Einstein, 1905b). In mathematical terms one can say that the description provided by eqn 11.5 is not invariant with respect to *Lorentz transformations* of the reference frame (Goldstein, 1950). In order to overcome this difficulty Dirac (1928) proposed a more general version of the Schrödinger equation, which has two important properties: (i) by construction the Dirac equation obeys the principle of special relativity, and (ii) when the velocity, v, of the particle is much smaller than the speed of light, c, the Dirac equation reduces (almost) to eqn 11.5. The Dirac equation reads:

$$i\hbar \frac{\partial}{\partial t}\Psi(\mathbf{r},t) = \left[c\boldsymbol{\alpha}\cdot(\mathbf{p} + e\mathbf{A}) - e\varphi + \beta\, m_e c^2\right]\Psi(\mathbf{r},t),\tag{11.6}$$

where the various new quantities are defined below. The reasoning which leads to the above equation will not be given here, but the interested reader will find a very clear explanation in the original work by Dirac (1928). In eqn 11.6 the wavefunction, $\Psi(\mathbf{r},t)$, has become an array of four functions:

$$\Psi(\mathbf{r},t) = \begin{bmatrix} \psi(\mathbf{r},t;1) \\ \psi(\mathbf{r},t;2) \\ \psi(\mathbf{r},t;3) \\ \psi(\mathbf{r},t;4) \end{bmatrix},$$

and is called a '4-spinor' (shorthand for 'spin-orbital'). An alternative notation, which is more convenient in relation to DFT calculations, consists of separating the 4-spinor into two '2-spinors' as follows:

$$\Psi(\mathbf{r},t) = \begin{bmatrix} \boldsymbol{\Psi}(\mathbf{r},t) \\ \boldsymbol{\psi}(\mathbf{r},t) \end{bmatrix}, \text{ with } \boldsymbol{\Psi}(\mathbf{r},t) = \begin{bmatrix} \psi(\mathbf{r},t;1) \\ \psi(\mathbf{r},t;2) \end{bmatrix} \text{ and } \boldsymbol{\psi}(\mathbf{r},t) = \begin{bmatrix} \psi(\mathbf{r},t;3) \\ \psi(\mathbf{r},t;4) \end{bmatrix}.\tag{11.7}$$

Since the wavefunction is now a 4-spinor, we can expect the quantities α and β to be matrices of size 4×4. In fact we have:

$$\alpha = \begin{pmatrix} 0 & \sigma \\ \sigma & 0 \end{pmatrix}, \qquad \beta = \begin{pmatrix} 1 & 0 \\ 0 & -1 \end{pmatrix}, \tag{11.8}$$

where 0 indicates a 2×2 matrix of zeros and 1 indicates the identity matrix of size 2×2. The quantity σ is a 2×2 'matrix of vectors':

$$\sigma = \sigma_x \mathbf{u}_x + \sigma_y \mathbf{u}_y + \sigma_z \mathbf{u}_z,$$

with σ_x, σ_y and σ_z being also 2×2 matrices, called *Pauli matrices*, and given by:

$$\sigma_x = \begin{pmatrix} 0 & 1 \\ 1 & 0 \end{pmatrix}, \qquad \sigma_y = \begin{pmatrix} 0 & -i \\ i & 0 \end{pmatrix}, \qquad \sigma_z = \begin{pmatrix} 1 & 0 \\ 0 & -1 \end{pmatrix}. \tag{11.9}$$

After getting over the oddity of having to deal with a 4-spinor instead of one wavefunction, it is worth noticing that eqn 11.6 differs from eqn 11.5 in that the $(\mathbf{p} + e\mathbf{A})$ term is not squared, and there appears one extra energy contribution, $m_e c^2$. This last term comes from the mass–energy equivalence in the theory of relativity.

Equation 11.6 is too complicated for practical DFT calculations. In fact we are still considering only one electron, but eventually what we will need is a many-body version of the Dirac equation, i.e. the counterpart of eqn 2.19.

In order to simplify the problem it is common to consider an approximate version of eqn 11.6, as we show in the following. In order to eliminate the time derivative we seek a stationary solution of the form $\Psi(\mathbf{r}, t) = \Psi(\mathbf{r}) \exp(-iEt/\hbar)$, similarly to what was done in Section 10.1.1. By making this replacement in eqn 11.6 and rewriting explicitly in terms of the 2-spinors, Ψ and ψ, we find rapidly:

$$(\ m_e c^2 - e\varphi - E)\Psi + c\sigma \cdot (\mathbf{p} + e\mathbf{A})\psi = 0, \tag{11.10}$$

$$(-m_e c^2 - e\varphi - E)\psi + c\sigma \cdot (\mathbf{p} + e\mathbf{A})\Psi = 0, \tag{11.11}$$

where the time dependence of the spinors has dropped out. Here each equation is truly an identity between 2×2 matrices. For example, we have:

$$\sigma \cdot \mathbf{p}\,\Psi = (\sigma_x p_x + \sigma_y p_y + \sigma_z p_z)\Psi = \begin{bmatrix} p_z & p_x - ip_y \\ p_x + ip_y & -p_z \end{bmatrix} \begin{bmatrix} \psi(\mathbf{r}; 1) \\ \psi(\mathbf{r}; 2) \end{bmatrix},$$

where in the last step we used eqn 11.9.

This system of equations (eqns 11.10 and 11.11) looks rather complicated; however, it can be simplified drastically if we proceed as follows: we express ψ in eqn 11.11 in terms of Ψ and replace it inside eqn 11.10. Furthermore, we refer the electron energy, E, to the rest energy, $m_e c^2$ (i.e. $E' = E - m_e c^2$ and relabel $E = E'$). This leads to one equation containing only Ψ and one containing only ψ:

$$\psi = \frac{c}{2m_e c^2 + e\varphi + E}\,\sigma \cdot (\mathbf{p} + e\mathbf{A})\Psi, \tag{11.12}$$

$$\sigma \cdot (\mathbf{p} + e\mathbf{A})\frac{c^2}{2m_e c^2 + e\varphi + E}\,\sigma \cdot (\mathbf{p} + e\mathbf{A})\Psi - e\varphi\Psi = E\,\Psi, . \tag{11.13}$$

At this point we observe that $m_e c^2 = 0.510$ MeV, while we expect the potential, $e\varphi$, and the electron energy levels, E, of valence electrons in materials to be of the order of a few to a few tens of eV. Therefore it is sensible to approximate the denominators appearing in these equations using a Taylor expansion as follows:

$$\frac{c^2}{2m_ec^2 + e\varphi + E} = \frac{1}{2m_e}\frac{1}{1+\eta} = \frac{1}{2m_e}(1 - \eta + \eta^2 + \ldots) \quad \text{with} \quad \eta = \frac{e\varphi + E}{2m_ec^2} \sim 10^{-4}.$$
(11.14)

If we retain only the 0th-order term in this expansion, we can rewrite eqns 11.12 and 11.13 as follows:

$$\psi = \frac{1}{2m_ec}\boldsymbol{\sigma} \cdot (\mathbf{p} + e\mathbf{A})\boldsymbol{\Psi},$$
(11.15)

$$\frac{1}{2m_e}[\boldsymbol{\sigma} \cdot (\mathbf{p} + e\mathbf{A})]^2\boldsymbol{\Psi} - e\varphi\boldsymbol{\Psi} = E\,\boldsymbol{\Psi}.$$
(11.16)

Here we see that the spinor ψ is smaller than $\boldsymbol{\Psi}$ by a factor of the order of p/m_ec, p being an 'average' momentum of the electron. If we replace p by the typical crystal momentum in a solid, i.e. $\hbar\pi/a$ with a the lattice parameter, we find:

$$\frac{p}{m_ec} \sim \frac{\pi}{a}\frac{\hbar c}{m_ec^2} \sim 0.002 \quad \text{for} \quad a = 5\,\text{Å}.$$

This simple estimate shows that for all practical purposes ψ is much smaller than $\boldsymbol{\Psi}$ and can be ignored. Accordingly, in the literature $\boldsymbol{\Psi}$ is referred to as the 'large component' and ψ as the 'small component' of the Dirac 4-spinor.

Equation 11.16 looks formally very similar to the Schrödinger equation reported at the beginning of this section (eqn 11.5). This similarity can be made more obvious by using the following identity, which is derived in Exercise 11.1:

$$[\boldsymbol{\sigma} \cdot (\mathbf{p} + e\mathbf{A})]^2 = (\mathbf{p} + e\mathbf{A})^2 + e\hbar\,\boldsymbol{\sigma} \cdot \mathbf{B}.$$
(11.17)

Using this identity eqn 11.16 becomes:

$$\left[\frac{1}{2m_e}(\mathbf{p} + e\mathbf{A})^2 - e\varphi + \frac{e\hbar}{2m_e}\boldsymbol{\sigma} \cdot \mathbf{B}\right]\boldsymbol{\Psi} = E\,\boldsymbol{\Psi}.$$
(11.18)

This last equation is traditionally written by introducing the *Bohr magneton*:

$$\mu_\text{B} = \frac{e\hbar}{2m_e} = 5.788 \cdot 10^{-5}\,\text{eV/T},$$
(11.19)

as well as the *spin operator*:

$$\mathbf{S} = \frac{\hbar}{2}\boldsymbol{\sigma},$$
(11.20)

which is simply a 2×2 matrix. Using these definitions eqn 11.18 can be written in the usual form:

$$\left[\frac{1}{2m_e}(\mathbf{p} + e\mathbf{A})^2 - e\varphi + \frac{2\mu_\text{B}}{\hbar}\mathbf{S} \cdot \mathbf{B}\right]\boldsymbol{\Psi} = E\,\boldsymbol{\Psi}.$$
(11.21)

spin term

This equation is referred to as the *Pauli equation* (Pauli, 1927). The most important aspect of this result is that the Hamiltonian carries an extra term with respect to the Schrödinger equation (eqn 11.5). Since this extra energy is similar to the magnetic coupling of eqns 11.1 and 11.2, we can say that the electron has an intrinsic degree of freedom which is somewhat similar to an angular momentum.

The similarity between the spin term in the Pauli equation and the angular momentum becomes more clear when we narrow our discussion to the case of a uniform and constant magnetic field. Indeed, as shown in Exercise 11.2, under these conditions eqn 11.21 further simplifies to:

$$\left[\frac{\mathbf{p}^2}{2m_e} - e\varphi + \frac{\mu_B}{\hbar}(\mathbf{L} + 2\mathbf{S}) \cdot \mathbf{B} + \frac{e}{4}\frac{\mu_B}{\hbar}(\mathbf{B} \times \mathbf{r})^2\right]\mathbf{\Psi} = E\,\mathbf{\Psi}. \tag{11.22}$$

By comparing the third term in eqn 11.22 with eqn 11.2 we see that the electron comes with an *effective* magnetic dipole moment corresponding to the operator:

$$\mathbf{m} = \frac{\mu_B}{\hbar}(\mathbf{L} + 2\mathbf{S}). \tag{11.23}$$

Therefore we can interpret the spin as an additional intrinsic angular momentum.

Exercise 11.1 In this exercise we want to verify the identity in eqn 11.17. The aim of this exercise is to give a glimpse of the complicated machinery behind Dirac's and Pauli's equations and the formalism to study magnetic fields. The reader already familiar with these aspects from introductory quantum mechanics courses may want to skip this exercise.

▶Using the definition of the Pauli matrices in eqn 11.9, show that the following relations hold:

$$\sigma_x\sigma_y = i\sigma_z, \qquad \sigma_y\sigma_x = -i\sigma_z, \qquad \sigma_x^2 = \mathbf{1}. \tag{11.24}$$

▶Extend these rules to any permutations of the indices x, y, z.

▶Now let us define $\boldsymbol{\pi} = \mathbf{p} + e\mathbf{A}$ to slim down the notation. In order to show that eqn 11.17 holds we need to evaluate $(\boldsymbol{\sigma} \cdot \boldsymbol{\pi})^2$. By using the definition of 'commutator' of two operators $[A, B] = AB - BA$ (as in eqn 10.58), show that:

$$(\boldsymbol{\sigma} \cdot \boldsymbol{\pi})^2 = \boldsymbol{\pi}^2\mathbf{1} + i[\pi_x, \pi_y]\sigma_z + i[\pi_y, \pi_z]\sigma_x + i[\pi_z, \pi_x]\sigma_y. \tag{11.25}$$

▶We now need to evaluate every commutator in the last expression. By writing down all the terms explicitly and noting that $\partial^2/\partial x\partial y = \partial^2/\partial y\partial x$ and similarly for x, z and y, z, show that:

$$\frac{1}{e}[\pi_x, \pi_y] = [A_x, p_y] + [p_x, A_y]. \tag{11.26}$$

▶Each term in the last equation can be evaluated explicitly by establishing the effect of the operator on a wavefunction, e.g. by evaluating $[A_x, p_y]\psi$. Using this expedient show that:

$$[A_x, p_y] = i\hbar\frac{\partial A_x}{\partial y}, \tag{11.27}$$

and similarly for the second term in eqn 11.26. ▶By combining together eqns 11.25–11.27 and remembering that $\mathbf{B} = \nabla \times \mathbf{A}$ from eqn 11.3 you should now obtain:

$$i[\pi_x, \pi_y] = e\hbar B_z, \tag{11.28}$$

and similar relations for the commutators involving y, z and z, x. By replacing this last result inside eqn 11.25 we finally obtain eqn 11.17.

Exercise 11.2 In this exercise we want to understand how, by starting from eqn 11.21, the angular momentum, **L**, of the electron appears in eqn 11.22. As in Exercise 11.1, here we only perform some formal derivations involving vector algebra, and the reader familiar with these aspects may want to skip this part.

▶By remembering that $\mathbf{p} = -i\hbar\nabla$ and the definition of Bohr magneton in eqn 11.19, show that:

$$\frac{1}{2m_e}(\mathbf{p} + e\mathbf{A})^2 = \frac{\mathbf{p}^2}{2m_e} - i\mu_B(\nabla \cdot \mathbf{A} + \mathbf{A} \cdot \nabla) + \frac{e}{\hbar}\mu_B\mathbf{A}^2. \tag{11.29}$$

▶By directly applying the operator $(\nabla \cdot \mathbf{A} + \mathbf{A} \cdot \nabla)$ to a wavefunction, ψ, show that

$$(\nabla \cdot \mathbf{A} + \mathbf{A} \cdot \nabla)\psi = (\nabla \cdot \mathbf{A})\psi + 2\mathbf{A} \cdot (\nabla\psi) \rightarrow \nabla \cdot \mathbf{A} + \mathbf{A} \cdot \nabla = \nabla \cdot \mathbf{A} + 2\mathbf{A} \cdot \nabla. \tag{11.30}$$

▶Now we consider the special case of a uniform and constant magnetic field, **B**. Show that the choice

$$\mathbf{A} = \frac{1}{2}\mathbf{B} \times \mathbf{r} \tag{11.31}$$

is legitimate since it satisfies eqn 11.3. ▶Finally, by replacing eqns 11.30 and 11.31 inside eqn 11.29, and using the permutation rules of the triple product between vectors, show that:

$$\frac{1}{2m_e}(\mathbf{p} + e\mathbf{A})^2 = \frac{\mathbf{p}^2}{2m_e} + \frac{\mu_B}{\hbar}\mathbf{L} \cdot \mathbf{B} + \frac{e}{4\hbar}\mu_B(\mathbf{B} \times \mathbf{r})^2.$$

This last result accounts for the various terms appearing in eqn 11.22.

The fact that the electron possesses an intrinsic angular momentum called spin is one of the cornerstones of quantum mechanics, and a thorough review of the physics of spin is well beyond the scope of this presentation. For the purposes of our simplified discussion there are only two key points worth remembering:

o The spin is a direct consequence of relativity and arises naturally within the framework of Dirac's equation. Whenever in doubt about how to interpret or calculate spin-related quantities in materials, our best bet is to go back to eqn 11.6 or its approximate version, eqn 11.21, and take it from there.

o From eqn 11.23 we are tempted to associate the spin with the idea that the electron is like a 'spinning top', with mass and charge distributed over a small but finite volume. This is a good *mnemonic expedient* to form an intuitive picture of spins in materials; however, the accepted view is that electrons are point-like charges. The *apparent* extended distribution of mass and charge is only an artefact. In this respect Messiah wrote (1965, page 948): '*In the non-relativistic limit, the Dirac electron appears not as a point charge, but as a distribution of charge and current extending over a domain of linear dimension $\hbar/m_e c$. This explains the appearance of interaction terms characteristic of the presence of a magnetic moment.*'

The derivations reported in this section are rather lengthy, but they are useful for making it clear that, even though magnetism is admittedly a complicated business, it is possible to formulate the theory of magnetic properties of materials while keeping the empiricism to a minimum. A very clear derivation of the equations presented here is offered by Bransden and Joachain (1983, Appendix 7). A more advanced presentation, which assumes knowledge of special relativity as a pre-requisite, can be found in Chapter 20 of Messiah (1965).

11.2   Charge density and spin density

After an introduction to Dirac's formulation of quantum mechanics in the previous section, it is now time to see how the concept of the 2-spinor Ψ can be used for practical calculations. In this section we continue with the case of *one isolated electron*, since this allows us to understand the key concepts without complicating unnecessarily the formalism. We will generalize our results to the case of many electrons in Section 11.5. For the sake of simplicity we are going to ignore two terms in eqn 11.22, namely $(\mathbf{B} \times \mathbf{r})^2$ and $\mathbf{L} \cdot \mathbf{B}$:

○   The term $e\mu_B(\mathbf{B} \times \mathbf{r})^2/4\hbar$ is responsible for *diamagnetic properties*. To see this, let us set for the time being $\mathbf{B} = B\mathbf{u}_z$. In this case we have:

$$\frac{e\mu_B}{4\hbar}(\mathbf{B} \times \mathbf{r})^2 = \mathbf{m}_{\mathrm{dia}} \cdot \mathbf{B} \quad \text{with} \quad \mathbf{m}_{\mathrm{dia}} = \frac{e\mu_B}{4\hbar}(x^2 + y^2)\mathbf{B}.$$

If we compare this term with the expression in eqn 11.2 we see that in this case the energy is minimized when the 'diamagnetic moment', $\mathbf{m}_{\mathrm{dia}}$, is antiparallel to the magnetic field. The origin of diamagnetic behaviour is precisely this tendency to counter the external magnetic field.

○   The term $\mu_B(\mathbf{L} \cdot \mathbf{B})/\hbar$ is responsible for *orbital paramagnetism*, i.e. the tendency of the electronic currents to produce a magnetic dipole moment aligned parallel to the external magnetic field. The mechanism of paramagnetism is essentially the one discussed in relation to eqns 11.1 and 11.2 at the beginning of Section 11.1.

The study of these two terms forms an entire research field in the area of computational materials science, and would deserve a thorough discussion. In particular, these terms are important for calculating nuclear magnetic resonance (NMR) parameters from first principles. The interested reader will find more details on such calculations in the original research papers and reviews, such as Mauri and Louie (1996) and Pickard and Mauri (2001) in the solid-state literature, and Schreckenbach and Ziegler (1998) and Bonhomme *et al.* (2012) in the chemistry literature.

In the following we concentrate on a simplified version of eqn 11.22, whereby we ignore diamagnetism and orbital paramagnetism. This approximation is justified in the study of the ground state of magnetically ordered systems (e.g. ferromagnets and antiferromagnets). The simplified version of eqn 11.22 reads:

$$\left[\frac{\mathbf{p}^2}{2m_e} - e\varphi + \frac{2\mu_B}{\hbar}\mathbf{S} \cdot \mathbf{B}\right]\Psi = E\,\Psi. \tag{11.32}$$

Now we want to understand how to calculate useful properties once a solution, Ψ, for this equation has been found. In particular, we are interested in the electron charge density and in the electron spin density, since these quantities are key to DFT calculations for magnetic materials.

In order to simplify the algebra it is convenient to write the 2-spinor as follows:

$$\Psi = \psi(\mathbf{r}; 1)\,\chi_\uparrow + \psi(\mathbf{r}; 2)\,\chi_\downarrow, \quad \text{with} \quad \chi_\uparrow = \begin{pmatrix} 1 \\ 0 \end{pmatrix} \quad \text{and} \quad \chi_\downarrow = \begin{pmatrix} 0 \\ 1 \end{pmatrix} \tag{11.33}$$

(the reason for using the labels ↑ and ↓ will become clear in Exercise 11.3). In the Schrödinger formulation of quantum mechanics the electron density associated with a wavefunction $\psi(\mathbf{r})$ is $n(\mathbf{r}) = |\psi(\mathbf{r})|^2$. In Dirac's formulation the prescription is only slightly different:

$$n(\mathbf{r}) = \mathbf{\Psi}^\dagger \mathbf{\Psi} \tag{11.34}$$

$$= \left[\psi(\mathbf{r};1)\,\chi_\uparrow + \psi(\mathbf{r};2)\,\chi_\downarrow\right]^\dagger \left[\psi(\mathbf{r};1)\,\chi_\uparrow + \psi(\mathbf{r};2)\,\chi_\downarrow\right]$$

$$= |\psi(\mathbf{r};1)|^2 + |\psi(\mathbf{r};2)|^2. \tag{11.35}$$

Here the symbol † stands for the Hermitian conjugate, i.e. complex and transpose, and we used the fact that $\chi_\uparrow^\dagger\chi_\uparrow = 1$, $\chi_\uparrow^\dagger\chi_\downarrow = 0$. As expected, the density given by eqn 11.34 depends only on the position variable. Since the spinor $\mathbf{\Psi}$ is meant to describe *one electron*, we need to use the normalization condition:

$$1 = \int d\mathbf{r}\, n(\mathbf{r}) = \int d\mathbf{r}\, \left[|\psi(\mathbf{r};1)|^2 + |\psi(\mathbf{r};2)|^2\right]. \tag{11.36}$$

Another useful property is the energy associated with the spinor $\mathbf{\Psi}$. In order to obtain E from eqn 11.32 we can proceed similarly to eqn 2.53. The only difference is that this time we multiply both sides by $\mathbf{\Psi}^\dagger$ in order to have a product of a 1×2 vector and a 2×1 vector yielding a number:

$$\langle\mathbf{\Psi}|\left[\frac{\mathbf{p}^2}{2m_e} - e\varphi + \frac{2\mu_\mathrm{B}}{\hbar}\mathbf{S}\cdot\mathbf{B}\right]|\mathbf{\Psi}\rangle = \langle\mathbf{\Psi}|E|\mathbf{\Psi}\rangle = E\langle\mathbf{\Psi}|\mathbf{\Psi}\rangle = E, \tag{11.37}$$

where the bracket between two spinors indicates the matrix product and the integral over space. For example, for an operator $\hat{A}$:

$$\langle\mathbf{\Psi}|\hat{A}|\mathbf{\Psi}\rangle = \int d\mathbf{r}\, \mathbf{\Psi}^\dagger(\mathbf{r})\hat{A}\,\mathbf{\Psi}(\mathbf{r}). \tag{11.38}$$

We are interested in the contribution to the electron energy in eqn 11.37 arising from the magnetic interaction. Let us call this contribution E_B, in analogy with eqn 11.2. We have:

$$E_\mathrm{B} = \langle\mathbf{\Psi}|\frac{2\mu_\mathrm{B}}{\hbar}\mathbf{S}\cdot\mathbf{B}|\mathbf{\Psi}\rangle = \frac{2\mu_\mathrm{B}}{\hbar}\int d\mathbf{r}\, \mathbf{\Psi}^\dagger\mathbf{S}\mathbf{\Psi}\cdot\mathbf{B}.$$

If we now define the *density of spin–angular momentum*, or in short *spin density*, as:

$$\mathbf{s}(\mathbf{r}) = \mathbf{\Psi}^\dagger(\mathbf{r})\mathbf{S}(\mathbf{r})\mathbf{\Psi}(\mathbf{r}), \tag{11.39}$$

and the associated density of spin–magnetic dipole, or *spin polarization*, as:

$$\mathbf{m}(\mathbf{r}) = -\frac{2\mu_\mathrm{B}}{\hbar}\mathbf{s}(\mathbf{r}), \tag{11.40}$$

then the energy E_B can be written compactly as:

$$E_\mathrm{B} = -\int d\mathbf{r}\, \mathbf{m}(\mathbf{r})\cdot\mathbf{B}(\mathbf{r}).$$

This result is entirely analogous to our classical model at the beginning of the chapter, and shows that the electron behaves like a collection of tiny magnetic dipoles, $\mathbf{m}(\mathbf{r})d\mathbf{r}$, spread over space, with spin–angular momentum $\mathbf{s}(\mathbf{r})d\mathbf{r}$ at each point.

Let us now take a closer look at the newly introduced spin density. If we replace the spinor components from eqn 11.33 and the definition of spin operator from eqns 11.20 and 11.9 and carry out the algebra, then we obtain the Cartesian components of the spin density as follows:

$$s_x(\mathbf{r}) = \frac{\hbar}{2} \, 2 \, \mathrm{Re}\Big[\psi^*(\mathbf{r}; 1)\psi(\mathbf{r}; 2)\Big], \tag{11.41}$$

$$s_y(\mathbf{r}) = \frac{\hbar}{2} \, 2 \, \mathrm{Im}\Big[\psi^*(\mathbf{r}; 1)\psi(\mathbf{r}; 2)\Big], \tag{11.42}$$

$$s_z(\mathbf{r}) = \frac{\hbar}{2}\Big[|\psi(\mathbf{r}; 1)|^2 - |\psi(\mathbf{r}; 2)|^2\Big]. \tag{11.43}$$

As we will see below, these expressions for the spin density constitute the starting point for DFT calculations of molecules and solids exhibiting magnetic order.

Before moving on to discuss the case of *many electrons* it is useful to spend a few words on the interpretation of eqns 11.41–11.43. If $s(\mathbf{r})d\mathbf{r}$ is supposed to represent an infinitesimal angular momentum, then we can picture it as being represented by a tiny arrow, whose length and orientation depend on the space coordinate $\mathbf{r}$. Therefore it should be possible to rewrite $s(\mathbf{r})$ in terms of its magnitude and two angles. This is indeed possible if we proceed as follows. We rewrite the spinor components in terms of the electron density, $n(\mathbf{r})$, and three real functions, $\theta(\mathbf{r})$, $\Phi(\mathbf{r})$ and $\gamma(\mathbf{r})$:

$$\psi(\mathbf{r}; 1) = \sqrt{n(\mathbf{r})}\, e^{-i\gamma(\mathbf{r})/2}\, e^{-i\Phi(\mathbf{r})/2} \cos\frac{\theta(\mathbf{r})}{2}, \tag{11.44}$$

$$\psi(\mathbf{r}; 2) = \sqrt{n(\mathbf{r})}\, e^{-i\gamma(\mathbf{r})/2}\, e^{+i\Phi(\mathbf{r})/2} \sin\frac{\theta(\mathbf{r})}{2}. \tag{11.45}$$

We note that this is only another way of expressing $\psi(\mathbf{r}; 1)$ and $\psi(\mathbf{r}; 2)$, and we are simply replacing four real functions (the real and imaginary components of the spinor) by another set of four real functions. By replacing these expressions inside eqns 11.41–11.43 we obtain:

$$\mathbf{s}(\mathbf{r}) = \frac{\hbar}{2} n(\mathbf{r})\, \hat{\mathbf{s}}(\mathbf{r}), \tag{11.46}$$

with

$$\hat{\mathbf{s}}(\mathbf{r}) = \sin\theta(\mathbf{r})\cos\Phi(\mathbf{r})\mathbf{u}_x + \sin\theta(\mathbf{r})\sin\Phi(\mathbf{r})\mathbf{u}_y + \cos\theta(\mathbf{r})\mathbf{u}_z. \tag{11.47}$$

This last result indicates that $\hat{\mathbf{s}}(\mathbf{r})$ is a unit vector whose orientation is defined by the polar angle, $\theta(\mathbf{r})$, and the azimutal angle, $\Phi(\mathbf{r})$, as shown in Figure 11.1. Since θ and Φ depend on the position, the infinitesimal spin $\mathbf{s}(\mathbf{r})d\mathbf{r}$ may be oriented along different directions at different points.

Using this strategy of expressing the spinor components in terms of directions in space, it is now easy to calculate the expectation value of the spin of the electron. We simply insert eqns 11.39 and 11.46 inside eqn 11.38 to find:

$$\langle\Psi|\mathbf{S}|\Psi\rangle = \int d\mathbf{r}\, \Psi^\dagger(\mathbf{r})\mathbf{S}\,\Psi(\mathbf{r}) = \int d\mathbf{r}\, \mathbf{s}(\mathbf{r}) = \frac{\hbar}{2}\int d\mathbf{r}\, n(\mathbf{r})\,\hat{\mathbf{s}}(\mathbf{r}) = \frac{\hbar}{2}\hat{\mathbf{s}}_{\mathrm{ave}}, \tag{11.48}$$

with $\hat{\mathbf{s}}_{\mathrm{ave}} = \int d\mathbf{r}\, n(\mathbf{r})\,\hat{\mathbf{s}}(\mathbf{r})/\int d\mathbf{r}\, n(\mathbf{r})$ being the spin direction averaged over the spatial distribution of the electron charge. We can interpret this result by saying that, on average, *the electron has a spin $\hbar/2$ pointing along the direction $\hat{\mathbf{s}}_{\mathrm{ave}}$.*

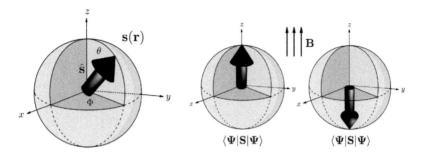

Fig. 11.1 Schematic representation of the direction of the spin angular momentum. Left panel: according to eqn 11.46 the spin density, $s(r)$, can take any orientation, $\hat{s}(r)$, as defined by the polar angle, $\theta(r)$, and the azimutal angle, $\Phi(r)$. These angles can be calculated at any point, r, once the spinor components $\psi(r; 1)$ and $\psi(r; 2)$ have been determined. Right panel: in the presence of a uniform magnetic field, B, the expectation value of the electron spin is either parallel (left) or antiparallel (right) to the magnetic field (see Exercise 11.3). In this case it is common to speak of spin-up and spin-down electrons. This representation of electron spin is usually referred to as the 'Bloch sphere' in the area of quantum computing (Nielsen and Chuang, 2010).

The result expressed by eqn 11.48 has profound consequences. In fact, owing to the *spin–statistics theorem* (Pauli, 1940), particles with half-integer spin (in units of $\hbar$) are *fermions*. As we have stated briefly in Section 2.6, fermions have the following properties:

○ They must obey Pauli's exclusion principle, which states that no two fermions can occupy the same quantum state;

○ The wavefunction of many fermions must change sign if we exchange any two particles (see Section 11.4).

The derivation of the spin–statistics theorem is rather challenging and will not be given here, but the curious reader can consult the original papers (e.g. Pauli, 1940; note that this requires a solid background in relativity and quantum field theory). This theorem is one of the pillars of quantum mechanics, and is connected with the invariance of the law of physics with respect to Lorentz transformations. The remarkable aspect of this theorem is that it establishes a connection between an intrinsic property of one electron (the spin) and collective behaviour of many of these electrons (the exclusion principle).

Given the importance of the spin, one may wonder why in Chapters 2–10 we avoided discussing the spin of the systems considered. The reason is that in the materials considered so far the magnetization density vanishes; that is, the number of electrons with spin along a certain direction equals the number of spins along the opposite direction. In these cases we say that the system is *non-spin-polarized*. As we might expect, the study of magnetism involves instead systems which are *spin-polarized*, either permanently (e.g. in ferromagnetism), or under the influence of an external magnetic field (paramagnetism).

Exercise 11.3 In this exercise we want to calculate the energy of a free electron immersed in a uniform and constant magnetic field, $\mathbf{B} = B\mathbf{u}_z$. In order to simplify the algebra it is convenient to choose a reference frame with the z axis parallel to the magnetic field. ▶Show that under these conditions eqn 11.32 separates into the following two equations for the spinor components:

$$-\frac{\hbar^2}{2m_e}\nabla^2\psi(\mathbf{r};1) + \mu_B B\psi(\mathbf{r};1) = E\,\psi(\mathbf{r};1),$$

$$-\frac{\hbar^2}{2m_e}\nabla^2\psi(\mathbf{r};2) - \mu_B B\psi(\mathbf{r};2) = E\,\psi(\mathbf{r};2).$$

☐It is important to observe that these equations must be satisfied *simultaneously* for the same energy, E, by the spinor components $\psi(\mathbf{r};1)$ and $\psi(\mathbf{r};2)$. ▶Show that the free electron wavefunctions

$$\psi(\mathbf{r};1) = \frac{c_1}{\sqrt{V}}\exp(i\mathbf{k}_1\cdot\mathbf{r}), \qquad \psi(\mathbf{r};2) = \frac{c_2}{\sqrt{V}}\exp(i\mathbf{k}_2\cdot\mathbf{r}),$$

with

$$|\mathbf{k}_1| = \sqrt{\frac{2m_e(E-\mu_B B)}{\hbar^2}}, \qquad |\mathbf{k}_2| = \sqrt{\frac{2m_e(E+\mu_B B)}{\hbar^2}},$$

are solutions of the above equations. Here V is the volume in which the electron is contained, and $|c_1|^2 + |c_2|^2 = 1$ to guarantee the normalization of the spinor. ☐In order to determine the constants c_1 and c_2 we need to use the following identity:

$$\int d\mathbf{r}\exp(i\mathbf{k}\cdot\mathbf{r}) = V\delta(\mathbf{k}),$$

with $\delta(\mathbf{k})$ the Dirac delta function. This property has already been derived in Exercise 5.4. ▶Show that this relation can also be obtained from Exercise 10.5 by replacing the wavefunctions with planewaves, and the momentum operator with a constant. ▶By making use of the above identity, show that the expectation values of the spin of this electron (eqns 11.41, 11.42 and 11.48) along x and y are:

$$S_x = S_y = 0.$$

☐From eqn 11.46 we know that there must be one direction in space along which the magnitude of the expectation value of the spin is $\hbar/2$. Since $S_x = S_y = 0$ the only possibility is:

$$S_z = \pm\frac{\hbar}{2}. \qquad (11.49)$$

▶Using eqn 11.43 and this last result, show that the constants c_1 and c_2 can take only the following values (apart from a complex factor of norm 1):

$$S_z = +\frac{\hbar}{2}: c_1 = 1,\ c_2 = 0, \qquad S_z = -\frac{\hbar}{2}: c_1 = 0,\ c_2 = 1.$$

☐Putting together the results of this exercise we can conclude that the stationary quantum states of a free electron in a uniform magnetic field are described by spinors of either one of the following forms:

$$\Psi = \frac{1}{\sqrt{V}}\exp(i\mathbf{k}\cdot\mathbf{r})\,\chi_\uparrow \quad \text{with } E = \hbar^2|\mathbf{k}|^2/2m_e - \mu_B B, \quad \text{or}$$

$$\Psi = \frac{1}{\sqrt{V}}\exp(i\mathbf{k}\cdot\mathbf{r})\,\chi_\downarrow \quad \text{with } E = \hbar^2|\mathbf{k}|^2/2m_e + \mu_B B.$$

The spinor with the lowest energy has the spin oriented parallel to the magnetic field, as we would expect for a tiny bar magnet. The spinor with the highest energy has the spin antiparallel to the field. The adoption of the subscripts ↑ and ↓ in eqn 11.33 was done precisely in order to distinguish these two possibilities.

☐So far we have focused on the expectation value of **S**; however, we may also ask what is the expectation value of the 'length' of the spin vector. This is simply evaluated as the expectation value of $|\mathbf{S}| = [S_x^2 + S_y^2 + S_z^2]^{1/2}$. ▶Show that the explicit expression for the 'spin norm' operator $|\mathbf{S}|$ is:

$$|\mathbf{S}| = \sqrt{3}\,\frac{\hbar}{2}\,\mathbf{1}.$$

This result seems trivial but we should note that the square root of a matrix is *not* the square root of its elements, unless the matrix is already diagonal. The calculation of the square root of a matrix can be performed immediately using the command 'sqrtm' of the program 'Octave'. ▶From this result deduce that the expectation value of $|\mathbf{S}|$ of *any* spinor **Ψ** is

$$\langle \mathbf{\Psi} | \, |\mathbf{S}| \, | \mathbf{\Psi} \rangle = \sqrt{3}\,\frac{\hbar}{2}.$$

It is important not to confuse this result with eqn 11.48, which instead states that $|\langle \mathbf{\Psi}|\mathbf{S}|\mathbf{\Psi}\rangle| = \hbar/2$. ☐The situation described in this exercise essentially corresponds to a simplified version of the Stern–Gerlach experiment (see for example Chapter 1 of Sakurai and Napolitano, 2011). As a word of caution, in this exercise we ignored the diamagnetic and paramagnetic terms of the Hamiltonian in eqn 11.22. Had we taken these terms into account we would have found the famous 'Landau levels' (see for example Chapter 15 of Landau and Lifshitz, 1965).

In Exercise 11.3 we have found that, in the presence of a uniform magnetic field, the spin of the electron is 'quantized', i.e. in stationary states electrons can have the expectation value of their spin only parallel or antiparallel to the magnetic field, and the component along the axis of the field can take only the values $\pm\hbar/2$. This is a general result which is not restricted to the simple case of a free electron. In fact the quantization of the spin angular momentum in atoms under magnetic fields is typically one of the starting points of textbooks on atomic physics (e.g. Bransden and Joachain, 1983).

Whenever we know that the spin is quantized along one axis, it is common practice to speak of 'spin-up' electrons and 'spin-down' electrons when referring to the spinors corresponding to $S_z = +\hbar/2$ and $-\hbar/2$, respectively. This terminology clearly draws from the pictorial representation given in Figure 11.1. In these cases we know that the spinors have only one non-zero component, and it is sometimes useful to label these components by the arrows ↑, ↓ as follows:

$$S_z = +\frac{\hbar}{2}: \quad \mathbf{\Psi} = \begin{bmatrix} \psi_\uparrow(\mathbf{r}) \\ 0 \end{bmatrix} = \psi_\uparrow(\mathbf{r})\,\chi_\uparrow, \tag{11.50}$$

$$S_z = -\frac{\hbar}{2}: \quad \mathbf{\Psi} = \begin{bmatrix} 0 \\ \psi_\downarrow(\mathbf{r}) \end{bmatrix} = \psi_\downarrow(\mathbf{r})\,\chi_\downarrow. \tag{11.51}$$

In Section 11.5 we will see that this naming convention is very common for describing molecules and solids for which electron spins are 'aligned' along a preferred direction. However, we need to keep in mind that the solutions of eqn 11.21 are truly 2-spinors consisting of two complex functions, and generally the direction of the spin can take any orientation as eqn 11.48 shows (see also Figure 11.1). As an example, a typical situation where we find many spin directions is provided by 'Bloch walls': in ferromagnets there exist domains of atoms with spin oriented along one direction adjacent to domains with spin oriented along a different direction. The Bloch wall is the transition region between these domains. In this case we have the gradual rotation of the electron spins from one domain to the next. The interested reader will find an introduction to magnetic domains and Bloch walls in Kittel (1976).

11.3   Spin in a system with many electrons

In the previous sections we have discussed the case of only *one* electron. The extension of the Dirac theory to the case of *many electrons* is very involved; therefore we will touch upon only the most basic concepts, without any pretence of rigour.

As we have seen in Chapter 2, the description of a many-electron system within the standard Schrödinger theory requires a collective many-body wavefunction. For example, in the case of two electrons we would have $\Psi(\mathbf{r}_1, \mathbf{r}_2)$. Accordingly we can imagine that in the Dirac theory introduced in Section 11.1 we might need a 4-spinor for two electrons: $\Psi(\mathbf{r}_1, \mathbf{r}_2)$. As it turns out this is not correct, since the complete theory requires a 16-component spinor of the two variables $\mathbf{r}_1$ and $\mathbf{r}_2$ (Breit, 1929). More generally, for N electrons the number of components in the many-body spinor becomes 4^N. This makes the complete theory enormously complex, and a certain number of approximations are necessary in order to move forward. Calculations using the whole machinery of Dirac spinors are now possible for small molecules, and there exists a specialized field of research on this aspect, broadly referred to as 'relativistic quantum chemistry' (Dyall and Faegri, 2007).

In the same way as the Dirac 4-spinor for one electron can be 'trimmed' down to a spinor with only two components, in the case of the 4^N-spinor for N electrons it is possible to lose several components which become very small in the limit of $v/c \ll 1$ (v being an average velocity of the electrons). After this trimming we end up with 2^N independent functions, which is four functions in the case of two electrons.

In eqn 11.33 we have seen that a 2-spinor for one electron can be written as a linear combination of the arrays $\chi_\uparrow$ and $\chi_\downarrow$. In Exercise 11.3 we have seen that, in the presence of a magnetic field, these arrays can be understood as describing electrons with spin magnetic moment pointing up or down. Intuitively we can extend these concepts to the case of two electrons by considering two completely independent electrons in the magnetic field of Exercise 11.3. In this case we expect four possibilities for the spins: up/up, up/down, down/up, and down/down. Therefore in order to describe this situation we need at least four functions depending on the coordinates $\mathbf{r}_1$ and $\mathbf{r}_2$.

These observations can be formalized by introducing the so-called 'Kronecker product' of the spin arrays. This product is indicated by the symbol '$\otimes$', and is defined as follows:

$$\chi_\uparrow \otimes \chi_\uparrow = \begin{pmatrix} 1 \\ 0 \end{pmatrix} \otimes \begin{pmatrix} 1 \\ 0 \end{pmatrix} = \begin{bmatrix} 1 \begin{pmatrix} 1 \\ 0 \end{pmatrix} \\ 0 \begin{pmatrix} 1 \\ 0 \end{pmatrix} \end{bmatrix} = \begin{bmatrix} 1 \\ 0 \\ 0 \\ 0 \end{bmatrix},$$

and similarly for the other possibilities, $\chi_{\uparrow,\downarrow} \otimes \chi_{\uparrow,\downarrow}$. In practice the Kronecker product is formed by 'attaching' the array on the right to each of the elements of the array on the left, which in turn act as multiplicative coefficients. In this operation the size of the final array is *not* the same as the size of the initial arrays. It is straightforward to check that the products $\chi_\uparrow \otimes \chi_\uparrow$, $\chi_\uparrow \otimes \chi_\downarrow$, $\chi_\downarrow \otimes \chi_\uparrow$ and $\chi_\downarrow \otimes \chi_\downarrow$ correspond to 4×1 column vectors with a 1 in the first, second, third, or fourth row, similarly to the above example.

Using this notation, the most general two-electron spinor can be written as:

$$\Psi(\mathbf{r}_1, \mathbf{r}_2) = \begin{bmatrix} f_1(\mathbf{r}_1, \mathbf{r}_2) \\ f_2(\mathbf{r}_1, \mathbf{r}_2) \\ f_3(\mathbf{r}_1, \mathbf{r}_2) \\ f_4(\mathbf{r}_1, \mathbf{r}_2) \end{bmatrix} = f_1 \chi_\uparrow \otimes \chi_\uparrow + f_2 \chi_\uparrow \otimes \chi_\downarrow + f_3 \chi_\downarrow \otimes \chi_\uparrow + f_4 \chi_\downarrow \otimes \chi_\downarrow, \quad (11.52)$$

with $f_1, \ldots f_4$ some functions of the space coordinates $\mathbf{r}_1$ and $\mathbf{r}_2$.

Since now we have a 4-spinor for two electrons, we need to generalize the spin operator of eqn 11.20 in such a way as to obtain the total spin of both electrons. This is done using again the Kronecker product, as follows:

$$\mathbf{S}_{\text{tot}} = \mathbf{S} \otimes \mathbb{1} + \mathbb{1} \otimes \mathbf{S}, \quad (11.53)$$

with $\mathbf{S}$ given in eqn 11.20. The purpose of this definition becomes clear if we consider the action of the total spin operator on any one of the spinor components in eqn 11.52. For example:

$$\mathbf{S}_{\text{tot}} f(\mathbf{r}_1, \mathbf{r}_2) \chi_\uparrow \otimes \chi_\downarrow = f(\mathbf{r}_1, \mathbf{r}_2)(\mathbf{S}\chi_\uparrow) \otimes \chi_\downarrow + f(\mathbf{r}_1, \mathbf{r}_2)\chi_\uparrow \otimes (\mathbf{S}\chi_\downarrow),$$

as can be verified by explicit evaluation of the matrix-vector products. This result shows that we should interpret the left- and right-hand side of $\chi_\uparrow \otimes \chi_\downarrow$ as referring to electron 1 and 2, respectively, and the total spin as the sum of two operators acting independently on such electrons.

Exercise 11.4 ►Using the definition of the Pauli matrices in eqn 11.9, evaluate the Kronecker products in eqn 11.53 and show that the x-, y- and z-components of the total spin of two electrons are given by:

$$S_{\text{tot},z} = \hbar \begin{bmatrix} 1 & 0 & 0 & 0 \\ 0 & 0 & 0 & 0 \\ 0 & 0 & 0 & 0 \\ 0 & 0 & 0 & -1 \end{bmatrix}, \quad S_{\text{tot},x} = \frac{\hbar}{2} \begin{bmatrix} 0 & 1 & 1 & 0 \\ 1 & 0 & 0 & 1 \\ 1 & 0 & 0 & 1 \\ 0 & 1 & 1 & 0 \end{bmatrix}, \quad S_{\text{tot},y} = i\frac{\hbar}{2} \begin{bmatrix} 0 & -1 & -1 & 0 \\ 1 & 0 & 0 & -1 \\ 1 & 0 & 0 & -1 \\ 0 & 1 & 1 & 0 \end{bmatrix}.$$

►Now calculate the matrix operator associated with the 'modulus' of the total spin of two electrons, defined by: $|\mathbf{S}_{\text{tot}}| = [S_{\text{tot},x}^2 + S_{\text{tot},y}^2 + S_{\text{tot},z}^2]^{1/2}$. In order to calculate the square root of a matrix you can use the 'sqrtm' command of 'Octave'. You should be able to find:

$$|\mathbf{S}_{\text{tot}}| = \sqrt{2}\frac{\hbar}{2} \begin{bmatrix} 2 & 0 & 0 & 0 \\ 0 & 1 & 1 & 0 \\ 0 & 1 & 1 & 0 \\ 0 & 0 & 0 & 2 \end{bmatrix}.$$

□We consider the two-electron spinor $\Psi = f(\mathbf{r}_1, \mathbf{r}_2) \chi_\uparrow \otimes \chi_\uparrow$, with f normalized such that $\int d\mathbf{r}_1 d\mathbf{r}_2 |f(\mathbf{r}_1, \mathbf{r}_2)|^2 = 1$. In analogy with eqn 11.48, the expectation value of the total spin of these two electrons is given by:

$$\langle \Psi | \mathbf{S}_{\text{tot}} | \Psi \rangle = \int d\mathbf{r}_1 d\mathbf{r}_2 \, \Psi^\dagger(\mathbf{r}_1, \mathbf{r}_2) \mathbf{S}_{\text{tot}} \Psi(\mathbf{r}_1, \mathbf{r}_2), \quad (11.54)$$

and a similar definition can be used for the expectation value of the total spin norm, $|\mathbf{S}_{\text{tot}}|$.
▶By explicitly performing the matrix-vector products, show that the expectation values of
the x, y and z components of the total spin and that of the spin norm for this state are as
follows:

$$\langle \Psi | S_{\text{tot},x} | \Psi \rangle = 0, \quad \langle \Psi | S_{\text{tot},y} | \Psi \rangle = 0, \quad \langle \Psi | S_{\text{tot},z} | \Psi \rangle = \hbar, \quad \langle \Psi | |\mathbf{S}_{\text{tot}}| | \Psi \rangle = \sqrt{2}\,\hbar.$$

▶In the same way as above verify the following expectation values:

if $\Psi = f(\mathbf{r}_1, \mathbf{r}_2)\,\chi_\uparrow \otimes \chi_\downarrow$ then $\langle \Psi | \mathbf{S}_{\text{tot}} | \Psi \rangle = \quad 0 \quad$ and $\langle \Psi | |\mathbf{S}_{\text{tot}}| | \Psi \rangle = \sqrt{2}\,\hbar/2,$

if $\Psi = f(\mathbf{r}_1, \mathbf{r}_2)\,\chi_\downarrow \otimes \chi_\uparrow$ then $\langle \Psi | \mathbf{S}_{\text{tot}} | \Psi \rangle = \quad 0 \quad$ and $\langle \Psi | |\mathbf{S}_{\text{tot}}| | \Psi \rangle = \sqrt{2}\,\hbar/2,$

if $\Psi = f(\mathbf{r}_1, \mathbf{r}_2)\,\chi_\downarrow \otimes \chi_\downarrow$ then $\langle \Psi | \mathbf{S}_{\text{tot}} | \Psi \rangle = -\hbar\mathbf{u}_z \quad$ and $\langle \Psi | |\mathbf{S}_{\text{tot}}| | \Psi \rangle = \sqrt{2}\,\hbar.$

☐We can summarize these findings by simply stating that, whenever the two electrons can
be described by a spinor that has *only one non-zero component*, the expectation value of the
total spin is along $\mathbf{u}_z$ and is obtained by adding $\hbar/2$ with the $+$ sign for $\uparrow$ and the $-$ sign for
$\downarrow$. It is important to keep in mind that this mnemonic rule is valid only when the 4-spinor
has only one component. In the most general case given by eqn 11.52 the total spin can take
any direction, similarly to the case of one electron in Figure 11.1.
▶Verify that the expectation value of the total spin along $\mathbf{u}_z$ for an arbitrary two-electron
spinor given by eqn 11.52 is:

$$\langle \Psi | S_{\text{tot},z} | \Psi \rangle = \int d\mathbf{r}_1 d\mathbf{r}_2 \Big[|f_1(\mathbf{r}_1, \mathbf{r}_2)|^2 - |f_4(\mathbf{r}_1, \mathbf{r}_2)|^2 \Big].$$

The corresponding expressions for $S_{\text{tot},x}$ and $S_{\text{tot},y}$ are similar to eqns 11.41 and 11.42 but
the evaluation is more tedious.

The formalism discussed in this section is admittedly a bit cumbersome, and one could
have presented the notion of spin in a completely heuristic fashion by simply stating a
set of rules. Here we have chosen to go through the mathematical framework in some
detail in order to make it clear that certain rules follow directly from a first-principles
theory of the electron. For instance, the use of Kronecker products to describe spins in
many-electron systems is simply a way to deal with the fact that the Dirac equation
is really a *matrix equation* involving many wavefunctions at once.
The reader interested in the algebra of electron spins will find more details in the
classic presentations by Bransden and Joachain (1983) or Merzbacher (1998), among
others. The use of Kronecker products for representing spin operators is discussed by
Cohen-Tannoudji *et al.* (1977), and the strictly mathematical aspects can be found in
Steeb and Shi (1997).

11.4   Spin and exchange energy

The discussion of the previous section was instrumental to introducing the connection
between the exchange energy of electrons and their spin configuration. This connection
is very important for understanding a number of magnetic properties of materials, such
as the microscopic origin of magnetic ordering in solids.
We have already stated on several occasions that, according to the Pauli exclusion
principle, the wavefunction of N electrons must change sign whenever we swap any two
electrons (the antisymmetry property). This principle applies generally to any spinor

wavefunction. In the simplest case of two electrons, Pauli's principle implies that the spinor in eqn 11.52 must be completely antisymmetric with respect to the exchange of the electron variables. Let us work out the consequences of this requirement.

The analysis is easier if we consider one spinor component at a time. If we have a two-electron spinor given by $\boldsymbol{\Psi}(\mathbf{r}_1, \mathbf{r}_2) = \Psi(\mathbf{r}_1, \mathbf{r}_2) \chi_\uparrow \otimes \chi_\uparrow$, when we swap the first and second electrons (i.e. the subscripts 1 and 2 and the order of the spinors) we find $\boldsymbol{\Psi}(\mathbf{r}_2, \mathbf{r}_1) = \Psi(\mathbf{r}_2, \mathbf{r}_1) \chi_\uparrow \otimes \chi_\uparrow$. Therefore, in order to have $\boldsymbol{\Psi}(\mathbf{r}_2, \mathbf{r}_1) = -\boldsymbol{\Psi}(\mathbf{r}_1, \mathbf{r}_2)$, the only possibility corresponds to having $\Psi(\mathbf{r}_2, \mathbf{r}_1) = -\Psi(\mathbf{r}_1, \mathbf{r}_2)$. In other words, if the spin part is symmetric with respect to the swap of the electrons, then the space part, Ψ, must be antisymmetric. The same observation applies to the case when we have only the $\chi_\downarrow \otimes \chi_\downarrow$ spinor component.

If we consider instead the second component of the spinor by setting $\boldsymbol{\Psi}(\mathbf{r}_1, \mathbf{r}_2) = \Psi(\mathbf{r}_1, \mathbf{r}_2) \chi_\uparrow \otimes \chi_\downarrow$, then the electron swap leads to $\boldsymbol{\Psi}(\mathbf{r}_2, \mathbf{r}_1) = \Psi(\mathbf{r}_2, \mathbf{r}_1) \chi_\downarrow \otimes \chi_\uparrow$. In this case there is no way of satisfying Pauli's principle. In fact, after the swap we end up with a spinor having a non-zero component in another row (remember $\chi_\uparrow \otimes \chi_\downarrow = [0\,1\,0\,0]^\top$ and $\chi_\downarrow \otimes \chi_\uparrow = [0\,0\,1\,0]^\top$). After a little thinking we realize that the only way to continue this reasoning is to consider spinors having both components, $\chi_\uparrow \otimes \chi_\downarrow$ and $\chi_\downarrow \otimes \chi_\uparrow$, non-zero, so that the swapping procedure is entirely contained within the same rows of the spinor. The two most obvious spinors that we can consider are: $\boldsymbol{\Psi}(\mathbf{r}_1, \mathbf{r}_2) = \Psi(\mathbf{r}_1, \mathbf{r}_2)(\chi_\uparrow \otimes \chi_\downarrow + \chi_\downarrow \otimes \chi_\uparrow)/\sqrt{2}$ and $\boldsymbol{\Psi}(\mathbf{r}_1, \mathbf{r}_2) = \Psi(\mathbf{r}_1, \mathbf{r}_2)(\chi_\uparrow \otimes \chi_\downarrow - \chi_\downarrow \otimes \chi_\uparrow)/\sqrt{2}$. The former choice is similar to the cases discussed in the previous paragraph: the spin part is symmetric, and therefore by the exclusion principle the space part must be antisymmetric. The latter choice leads to a different situation: since the swap changes the sign of the spin part, the space part must be symmetric with respect to the exchange of the two electrons.

These considerations can be summarized as follows. The 'most elementary' two-electron spinors satisfying Pauli's principle are given by:

$$\boldsymbol{\Psi}_t(\mathbf{r}_1, \mathbf{r}_2) = \Psi_t(\mathbf{r}_1, \mathbf{r}_2) \times \begin{cases} \chi_\uparrow \otimes \chi_\uparrow \\ \chi_\downarrow \otimes \chi_\downarrow \qquad\qquad\qquad \text{with } \boldsymbol{\Psi}_t(\mathbf{r}_2, \mathbf{r}_1) = -\boldsymbol{\Psi}_t(\mathbf{r}_1, \mathbf{r}_2), \\ (\chi_\uparrow \otimes \chi_\downarrow + \chi_\downarrow \otimes \chi_\uparrow)/\sqrt{2} \end{cases}$$

$$\boldsymbol{\Psi}_s(\mathbf{r}_1, \mathbf{r}_2) = \Psi_s(\mathbf{r}_1, \mathbf{r}_2) \times \quad (\chi_\uparrow \otimes \chi_\downarrow - \chi_\downarrow \otimes \chi_\uparrow)/\sqrt{2} \quad \text{with } \boldsymbol{\Psi}_s(\mathbf{r}_2, \mathbf{r}_1) = +\boldsymbol{\Psi}_s(\mathbf{r}_1, \mathbf{r}_2).$$

$$(11.55)$$

Here Ψ_t and Ψ_s are two normalized functions of the electron variables, respectively antisymmetric or symmetric in the electron exchange. Since in the antisymmetric case we have three options, we call the spinors $\boldsymbol{\Psi}_t$ *triplet states*. Similarly we call the spinor $\boldsymbol{\Psi}_s$ a *singlet state*. Table 11.1 shows that the triplets correspond to a norm of the total spin of $\sqrt{2}\hbar$, while the singlet corresponds to a total spin norm of 0.

Exercise 11.5 In Exercise 4.5 we studied the H_2 molecule using a simple approximation for the two-electron wavefunction, $\Psi(\mathbf{r}_1, \mathbf{r}_2)$. Here we want to build on the same example in order to understand the consequences of the connection between the exclusion principle and the spin. All the physical quantities in the equations are in Hartree atomic units for consistency with Exercise 4.5. The electronic Hamiltonian of the molecule when the protons are fixed at the positions $\pm d\mathbf{u}_x/2$ is given by eqns 4.25–4.27:

Table 11.1 Expectation values of the total spin and the spin norm for the singlet and triplet states in eqn 11.55. We indicate the states using a shorthand notation for convenience. The value of S is obtained from $|\mathbf{S}_{\text{tot}}|^2 = S(S+1)\hbar^2$, in analogy with the standard rules for the orbital angular momentum (Merzbacher, 1998). These values can be obtained directly from the results of Exercise 11.4. The 'family' label and the symmetry properties of the corresponding wavefunctions are also indicated for completeness. Atkins and de Paula (2006) provide a very appealing visualization of the notion of spin singlet and triplets (pages 337 and 347).

State	Expt. value $\mathbf{S}_{\text{tot}}$	Expt. value $|\mathbf{S}_{\text{tot}}|$	S	Family	Wavefun.
↑↑	$+\hbar\mathbf{u}_z$	$\sqrt{2}\hbar$	1	triplet	antisymm.
↓↓	$-\hbar\mathbf{u}_z$	$\sqrt{2}\hbar$	1	triplet	antisymm.
$(\uparrow\downarrow + \downarrow\uparrow)/\sqrt{2}$	0	$\sqrt{2}\hbar$	1	triplet	antisymm.
$(\uparrow\downarrow - \downarrow\uparrow)/\sqrt{2}$	0	0	0	singlet	symm.

$$\hat{H} = \hat{H}_1 + \hat{H}_2 + \Delta\hat{H},$$

with $\hat{H}_1$, $\hat{H}_2$ and $\Delta\hat{H}$ given in Exercise 4.5. In Exercise 4.5 it was observed that in a first approximation we can treat $\Delta\hat{H}$ as a small perturbation, and write the two-electron wavefunction as a product of the ground-state wavefunctions of $\hat{H}_1$ and $\hat{H}_2$ (see eqn 4.28):

$$\Psi_0(\mathbf{r}_1, \mathbf{r}_2) = \phi_{1s}(\mathbf{r}_1 + d\mathbf{u}_x/2)\phi_{1s}(\mathbf{r}_2 - d\mathbf{u}_x/2).$$

For the present exercise it is convenient to introduce the simplified notation:

$$\phi_A(\mathbf{r}) = \phi_{1s}(\mathbf{r} + d\mathbf{u}_x/2), \qquad \phi_B(\mathbf{r}) = \phi_{1s}(\mathbf{r} - d\mathbf{u}_x/2),$$

so that $\Psi_0(\mathbf{r}_1, \mathbf{r}_2) = \phi_A(\mathbf{r}_1)\phi_B(\mathbf{r}_2)$. In writing the wavefunction Ψ_0, the spin part and the normalization condition were completely ignored in Exercise 4.5. Now that we are more comfortable with the notion of spin, we can improve our attempt by constructing two new wavefunctions, one for the singlet and one for the triplet:

$$\Psi_s(\mathbf{r}_1, \mathbf{r}_2) = \frac{1}{\sqrt{2(1+\kappa^2)}}[\phi_A(\mathbf{r}_1)\phi_B(\mathbf{r}_2) + \phi_A(\mathbf{r}_2)\phi_B(\mathbf{r}_1)], \tag{11.56}$$

$$\Psi_t(\mathbf{r}_1, \mathbf{r}_2) = \frac{1}{\sqrt{2(1-\kappa^2)}}[\phi_A(\mathbf{r}_1)\phi_B(\mathbf{r}_2) - \phi_A(\mathbf{r}_2)\phi_B(\mathbf{r}_1)], \tag{11.57}$$

where κ is the overlap integral:

$$\kappa = \int d\mathbf{r}\, \phi_A(\mathbf{r})\phi_B(\mathbf{r}).$$

▶Verify that the wavefunctions $\Psi_s(\mathbf{r}_1, \mathbf{r}_2)$ and $\Psi_t(\mathbf{r}_1, \mathbf{r}_2)$ are correctly normalized to unity.
▶Verify that the following spinors all satisfy Pauli's principle, i.e. they change sign upon exchanging the two electrons. You need to remember that the electron swap is also performed on the spin part.

$$\Psi(\mathbf{r}_1, \mathbf{r}_2) = \Psi_t(\mathbf{r}_1, \mathbf{r}_2)\, \chi_\uparrow \otimes \chi_\uparrow,$$
$$\Psi(\mathbf{r}_1, \mathbf{r}_2) = \Psi_t(\mathbf{r}_1, \mathbf{r}_2)\, \chi_\downarrow \otimes \chi_\downarrow,$$
$$\Psi(\mathbf{r}_1, \mathbf{r}_2) = \Psi_t(\mathbf{r}_1, \mathbf{r}_2)\, (\chi_\uparrow \otimes \chi_\downarrow + \chi_\downarrow \otimes \chi_\uparrow)/\sqrt{2},$$
$$\Psi(\mathbf{r}_1, \mathbf{r}_2) = \Psi_s(\mathbf{r}_1, \mathbf{r}_2)\, (\chi_\uparrow \otimes \chi_\downarrow - \chi_\downarrow \otimes \chi_\uparrow)/\sqrt{2}.$$

☐Now we want to calculate the expectation values of the total energy, $E = \langle \Psi | \hat{H} | \Psi \rangle$, for each of the above states. ▶Evaluate the energy expectation values by replacing eqns 11.56 and 11.57 inside the brackets, and by using the explicit expression for the Hamiltonian, $\hat{H}$, given in Exercise 4.5. Note that this derivation is very lengthy and requires handling a dozen matrix elements. ▶Show that, if we neglect the overlap, κ, and the crystal field term, γ (defined in eqn 4.23), then the expectation value, E_s, for the singlet and the expectation value, E_t, for all the triplets reduce to:

$$E_s = 2E_{1s} + W + J, \qquad (11.58)$$
$$E_t = 2E_{1s} + W - J, \qquad (11.59)$$

with

$$J = \int \left\{ \frac{1}{|\mathbf{r}_1 - \mathbf{r}_2|} - \frac{1}{2} \left[\frac{1}{|\mathbf{r}_1 - d\mathbf{u}_x/2|} + \frac{1}{|\mathbf{r}_1 + d\mathbf{u}_x/2|} + \frac{1}{|\mathbf{r}_2 - d\mathbf{u}_x/2|} + \frac{1}{|\mathbf{r}_2 + d\mathbf{u}_x/2|} \right] \right\}$$
$$\times \phi_A(\mathbf{r}_1)\phi_B(\mathbf{r}_2)\phi_A(\mathbf{r}_2)\phi_B(\mathbf{r}_1)\, d\mathbf{r}_1 d\mathbf{r}_2, \qquad (11.60)$$

and

$$W = \int d\mathbf{r}_1 d\mathbf{r}_2 \, \frac{|\phi_A(\mathbf{r}_1)|^2 |\phi_B(\mathbf{r}_2)|^2}{|\mathbf{r}_1 - \mathbf{r}_2|}.$$

☐The quantity J is called the *exchange integral*, since it contains both $\phi_A(\mathbf{r}_1)\phi_B(\mathbf{r}_2)$ and the same product with the electron variables exchanged, $\phi_A(\mathbf{r}_2)\phi_B(\mathbf{r}_1)$. An approximate calculation of the exchange integral for H_2 performed by Heitler and London (1927) yields:

$$J = e^{-2d} \left[\frac{5}{8} \left(1 + d + \frac{1}{3}d^2 \right)^2 - 2 \left(1 + 2d + \frac{4}{3}d^2 + \frac{1}{3}d^3 \right) \right].$$

▶Plot the exchange integral as a function of the H–H separation, d. ☐From the plot we observe that the exchange integral is a negative quantity for any d; therefore the singlet energy in eqn 11.58 is always smaller than the triplet energy in eqn 11.59. This indicates that the *singlet spinor* is the ground state of the H_2 molecule.
▶By using eqns 11.56 and 11.57 inside eqn 2.10 show that the electron charge density in the singlet and triple states are:

$$\text{singlet:} \quad n(\mathbf{r}) = \frac{1}{1 + \kappa^2} \left[|\phi_A(\mathbf{r})|^2 + |\phi_B(\mathbf{r})|^2 + 2\kappa\, \phi_A(\mathbf{r})\phi_B(\mathbf{r}) \right],$$

$$\text{triplet:} \quad n(\mathbf{r}) = \frac{1}{1 + \kappa^2} \left[|\phi_A(\mathbf{r})|^2 + |\phi_B(\mathbf{r})|^2 - 2\kappa\, \phi_A(\mathbf{r})\phi_B(\mathbf{r}) \right].$$

☐In the case of the singlet, since the wavefunction is symmetric in $\mathbf{r}_1$ and $\mathbf{r}_2$, we have an accumulation of charge (with respect to the free atoms) in between the protons. This stabilizes the molecule by reducing the internuclear repulsion. In the case of the triplet we have instead a depletion of charge, leading to the opposite effect. In practice the spin configuration determines the symmetry of the wavefunction, thereby altering the charge density and the electrostatics.
▶The difference between E_s and E_t provides a measure of the stability of the singlet state relative to the triplet. Using eqns 11.58 and 11.59 and the Heitler–London expression for J, calculate the singlet–triplet splitting in the H_2 molecule (the equilibrium bond length is 0.74 Å). ☐The splitting calculated by neglecting the overlap, κ, turns out to be rather inaccurate, since the 1s orbitals have substantial overlap when the protons are 0.74 Å apart ($\kappa = 0.57$). The complete calculation performed by Heitler and London (1927) gives $E_t - E_s = 10.5$ eV.

□If the two protons are far apart from each other it is also conceivable that two electrons may remain bound to the same proton. This is a highly unlikely situation due to the unfavourable energetics, but the hydrogen anion does H^- exist and is present in the atmosphere of the sun and other stars (Chandrasekhar, 1944). If we forget for a moment the presence of two protons, we may ask what the spinor of two electrons bound to the same proton would look like. Alternatively we could consider the case of the He atom studied in Exercise 2.6. The analogue of eqns 11.56 and 11.57 in this case is:

$$\Psi(\mathbf{r}_1, \mathbf{r}_2) = \phi_A(\mathbf{r}_1)\phi_A(\mathbf{r}_2).$$

►Show that any spin–triplet spinor with this spatial component must have $\phi_A(\mathbf{r}) = 0$. □This simple result shows that the only admissible spinor with both electrons occupying the same wavefunction is the spin-singlet. In other words, *two electrons described by the same wavefunction cannot have parallel spins*. This observation corresponds to the 'intuitive' notion of Pauli's exclusion principle as it is usually taught in introductory quantum mechanics courses.

In Exercise 11.5 we have learned that the spin configuration of a many-electron system plays a crucial role in the energetics. In the case of the H_2 molecule the ground state is a spin-singlet state, since this is much more stable than the spin-triplet state. It is very important to observe that the energy difference between singlet and triplet arises *even if the Hamiltonian does not contain any magnetic interactions*. In fact, the effects of the spin manifest themselves through the exclusion principle, and hence the spatial symmetry of the electron wavefunction. Since this symmetry has a direct effect on the electrostatics, we conclude that there is a *direct link between spin and Coulomb interactions in many-electron systems*. This observation is very general and provides the key to understanding magnetic order in solids.

In order to see the connection between spin and energetics more clearly it is common to construct a *reduced* Hamiltonian whose expectation values are precisely the singlet and triplet energies. We can do this by observing that the expectation values of the spin-product operator $\mathbf{S} \otimes \mathbf{S}$ are $\hbar^2/4$ and $-3\hbar^2/4$ in the cases of the triplet and singlet, respectively, as can be seen by an explicit evaluation. This suggests that we may construct the reduced Hamiltonian in the following way:

$$\hat{H}_{\text{red}} = E_0 - 2J\frac{\mathbf{S}}{\hbar} \otimes \frac{\mathbf{S}}{\hbar}, \qquad \text{with } E_0 = 2E_{1s} + W - J/2. \tag{11.61}$$

It is immediately verifiable that, when we take the expectation values of this Hamiltonian on the singlet or triplet spin states of H_2, we find precisely the energies E_s and E_t in eqns 11.58 and 11.59. Obviously, the reduced Hamiltonian does not contain the full wealth of information as the complete Hamiltonian in eqn 4.25 does, but it serves only the purpose of distinguishing singlet from triplets.

We can interpret eqn 11.61 heuristically as follows. Since J is negative in the case of H_2, the lowest energy is obtained when the product $\mathbf{S} \otimes \mathbf{S}$ is negative, i.e. when the spins are antiparallel (↑↓). In the language of solid-state physics we would say that the spins of H_2 are aligned in an 'antiferromagnetic' configuration.

Clearly, the *sign and magnitude of the exchange integral*, J, determines what kind of 'magnetic order' is the most stable in a system. For example, in the case of the O_2 molecule the situation is reversed. In fact in O_2 we have $J > 0$ and the ground-state configuration is a spin triplet. Once again, in the language of solid-state physics we would say that the spins in the ground state of O_2 are in a 'ferromagnetic' configuration (note this is an abuse of language since the spins of a gas of O_2 molecules are randomly oriented and the system is paramagnetic).

Reduced Hamiltonians like that in eqn 11.61 are not very useful for simple systems such as H_2, but can be very powerful tools when generalized to solids. In fact it is common to study the magnetic order in complex solids by using an extension of eqn 11.61 which reads as follows:

$$\hat{H} = -\frac{2}{\hbar^2} \sum_{\langle ij \rangle} J_{ij}\, \mathbf{S}_i \cdot \mathbf{S}_j, \qquad (11.62)$$

where it is intended that the operator $\mathbf{S}_i$ (or $\mathbf{S}_j$) acts only on the spin of atom i (or j). The sum is performed only on the nearest-neighbour atoms (this is indicated by the notation '$\langle ij \rangle$'), and the exchange integrals, J_{ij}, refer to pairs of atoms. Similarly to the previous discussion for the H_2 molecule, in the simplest situation of $J_{ij} = J_0 > 0$ the Hamiltonian will favour the parallel alignment of spins (ferromagnetic configuration). Conversely, if $J_{ij} = J_0 < 0$ then the Hamiltonian will favour the antiparallel alignment of spins (e.g. antiferromagnets). The Hamiltonian in eqn 11.62 is called the *Heisenberg Hamiltonian* (Ashcroft and Mermin, 1976) and forms the basis for a large body of current research on magnetism in insulators, e.g. in multiferroic compounds.

11.5 Spin in density functional theory

After a lengthy introduction to the notion of spin in Sections 11.1–11.4, it is now time to move towards the extension of DFT to magnetic materials. As we already have seen in Chapter 3, the key advantage of DFT is that we can replace extremely difficult calculations of the total energy via the many-body Schrödinger equation (eqn 2.24) by affordable calculations based on the electron density.

In Section 3.1 we have seen how the Hohenberg–Kohn theorem allows us to write the total energy, E, of the electrons in their ground state as a functional of the electron density, $n(\mathbf{r})$. Subsequently in Section 3.2 we discussed how to write single-particle equations to determine Kohn–Sham wavefunctions (eqn 3.12) and the electron density (eqn 2.44).

The generalization of DFT to include magnetism goes through the inclusion of special relativity, and proceeds along the same lines as described in Sections 11.1 and 11.2 for one electron only. Since the Dirac equation for one electron (eqn 11.6) is really a matrix differential equation in the four components of the Dirac spinor, it is natural to expect that the (scalar) electron density of standard DFT will be replaced in relativistic DFT by a four-component function.

This is indeed what was found by Rajagopal and Callaway (1973). These authors formulated DFT starting from the Dirac equation, and were able to prove that the *total energy of a system of electrons in their ground state is a unique functional of the 'relativistic 4-current $J_\mu(\mathbf{r})$'*. The details of how this '4-current' is constructed are not very important in our present discussion. The key point is that this new character in the plot, $J_\mu(\mathbf{r})$, conceals within itself three quantities, namely the electron density, $n(\mathbf{r})$, the electron spin density, $\mathbf{s}(\mathbf{r})$, and also the electron current density, mentioned in Section 5.5.

In contemporary DFT calculations for magnetic materials it is common to neglect the electron current density, which can be dealt with separately when one is interested in diamagnetic effects or electric polarization. The interested reader can find details on how to deal with the electron current density in the work by Vignale and Rasolt (1988). When we restrict ourselves to considering only the electron density and the spin density, by neglecting the current and the small components of the Dirac spinors (eqn 11.15), we usually speak of 'spin-density functional theory' (spin-DFT).

The extension of the Hohenberg–Kohn theorem to spin-DFT can be summarized schematically as follows (Kohn and Sham, 1965; von Barth and Hedin, 1972):

without magnetism: $\qquad n(\mathbf{r}) \xrightarrow{\ F\ } E \qquad E = F[n(\mathbf{r})]$

with magnetism: $\qquad n(\mathbf{r}), \mathbf{s}(\mathbf{r}) \xrightarrow{\ G\ } E \qquad E = G[n(\mathbf{r}), \mathbf{s}(\mathbf{r})]$

In other words, the total energy is now a functional of the electron density (as in the non-magnetic case), and of the spin density. Similarly to the case of non-magnetic DFT, where we construct the electron density as the sum of the densities of independent electrons (see eqn 2.44), in spin-DFT we generalize this concept by considering the sum of the charge and spin densities of *independent 2-spinors*, as given by eqns 11.34 and 11.39:

$$n(\mathbf{r}) = \sum_i \mathbf{\Psi}_i^\dagger(\mathbf{r})\mathbf{\Psi}_i(\mathbf{r}), \tag{11.63}$$

$$\mathbf{s}(\mathbf{r}) = \sum_i \mathbf{\Psi}_i^\dagger(\mathbf{r})\mathbf{S}(\mathbf{r})\mathbf{\Psi}_i(\mathbf{r}), \tag{11.64}$$

where the sum runs over the N lowest-energy spinors of the system in the case of N electrons. If we represent the 2-spinors using the scalar/vector notation introduced in eqn 11.33,

$$\mathbf{\Psi}_i(\mathbf{r}) = \phi_i(\mathbf{r}; 1)\,\chi_\uparrow + \phi_i(\mathbf{r}; 2)\,\chi_\downarrow,$$

then the previous equations yield the following expressions for charge and spin densities:

$$n(\mathbf{r}) = \sum_i |\phi_i(\mathbf{r}; 1)|^2 + \sum_i |\phi_i(\mathbf{r}; 2)|^2, \tag{11.65}$$

$$s_x(\mathbf{r}) = \frac{\hbar}{2}\, 2\,\mathrm{Re} \sum_i \phi_i^*(\mathbf{r}; 1)\phi_i(\mathbf{r}; 2), \tag{11.66}$$

$$s_y(\mathbf{r}) = \frac{\hbar}{2}\, 2\,\mathrm{Im} \sum_i \phi_i^*(\mathbf{r}; 1)\phi_i(\mathbf{r}; 2), \tag{11.67}$$

$$s_z(\mathbf{r}) = \frac{\hbar}{2} \sum_i \left[|\phi_i(\mathbf{r}; 1)|^2 - |\phi_i(\mathbf{r}; 2)|^2 \right]. \tag{11.68}$$

These expressions parallel those that were derived in eqns 11.41–11.43 for an isolated electron, the only difference being that now we sum over all the electrons. In particular, the notion of spin density, $\mathbf{s}(\mathbf{r})$, carries precisely the same meaning as illustrated in

Figure 11.1; that is, in spin-DFT we can associate an infinitesimal magnetic dipole moment, $s(\mathbf{r})d\mathbf{r}$, to every point in space where the electron density is non-zero.

In the standard DFT literature the quantities $n(\mathbf{r})$ and $s(\mathbf{r})$ are commonly written in terms of a compact object called the *density matrix*. As the name suggests, this is indeed a matrix of functions, and is defined as follows:

$$n_{\alpha\beta}(\mathbf{r}) = \sum_i \phi_i^*(\mathbf{r}; \alpha)\phi_i(\mathbf{r}; \beta),$$

with α and β integers taking the values 1 or 2 in order to identify the spinor components $\chi_\uparrow$ and $\chi_\downarrow$. Some quick algebra using the Pauli matrices in eqn 11.9 shows that the charge and spin densities are related to the density matrix as follows:

$$n(\mathbf{r}) = \sum_\alpha n_{\alpha\alpha}(\mathbf{r}), \qquad s(\mathbf{r}) = \frac{\hbar}{2}\sum_{\alpha\beta} n_{\alpha\beta}(\mathbf{r})\boldsymbol{\sigma}_{\alpha\beta},$$

where $\boldsymbol{\sigma}_{\alpha\beta}$ indicate the (α, β) elements of the Pauli matrices. Using the newly defined density matrix, the most general statement of the spin-DFT theory is that the *total energy of the electrons in their ground state is a functional of the density matrix*, i.e. $E = G[n_{\alpha\beta}(\mathbf{r})]$.

At this point the entire machinery of the Kohn–Sham formulation of DFT discussed in Section 3.2 can be adapted to the case of spin-DFT. In particular, one can derive Kohn–Sham equations using the property that the density matrix in the ground state, $n_{\alpha\beta}^0$, minimizes the energy functional, in complete analogy with eqn 3.11:

$$\left.\frac{\delta G[n_{\alpha\beta}]}{\delta n_{\alpha\beta}}\right|_{n_{\alpha\beta}^0} = 0. \tag{11.69}$$

As shown by von Barth and Hedin (1972), this minimum principle (together with the condition that we have exactly N electrons and that the spinors are orthonormal) leads to the Kohn–Sham equations in spin-DFT. If we focus on the simplest situation where there is no external magnetic field, then the spin-DFT analogue of eqn 3.12 reads:

$$\left[-\frac{1}{2}\nabla^2 + V_n(\mathbf{r}) + V_H(\mathbf{r})\right]\phi_i(\mathbf{r}; \alpha) + \sum_\beta v_{\alpha\beta}^{xc}(\mathbf{r})\phi_i(\mathbf{r}; \beta) = \varepsilon_i\phi_i(\mathbf{r}; \alpha), \quad \text{with } \alpha = 1, 2,$$

$$\tag{11.70}$$

where we used atomic Hartree units. The exchange and correlation potential, $v_{\alpha\beta}^{xc}(\mathbf{r})$, in this case is a matrix of functions, and is defined as in eqn 3.13:

$$v_{\alpha\beta}^{xc}(\mathbf{r}) = \left.\frac{\delta E_{xc}}{\delta n_{\alpha\beta}}\right|_{n_{\alpha\beta}(\mathbf{r})}. \tag{11.71}$$

Various forms and properties of exchange and correlation functionals for spin-DFT are analysed by Vosko *et al.* (1980), and will not be discussed here, in order to keep the presentation as light as possible.

The Kohn–Sham equations (eqn 11.70) for spin-DFT look rather abstract, and the physical meaning of the newly introduced $v_{\alpha\beta}^{xc}(\mathbf{r})$ is not immediately apparent. In order to shed some light on the nature of this quantity it is helpful to proceed as follows. We define:

$$V_{xc} = \frac{v_{11}^{xc} + v_{22}^{xc}}{2}, \quad B_x^{xc} = \frac{v_{12}^{xc} + v_{21}^{xc}}{2\mu_{\mathrm{B}}}, \quad B_y^{xc} = i\frac{v_{12}^{xc} - v_{21}^{xc}}{2\mu_{\mathrm{B}}}, \quad B_z^{xc} = \frac{v_{11}^{xc} - v_{22}^{xc}}{2\mu_{\mathrm{B}}},$$

so that the exchange and correlation matrix can be written as (Kübler *et al.*, 1988):

$$v_{\alpha\beta}^{xc}(\mathbf{r}) = V_{xc}(\mathbf{r})\mathbf{1} + \mu_{\mathrm{B}}\boldsymbol{\sigma} \cdot \mathbf{B}_{xc}(\mathbf{r})$$

(note that in Hartree units $\mu_{\mathrm{B}} = 1/2$). This apparently complicated manipulation allows us to recast eqn 11.70 into the elegant and compact form:

$$\left[-\frac{1}{2}\nabla^2 + V_{\mathrm{n}}(\mathbf{r}) + V_{\mathrm{H}}(\mathbf{r}) + V_{xc}(\mathbf{r}) + \mu_{\mathrm{B}}\boldsymbol{\sigma} \cdot \mathbf{B}_{xc}(\mathbf{r}) \right] \mathbf{\Psi}_i(\mathbf{r}) = \varepsilon_i \mathbf{\Psi}_i(\mathbf{r}). \qquad (11.72)$$

This expression is entirely analogous to eqn 11.32, which was derived for an isolated electron. Here the presence of many electrons generates an *effective* magnetic field, $\mathbf{B}_{xc}(\mathbf{r})$, called the 'exchange and correlation magnetic field'. This extra field tends to align the spin of the electrons and may drive magnetic order. We could think of this additional interaction as a conceptual tool for including, within the framework of DFT, the notion of spin and the exclusion principle discussed in Sections 11.1–11.4.

11.6   Examples of spin-DFT calculations

Practical spin-DFT calculations can be grouped in two broad categories, usually referred to as *collinear* and *non-collinear* calculations. The formalism discussed in the previous section refers to the most general case of non-collinear spin-DFT, since the Kohn–Sham equations (eqn 11.72) involve spinors, $\mathbf{\Psi}_i(\mathbf{r})$, with two components each, and the spin density, $\mathbf{s}(\mathbf{r})$, can take any direction.

In collinear spin-DFT calculations one makes the *approximation* that the spin density can be oriented along only one direction. This situation is formally similar to that of the free electron in Exercise 11.3; therefore it is possible to choose the Kohn–Sham spinors in such a way that there is only one non-zero component. Similarly to Exercise 11.3 we can label the spinors as:

$$\mathbf{\Psi}_{i\uparrow} = \begin{bmatrix} \psi_{i\uparrow}(\mathbf{r}) \\ 0 \end{bmatrix} \quad \text{and} \quad \mathbf{\Psi}_{i\downarrow} = \begin{bmatrix} 0 \\ \psi_{i\downarrow}(\mathbf{r}) \end{bmatrix},$$

and define the 'up' and 'down' electron densities:

$$n_\uparrow(\mathbf{r}) = \sum_{i\uparrow} |\phi_{i\uparrow}(\mathbf{r})|^2, \quad n_\downarrow(\mathbf{r}) = \sum_{i\downarrow} |\phi_{i\downarrow}(\mathbf{r})|^2.$$

With these definitions the spin polarization of eqn 11.40 is directed along the z axis and its magnitude is simply given by $n_\uparrow(\mathbf{r}) - n_\downarrow(\mathbf{r})$ (in units of μ_{B}).

From a historical point of view it is interesting to note that collinear spin-DFT calculations pre-date the advent of non-collinear DFT (Kübler *et al.*, 1988). The slower

Table 11.2 Number of complex scalar functions per electron (e.g. $f(\mathbf{r})$ with f complex-valued) that need to be evaluated in the case of spin-unpolarized DFT calculations, spin-DFT calculations with collinear spins, and in non-collinear spin-DFT calculations.

	spin-unpolarized	collinear spins	non-collinear spins
functions per electron	1/2	1	2

development of non-collinear DFT is connected with the fact that these calculations are significantly more demanding in terms of computational resources.

In order to appreciate the increasing complexity of DFT calculations as we move from non-magnetic or 'spin-unpolarized' DFT to collinear spin-DFT, up to non-collinear spin-DFT, we may consider Table 11.2. In all the spin-unpolarized calculations discussed in Chapters 3–10 we implicitly assumed that two electrons with opposite spin are described by the same spatial wavefunction. In collinear spin-DFT calculations we need one spatial function for spin-up electrons and one for spin-down electrons, and hence one spatial function per electron. In non-collinear spin-DFT each electron is described by two functions. As the computational cost of standard DFT calculations is roughly proportional to the cube of the number of such functions, we can expect the execution time of a non-collinear DFT calculation to be $\sim 4^3 = 64$ times longer than the corresponding non-magnetic calculation.

The vast majority of contemporary DFT calculations for magnetic materials is performed under the approximation of collinear spins. This approximation leads to a good description of many basic properties of standard ferro- or antiferromagnetic materials, such as permanent magnetic moments and phase diagrams (Kübler, 2000). Non-collinear DFT calculations have become increasingly popular during the past two decades. These calculations are necessary for describing more exotic phases such as spin spirals, spin glasses and magnetic excitations such as spin waves. Non-collinear spins are also a key ingredient for investigating 'spin-orbit coupling', i.e. the interaction between the electron spins and the atomic orbits in magnetic atoms. This is important, for example, in the study of magnetic anisotropy.

Exercise 11.6 One of the simplest examples of collinear spin-DFT calculations is the study of the energetics of the H_2 molecule in various spin configurations. The results of calculations within DFT/LDA for H_2 are shown in Figure 11.2. ▶With reference to Table 11.2, explain why a spin-unpolarized DFT calculation of H_2 must correspond to the spin-singlet configuration. □From the left panel of Figure 11.2 we see that the spin-singlet configuration of H_2 is always lower in energy than the spin-triplet. From this we conclude that the ground state of this molecule is a singlet, while the triplet is an excited state. However, we know that DFT addresses only the energy of the ground state; therefore it seems surprising that we can calculate the DFT total energy in the triplet state. ▶Read paragraph II.C of Gunnarsson and Lundqvist (1976), and then explain why DFT calculations of the triplet state of H_2 are meaningful and can be expected to be as accurate as for the singlet. ▶By reading the binding energies for the singlet and triplet states obtained using DFT/LDA from Figure 11.2, calculate the exchange integral, J, for a few values of the H–H interatomic separation using the Heisenberg Hamiltonian (eqn 11.61). ▶Now compare your results with the values of J shown in the right panel of the same figure.

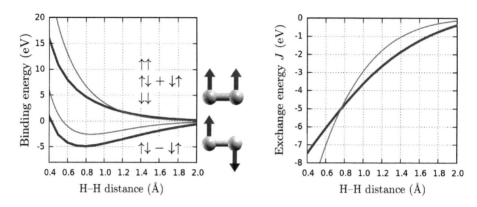

Fig. 11.2 Left panel: the binding energy of H_2 calculated using DFT/LDA as a function of H–H distance (thick blue lines). The top curves refer to spin-polarized calculations (triplet ground state); the bottom curves refer to spin-unpolarized calculations (singlet ground state). The thin red lines indicate the results of Heitler and London (1927) for comparison. The ball-and-stick models give an intuitive representation of the spin configurations in these calculations. Right panel: the exchange integral, J (see Section 11.4), for the two electrons in the H_2 molecule as a function of H–H distance (thick blue line), calculated within DFT/LDA. The thin red line is the corresponding result from Heitler and London (1927).

DFT calculations for a given spin configuration can be performed by minimizing the total energy under the constraint that the total magnetization takes a certain value (Dederichs *et al.*, 1984). In this case the calculations in the triplet state correspond to a fixed total magnetization of $2\mu_B$.

▶Extract the equilibrium bond length and the binding energy of H_2 from the left panel of Figure 11.2, and compare these values with experimental data. ▶From the results shown in Figure 11.2 establish whether we should expect a gas of hydrogen molecules at ambient conditions to be paramagnetic or non-magnetic.

11.6.1   Itinerant spins: bcc Fe

As an example of of spin-DFT calculations for solids we consider α-Fe. Iron is the dominant component of the Earth's core (Section 1.2.3) and one of the most common elements in the crust, and has been in use since about 1,000 BC. The most common uses of iron today are in the steel industry. Indeed, the vast majority of steels are iron alloys containing small amounts (around 1%) of carbon, and are essentially ubiquitous, being used in virtually every building either as a skeleton or in reinforced concrete. The stable form of iron at ambient condition is α-Fe, a body-centred cubic (bcc) crystal with a lattice parameter $a = 2.86$ Å. At the microscopic level α-Fe is a ferromagnet, i.e. it possesses a permanent magnetic moment in each of its magnetic domains. However, at the macroscopic level ferromagnetism manifests itself only below the Curie temperature, $T_C = 1,043$ K, when the magnetic moments of each domain tend to orient along the same direction. Within this context DFT calculations are most useful to address the microscopic mechanisms of the ferromagnetic order, as opposed to the interdomain interactions which happen on length scales of the order of many microns.

Figure 11.3 shows the results of DFT/LDA calculations for bcc Fe. The electronic configuration of Fe is $[\text{Ar}]4s^23d^6$; therefore we have eight valence electrons in the unit cell, and the calculations take only a fraction of a minute on a standard laptop computer. The calculations indicate that bcc Fe has a lattice parameter of 2.76 Å, and that the ferromagnetic phase with all the spins aligned is more stable than the non-magnetic phase by 0.39 eV/atom. From this we deduce that the ground state of bcc Fe is ferromagnetic, as shown schematically in the ball-and-stick model in Figure 11.3. The calculated magnetic moment corresponds to 2.14 μ_B per Fe atom, in good agreement with the experimental value of 2.22 μ_B extracted from measurements of the hysteresis cycle (Kittel, 1976).

From the electronic density of states shown in Figure 11.3 we see that in the ferromagnetic configuration there are more electrons with 'spin up' (blue) than

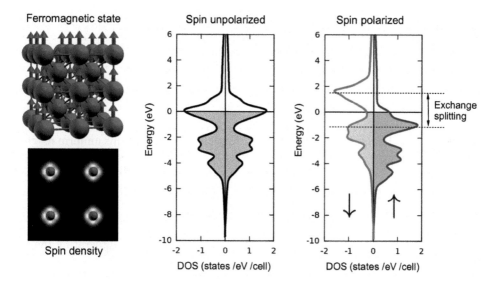

Fig. 11.3 The top left panel shows a schematic representation of the spins in the ferromagnetic configuration of iron: all the spins are oriented in the same direction. The bottom left panel shows the magnetization density obtained from DFT/LDA calculations, plotted on a plane cutting through one face of the cubic structure. In this plot we can recognize the contribution of the $3d$ states of Fe. The DOS plots correspond to a non-magnetic calculation (centre) and to a spin-polarized calculation (right). In spin-DFT calculations it is common to compare the DOS for spin-up and spin-down electrons in the same plot, as in this figure. In the spin-polarized DOS we see that the exchange and correlation magnetic field stabilizes the spin-up electrons by lowering their energy, while the spin-down electrons become more energetic. This is similar to what we have seen in Exercise 11.3 for the case of a free electron. More details on DFT calculations of the magnetic properties of Fe can be found in Moruzzi *et al.* (1986).

electrons with 'spin down' (red). It is precisely such majority of spin-up electrons (spin polarization) which gives Fe its permanent magnetic moment. Since the spin polarization is associated with conduction electrons we speak of *itinerant magnetism* in Fe.

The spin-polarized density of states shown in Figure 11.3 suggests a very simple model of ferromagnetism in iron. In fact from the figure we realize that, upon the onset of spin polarization, the DOS for the spin-up electrons has been shifted downwards with respect to the unpolarized case. Similarly, the DOS for the spin-down electrons has moved upwards. The splitting between the DOS of spin-up and spin-down electrons is usually referred to as *exchange splitting*, although as we will see below both exchange and correlation effects are involved.

From eqn 11.72 we can see that, in collinear DFT calculations, the only difference between the Kohn–Sham Hamiltonians for spin-up or spin-down electrons is that the exchange and correlation magnetic field appears with different signs, $\pm\mu_B B_{xc}(\mathbf{r})$. Within a very crude approximation we can think of the electrons in iron as if described by two uniform electron gases, with densities $n_\uparrow$ and $n_\downarrow$. In this case B_{xc} is just a constant depending on the value of the spin polarization, $m = n_\uparrow - n_\downarrow$: $B_{xc} = B_{xc}(m)$. For example, if we only consider the LDA exchange as in eqn 3.23, it can be shown that (von Barth and Hedin, 1972):

$$\mu_B B_x = -\left(\frac{3}{16\pi}\right)^{\frac{1}{3}} n^{\frac{1}{3}} \left[\left(1 + \frac{m}{n}\right)^{\frac{1}{3}} - \left(1 - \frac{m}{n}\right)^{\frac{1}{3}}\right], \qquad (11.73)$$

with $n = n_\uparrow + n_\downarrow$ being the uniform electron density. The effect of this uniform magnetic field on the energy of the electrons is identical to what we have seen in Exercise 11.3. As a result the DOS of spin-up and spin-down electrons can be obtained from the DOS of the non-magnetic calculation, $\rho(E)$, by shifting the energy axis:

$$\rho_\uparrow(E) = \rho[E + \mu_B B_{xc}(m)], \qquad \rho_\downarrow(E) = \rho[E - \mu_B B_{xc}(m)].$$

Using this approximation we can express the spin polarization by integrating the DOS up to the Fermi level:

$$m = \int_{-\infty}^{\epsilon_F} [\rho_\uparrow(E) - \rho_\downarrow(E)] dE = -\int_{\epsilon_F - \mu_B B_{xc}(m)}^{\epsilon_F + \mu_B B_{xc}(m)} \rho(E) dE.$$

If we approximate the density of states inside the integral using its value at the Fermi level, ρ_F, we obtain a non-linear equation for the spin polarization:

$$m = -2\rho_F \mu_B B_{xc}(m). \qquad (11.74)$$

This equation admits the solution $m = 0$, which corresponds to the non-magnetic DFT calculation. Possible solutions with $m \neq 0$ (which may exist since B_{xc} is a non-linear function of m, see eqn 11.73) correspond instead to ferromagnetic states. This simple analysis shows that the magnetization of ferromagnets depends crucially on (i) the density of states at the Fermi level and (ii) the exchange and correlation magnetic field (Gunnarsson, 1976; Zeller, 2006).

Equation 11.74 is a simplified version of what is commonly known as the *Stoner criterion* (Stoner, 1938), which provides a very general framework for understanding itinerant magnetism (see for instance Blundell, 2001).

Exercise 11.7 ►By reading the exchange splitting in bcc Fe from the spin-polarized DOS in Figure 11.3, calculate the mean value of the exchange and correlation magnetic field, B_{xc}. ►Calculate the spontaneous magnetization in bcc Fe by using eqn 11.74 and by reading the density of states at the Fermi level from Figure 11.3. Show that this result is rather inaccurate when compared with the complete DFT calculation (yielding 2.14 μ_B), and explain the reason for such inaccuracy.

11.6.2 Localized spins: MnO

As another example of spin-DFT calculations we consider the case of manganous oxide, MnO. MnO is commonly used for the production of Mn salts and as an additive in fertilizers, and is naturally occurring as the mineral manganosite.

To a first approximation the structure of MnO is essentially the same as that of rock-salt, NaCl, as shown in Figure 11.4. The magnetic element in this oxide is Mn, owing to its electronic configuration, $[Ar]4s^2 3d^5$. In fact, according to Hund's rules (Kittel, 1976), an isolated Mn atom will have all the spins of its five $3d$ electrons aligned, and hence a magnetic moment of $5\mu_B$. Within the simplest intuitive picture, we can imagine the Mn atoms to be isolated from each other by the O atoms in between (Figure 11.4); therefore we may expect a maximum magnetic moment in the solid corresponding to $5\mu_B$ per formula unit.

However, at variance with the case of α-Fe discussed in the previous section, MnO does not become ferromagnetic at low temperature. Quite on the contrary, below the Néel temperature, $T_N = 118$ K, MnO becomes *antiferromagnetic* (Kittel, 1976). In this configuration, for each Mn atom with total spin in a given direction, there is another Mn atom with total spin in the opposite direction, in such a way that the total magnetization vanishes.

The identification of the magnetic structure of antiferromagnets can be performed using neutron diffraction, exploiting the fact that neutrons have a spin magnetic moment (Section 8.1.2). The magnetic phase of MnO observed in experiment below the Néel temperature is referred to as type-II structure, and consists of spins aligned parallel to each other within (111) planes, and antiparallel between each plane, as shown in Figure 11.4.

By performing standard DFT calculations for MnO, one obtains a type-II antiferromagnetic ground state as in experiment. The calculated magnetic moment per Mn atom, 3.72–4.15 μ_B, depending on the exchange and correlation functional adopted (Dufek *et al.*, 1994), is also in reasonable agreement with the experimental value of 4.58 μ_B (Cheetham and Hope, 1983).

An intuitive understanding of the magnetic properties of MnO and similar systems requires the development of atomistic models which are rather different from the Stoner's model discussed for iron in Section 11.6.1. In fact, in this case we are dealing with an *insulator* (Figure 11.4); therefore the very concept of DOS at the Fermi level is not meaningful.

In order to rationalize the magnetic properties of oxides it is common to construct simplified models which relate the energetics of the system to the underlying spin configuration. For example, Pask *et al.* (2001) adapted the Heisenberg Hamiltonian in eqn 11.62 to MnO by including exchange interactions between both nearest-neighbour and next-nearest-neighbour Mn atoms:

$$\hat{H}_{\mathrm{red}} = E_0 + \frac{1}{\hbar^2} \sum_{\langle ij \rangle} J_1 \, \mathbf{S}_i \cdot \mathbf{S}_j + \frac{1}{\hbar^2} \sum_{[ij]} J_2 \, \mathbf{S}_i \cdot \mathbf{S}_j. \tag{11.75}$$

In this expression the sums run over pairs of nearest-neighbour (indicated by '$\langle ij \rangle$') or next-nearest-neighbour (indicated by '$[ij]$') Mn atoms, E_0, J_1 and J_2 are parameters to be determined, and the spins, $\mathbf{S}_i$, $\mathbf{S}_j$, are allowed to point up or down in a common direction. Note that, in eqn 11.75, Pask *et al.* use a different sign convention with respect to eqn 11.62. By requiring that $\hat{H}_{\mathrm{red}}$ reproduces the total energy obtained from DFT in various spin configurations, it is possible to determine the exchange parameters J_1 and J_2. These parameters are subsequently used to develop an atomic-scale understanding of magnetism in MnO.

Exercise 11.8 We want to extract the exchange parameters, J_1 and J_2, of the reduced Hamiltonian in eqn 11.75 by using the total energies calculated using spin-DFT by Pask *et al.* (2001). Since there are three unknown parameters, we need three total-energy calculations in inequivalent spin configurations. As shown in Figure 11.4, the simplest spin textures that one can consider for this purpose are the ferromagnetic phase (FM), the type-I antiferromagnetic phase (AF-I) and the type-II antiferromagnetic phase (AF-II). In the FM phase all the Mn

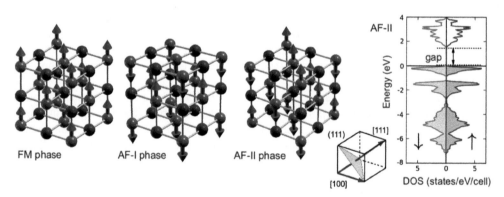

Fig. 11.4 Possible magnetic phases of MnO. In the ferromagnetic phase (FM) all the spins are parallel. In the type-I antiferromagnetic phase we have (001) planes where the Mn atoms have parallel spins, and adjacent planes have antiparallel spins. The type-II antiferromagnetic phase exhibits parallel spins within (111) planes, and antiparallel spins in between the planes. Here purple and red denote Mn and O atoms, respectively. The (111) plane is shown in the small inset. The plot on the right is the DOS of the AF-II phase. Contrary to Figure 11.3 here the spin-up and spin-down DOS are identical (although electrons with opposite spins are localized on different atoms). The shading indicates the occupied electronic states. We can see that MnO exhibits a band gap; therefore this system is an insulator.

spins point along the same direction, similarly to the case of iron in Section 11.6.1. In the AF-I phase, we see from Figure 11.4 that (001) planes of spin-up electrons alternate with (001) planes of spin-down electrons. Similarly, in the AF-II phase the (111) planes contain same-spin Mn atoms, and adjacent planes have antiparallel spins.

The DFT total energies calculated by Pask *et al.* (2001) for each of these three phases are:

	FM	AF-I	AF-II
Energy (meV/MnO formula unit)	167.4	86.4	0.0

(here the total energy is given relative to that of the AF-II phase).

▶Show that in the MnO structure in Figure 11.4 there are six pairs of nearest-neighbour Mn atoms per formula unit. □Hint: One possible way to proceed is to count how many lines connect nearest-neighbour Mn atoms in one of the ball-and-stick models in the figure. Lines on the faces of the cube are shared between adjacent cubes and should be counted 1/2. Then we divide this number by the number of Mn atoms in the structure, with each atom weighted 1/2 if on the surface, and 1/8 if on the corner. ▶In a similar spirit as above, show that there are three pairs of next-nearest-neighbour Mn atoms per formula unit. ▶Using the previous results, show that in the FM phase (where all the Mn spins are along the same direction and have magnitude S), eqn 11.75 gives the total energy:

$$E_{\text{FM}} = E_0 + (6J_1 + 3J_2)(S/\hbar)^2.$$

▶Now we consider the AF-II phase in Figure 11.4. Show that in this case, since half of the nearest-neighbour pairs have parallel spins, and half have antiparallel spins, only next-nearest-neighbour pairs contribute to the energy and we have:

$$E_{\text{AF-II}} = E_0 - 3J_2\,(S/\hbar)^2.$$

▶Using similar reasoning as above, show that for the antiferromagnetic phase, AF-I, eqn 11.75 yields:

$$E_{\text{AF-I}} = E_0 + (-2J_1 + 3J_2)(S/\hbar)^2.$$

▶Calculate the values of J_1 and J_2 using the expressions for E_{FM}, $E_{\text{AF-II}}$ and $E_{\text{AF-I}}$ just obtained, and the DFT total energies in the table. For definiteness we can use $S = 5\hbar/2$ as for isolated Mn atoms. You should find $J_1 = 1.6$ meV and $J_2 = 2.8$ meV. Since J_1 and J_2 are both positive, the energy of the system is minimized by an antiferromagnetic arrangement of the spins, in agreement with the table. ▶Try to explain why the calculated exchange parameters are so much smaller than in the case of the H_2 molecule discussed in Exercise 11.6.

The exchange parameters, J_1 and J_2, determined in Exercise 11.8 are referred to as *direct exchange* coupling and *super-exchange* coupling. The former has to do with the 'direct' exchange interaction between nearest-neighbour Mn atoms, similarly to the H_2 molecule of Exercise 11.6. The latter has to do with the interaction of Mn spins when there is an O atom in the way. In this case the Mn atoms are too far apart for the standard exchange integral to be significant, and the interaction takes place via the p orbitals of the bridging O atoms. As a curiosity, the notion of super-exchange was developed precisely in order to explain the antiferromagnetic phase of MnO (Kramers, 1934; Anderson, 1950).

The strategy of *mapping* the results of spin-DFT calculations onto simpler model Hamiltonians such as in eqn 11.75 is very useful for the study of *magnetic phase diagrams*. In fact, since such diagrams describe statistical averages over large ensembles, it is practically hopeless to address these aspects by brute-force DFT calculations. In these cases it is advantageous to use DFT as a 'seed' to generate the parameters of a model Hamiltonian. Subsequently one studies such model Hamiltonian using specialized methods of computational statistical mechanics (see for example Ferrenberg and Swendsen, 1988).

The study of the phase diagram corresponding to the fcc Heisenberg model with first and second neighbours in eqn 11.75 has been carried out by Phani *et al.* (1980). In this case Monte Carlo calculations indicate that the Néel temperature for the transition to the AF-II phase corresponds to $k_B T_N \approx 8J_1(S/\hbar)^2$. If we use the exchange parameter $J_1 = 1.6$ meV from Exercise 11.8, we find a Néel temperature of 930 K.

This result is clearly in disagreement with the measured transition temperature of 118 K; however, we should not be too surprised. In fact, the relevant interaction energies are so small ($\sim$1 meV) that even the smallest details can make a difference. For example, tiny changes in the lattice parameters, possible spin–orbit interactions, the choice of the DFT exchange and correlation functional, the inclusion of an additional term in the Heisenberg Hamiltonian: all these factors may play a role at this energy scale. As a qualitative indication of the sensitivity of T_N to the exchange parameters, we note that using $J_1 = J_2 = 0.5$ meV we would obtain $T_N \approx 130$ K, in much better agreement with experiment.

These considerations indicate that the prediction of critical temperatures in magnetic systems from first principles is very challenging, precisely because very small energies are involved. Much work is currently underway in order to achieve predictive power in the study of magnetic phase diagrams starting from DFT calculations.

As a closing remark we point out that, within the research community specializing in condensed matter, MnO and similar transition metal monoxides, such as FeO, NiO and CoO, are important since they constitute the prototypes of so-called *Mott–Hubbard* insulators. In this class of systems the localized nature of the $3d$ orbitals renders DFT/LDA and similar approaches inadequate. This can intuitively be understood by recalling that the LDA is based on the assumption that the electron density is a slowly varying function of the space coordinate (Section 3.3.2). Clearly, this assumption breaks down in the presence of localized electronic states such as the d electrons in transition metal oxides.

The development of computational methods for extending the scope of DFT calculations to Mott–Hubbard systems forms an entire research area on its own right, and is beyond the scope of this presentation. The interested reader will find more details in the vast literature on this subject, from the 'DFT+U' method (Anisimov *et al.*, 1991) to the 'dynamical mean-field theory' (Georges *et al.*, 1996; Kotliar *et al.*, 2006).

Appendix A
Derivation of the Hartree–Fock equations

In this appendix we go through the derivation of the Hartree–Fock equations (eqns 2.57 and 2.60 of Section 2.8). In order to keep the formalism as simple as possible we consider the case of two electrons, and for the time being we ignore the spin. The generalization of the following derivation to an arbitrary number of electrons carrying a spin degree of freedom will be indicated at the end.

We start by rewriting the many-body Schrödinger equation, eqn 2.28, for the particular case of $N = 2$ electrons:

$$\left[\hat{H}_0(\mathbf{r}_1) + \hat{H}_0(\mathbf{r}_2) + \frac{1}{|\mathbf{r}_1 - \mathbf{r}_2|}\right]\Psi = E\,\Psi, \tag{A.1}$$

where the single-particle Hamiltonian, $\hat{H}_0$, is the same as in eqn 2.31. The central idea of the Hartree–Fock method is to look for solutions, Ψ, of this equation in the form of a Slater determinant. Below we report the definition of a Slater determinant given in eqn 2.41, and for convenience we indicate the single-particle wavefunctions by ψ_1 and ψ_2:

$$\Psi(\mathbf{r}_1, \mathbf{r}_2) = \frac{1}{\sqrt{2}}\left[\psi_1(\mathbf{r}_1)\psi_2(\mathbf{r}_2) - \psi_1(\mathbf{r}_2)\psi_2(\mathbf{r}_1)\right]. \tag{A.2}$$

At this stage we do not know ψ_1 and ψ_2, and in order to proceed we need additional information. One piece of information that we can use is that the two-body wavefunction Ψ in eqn A.2 is not just any solution of eqn A.1, but the solution with the lowest energy, E, i.e. the electronic ground state.

In order to find the functions ψ_1 and ψ_2 which minimize the total energy, the first step is to write E as an *explicit functional* of the wavefunctions. This is achieved by combining eqns 2.53, A.1 and A.2. After a few tedious but straightforward manipulations we obtain:

$$E = \langle\psi_1|\hat{H}_0|\psi_1\rangle\langle\psi_2|\psi_2\rangle + \langle\psi_2|\hat{H}_0|\psi_2\rangle\langle\psi_1|\psi_1\rangle - \langle\psi_1|\hat{H}_0|\psi_2\rangle\langle\psi_2|\psi_1\rangle - \langle\psi_2|\hat{H}_0|\psi_1\rangle\langle\psi_1|\psi_2\rangle$$
$$+ \int d\mathbf{r}_1 d\mathbf{r}_2 \frac{|\psi_1(\mathbf{r}_1)|^2|\psi_2(\mathbf{r}_2)|^2}{|\mathbf{r}_1 - \mathbf{r}_2|} - \int d\mathbf{r}_1 d\mathbf{r}_2 \frac{\psi_1^*(\mathbf{r}_1)\psi_2^*(\mathbf{r}_2)\psi_1(\mathbf{r}_2)\psi_2(\mathbf{r}_1)}{|\mathbf{r}_1 - \mathbf{r}_2|}, \tag{A.3}$$

where the Dirac notation introduced in Chapter 2 has been used.

Equation A.3 simplifies considerably if we require that the functions $\psi_1(\mathbf{r})$ and $\psi_2(\mathbf{r})$ are orthonormal, i.e. normalized and orthogonal. This requirement corresponds to setting:

$$\langle\psi_1|\psi_1\rangle = \langle\psi_2|\psi_2\rangle = 1, \text{ and } \langle\psi_1|\psi_2\rangle = \langle\psi_2|\psi_1\rangle = 0. \tag{A.4}$$

These conditions automatically imply that our trial solution, Ψ, is correctly normalized, as discussed in Section 2.6. Using eqn A.4 inside eqn A.3 we find immediately:

$$E = \int d\mathbf{r}\,\psi_1^*(\mathbf{r})\hat{H}_0(\mathbf{r})\psi_1(\mathbf{r}) + \int d\mathbf{r}\,\psi_2^*(\mathbf{r})\hat{H}_0(\mathbf{r})\psi_2(\mathbf{r})$$
$$+ \int d\mathbf{r}_1 d\mathbf{r}_2 \frac{\psi_1^*(\mathbf{r}_1)\psi_2^*(\mathbf{r}_2)\psi_1(\mathbf{r}_1)\psi_2(\mathbf{r}_2)}{|\mathbf{r}_1 - \mathbf{r}_2|} - \int d\mathbf{r}_1 d\mathbf{r}_2 \frac{\psi_1^*(\mathbf{r}_1)\psi_2^*(\mathbf{r}_2)\psi_1(\mathbf{r}_2)\psi_2(\mathbf{r}_1)}{|\mathbf{r}_1 - \mathbf{r}_2|}. \tag{A.5}$$

Equation A.5 shows how the energy E is a *functional* of ψ_1 and ψ_2, in short $E = E[\psi_1, \psi_2]$. Now we can search for the functions ψ_1 and ψ_2 which minimize this functional. In principle we could do so by requiring that the functional derivatives of E with respect to ψ_1 and ψ_2 are equal to zero (see Exercise 3.4 for the definition of functional derivative):

$$\frac{\delta E}{\delta \psi_1} = 0, \quad \frac{\delta E}{\delta \psi_2} = 0. \tag{A.6}$$

This is completely analogous to the calculation of the minima and maxima of a function of several variables in standard calculus. However, here the minimization procedure is complicated by the fact that our functions need to satisfy the constraints in eqn A.4. In order to deal effectively with those constraints it is convenient to use the method of Lagrange multipliers (Nocedal and Wright, 1999). In this method one introduces a new functional which automatically incorporates the constraints:

$$L[\psi_1, \psi_2, \lambda_{11}, \ldots, \lambda_{22}] = E[\psi_1, \psi_2] - \sum_{ij} \lambda_{ij}\left[\langle\psi_i|\psi_j\rangle - \delta_{ij}\right], \tag{A.7}$$

where the constants $\lambda_{11}, \lambda_{12}, \lambda_{21}$ and λ_{22} are unknown. The replacement of E by L is sensible since the term in the square brackets is required to vanish by eqn A.4. The constants λ_{ij} are called *Lagrange multipliers*, and can be considered new independent variables. In Lagrange's method the *constrained* minimization problem defined by eqns A.6 and A.4 is replaced by the following *unconstrained* minimization problem:

$$\frac{\delta L}{\delta \psi_i} = 0, \quad i = 1, 2, \qquad \frac{\delta L}{\delta \lambda_{ij}} = 0, \quad i, j = 1, 2. \tag{A.8}$$

It can immediately be verified that the functional derivatives with respect to the Lagrange multipliers yield precisely the constraints in eqn A.4.

Before proceeding we note that the derivatives with respect to ψ_i and ψ_i^* are not independent; however, they can effectively be treated as such. This is a subtle point which has to do with the fact that the Hamiltonian is Hermitian, and a clarification on this aspect can be found in Chapter 2.8 of Bransden and Joachain (1983). In the following, as a matter of convenience, we evaluate the derivatives with respect to ψ_i^*. Let us calculate as an example the functional derivative of the first term in eqn A.5, which we call G for definiteness:

$$G = \int d\mathbf{r}\,\psi_1^*(\mathbf{r})\hat{H}_0(\mathbf{r})\psi_1(\mathbf{r}).$$

Using the definition of functional derivative given in eqn 3.24, we find:

$$\int d\mathbf{r}\, h(\mathbf{r}) \frac{\delta G}{\delta \psi_1^*} = \frac{d}{d\epsilon} \int d\mathbf{r}\, [\psi_1^*(\mathbf{r}) + \epsilon h(\mathbf{r})]\, \hat{H}_0(\mathbf{r})\psi_1(\mathbf{r}) = \int d\mathbf{r}\, h(\mathbf{r})\hat{H}_0(\mathbf{r})\psi_1(\mathbf{r}).$$

Therefore we have:

$$\frac{\delta G}{\delta \psi_1^*} = \hat{H}_0(\mathbf{r})\psi_1(\mathbf{r}).$$

The functional derivatives of all the other terms of eqn A.5 and of the orthonormality constraints in eqn A.7 can be obtained similarly. We find:

$$\frac{\delta L}{\delta \psi_1^*} = \hat{H}_0(\mathbf{r})\psi_1(\mathbf{r}) + \int d\mathbf{r}' \frac{|\psi_2(\mathbf{r}')|^2}{|\mathbf{r} - \mathbf{r}'|}\psi_1(\mathbf{r}) - \int d\mathbf{r}' \frac{\psi_2^*(\mathbf{r}')\psi_2(\mathbf{r})}{|\mathbf{r} - \mathbf{r}'|}\psi_1(\mathbf{r}') - \lambda_{11}\psi_1(\mathbf{r}) - \lambda_{12}\psi_2(\mathbf{r}).$$

$$\frac{\delta L}{\delta \psi_2^*} = \hat{H}_0(\mathbf{r})\psi_2(\mathbf{r}) + \int d\mathbf{r}' \frac{|\psi_1(\mathbf{r}')|^2}{|\mathbf{r} - \mathbf{r}'|}\psi_2(\mathbf{r}) - \int d\mathbf{r}' \frac{\psi_1^*(\mathbf{r}')\psi_1(\mathbf{r})}{|\mathbf{r} - \mathbf{r}'|}\psi_2(\mathbf{r}') - \lambda_{21}\psi_1(\mathbf{r}) - \lambda_{22}\psi_2(\mathbf{r}).$$

$$\frac{\delta L}{\delta \lambda_{ij}} = \delta_{ij} - \langle \psi_i | \psi_j \rangle \quad \text{for any} \quad i,j = 1,2.$$

By setting all these derivatives to zero we obtain the conditions that the functions ψ_1 and ψ_2 have to satisfy in order to minimize the energy:

$$\hat{H}_0(\mathbf{r})\psi_1(\mathbf{r}) + \int d\mathbf{r}' \frac{|\psi_2(\mathbf{r}')|^2}{|\mathbf{r} - \mathbf{r}'|}\psi_1(\mathbf{r}) - \int d\mathbf{r}' \frac{\psi_2^*(\mathbf{r}')\psi_2(\mathbf{r})}{|\mathbf{r} - \mathbf{r}'|}\psi_1(\mathbf{r}') = \lambda_{11}\psi_1(\mathbf{r}) + \lambda_{12}\psi_2(\mathbf{r}).$$

$$\text{(A.9)}$$

$$\hat{H}_0(\mathbf{r})\psi_2(\mathbf{r}) + \int d\mathbf{r}' \frac{|\psi_1(\mathbf{r}')|^2}{|\mathbf{r} - \mathbf{r}'|}\psi_2(\mathbf{r}) - \int d\mathbf{r}' \frac{\psi_1^*(\mathbf{r}')\psi_1(\mathbf{r})}{|\mathbf{r} - \mathbf{r}'|}\psi_2(\mathbf{r}') = \lambda_{21}\psi_1(\mathbf{r}) + \lambda_{22}\psi_2(\mathbf{r}).$$

$$\text{(A.10)}$$

$$\int d\mathbf{r}\psi_i^*(\mathbf{r})\psi_j(\mathbf{r}) = \delta_{ij} \quad \text{for any} \quad i,j = 1,2. \tag{A.11}$$

Equations A.9 and A.10 can be recast in a form which is reminiscent of the Hartree–Fock equations in Section 2.8. To this end we use the expression for the Hartree potential in eqn 2.47 and that for the exchange potential in eqn 2.60:

$$V_{\mathrm{H}}(\mathbf{r}) = \sum_j \int d\mathbf{r}' \frac{|\psi_j(\mathbf{r}')|^2}{|\mathbf{r} - \mathbf{r}'|}, \qquad V_{\mathrm{X}}(\mathbf{r}, \mathbf{r}') = -\sum_j \frac{\psi_j^*(\mathbf{r}')\psi_j(\mathbf{r})}{|\mathbf{r} - \mathbf{r}'|}, \tag{A.12}$$

with $j = 1,2$. By replacing these expressions inside eqns A.9 and A.10 we find immediately:

$$\left[-\frac{\nabla^2}{2} + V_{\mathrm{n}}(\mathbf{r}) + V_{\mathrm{H}}(\mathbf{r})\right]\psi_1(\mathbf{r}) + \int d\mathbf{r}'\, V_{\mathrm{X}}(\mathbf{r},\mathbf{r}')\,\psi_1(\mathbf{r}') = \lambda_{11}\psi_1(\mathbf{r}) + \lambda_{12}\psi_2(\mathbf{r}), \quad \text{(A.13)}$$

$$\left[-\frac{\nabla^2}{2} + V_{\mathrm{n}}(\mathbf{r}) + V_{\mathrm{H}}(\mathbf{r})\right]\psi_2(\mathbf{r}) + \int d\mathbf{r}'\, V_{\mathrm{X}}(\mathbf{r},\mathbf{r}')\,\psi_2(\mathbf{r}') = \lambda_{21}\psi_1(\mathbf{r}) + \lambda_{22}\psi_2(\mathbf{r}). \quad \text{(A.14)}$$

The final step is to eliminate one of the two functions on the right-hand side of both equations. This can be achieved by introducing a 2×2 matrix, S, which diagonalizes the Lagrange multipliers:

$$S \begin{pmatrix} \lambda_{11} & \lambda_{12} \\ \lambda_{21} & \lambda_{22} \end{pmatrix} S^{-1} = \begin{pmatrix} \varepsilon_1 & 0 \\ 0 & \varepsilon_2 \end{pmatrix}. \tag{A.15}$$

Since the Hamiltonian, $\hat{H}$, is Hermitian, the Lagrange multipliers are also Hermitian, i.e. $\lambda_{ij}^* = \lambda_{ji}$ (Messiah, 1965, Section II.9); therefore S is a unitary matrix and the eigenvalues ε_1, ε_2 are real numbers.

If we now define new wavefunctions, ϕ_1 and ϕ_2, as follows:

$$\phi_i = \sum_j S_{ij} \psi_j,$$

we can rewrite eqns A.9 and A.10 as:

$$\left[-\frac{\nabla^2}{2} + V_n(\mathbf{r}) + V_H(\mathbf{r}) \right] \phi_1(\mathbf{r}) + \int d\mathbf{r}' \, V_X(\mathbf{r}, \mathbf{r}') \, \phi_1(\mathbf{r}') = \varepsilon_1 \, \phi_1(\mathbf{r}), \tag{A.16}$$

$$\left[-\frac{\nabla^2}{2} + V_n(\mathbf{r}) + V_H(\mathbf{r}) \right] \phi_2(\mathbf{r}) + \int d\mathbf{r}' \, V_X(\mathbf{r}, \mathbf{r}') \, \phi_2(\mathbf{r}') = \varepsilon_2 \, \phi_2(\mathbf{r}). \tag{A.17}$$

In this form we can see that the constants introduced as Lagrange multipliers are effectively proxies for the Hartree–Fock eigenvalues.

Using the fact that S is a unitary matrix it can also be verified that, in terms of the new functions, ϕ_1 and ϕ_2, the orthonormality conditions in eqn A.11 and the Hartree and exchange potentials in eqn A.12 become:

$$\int d\mathbf{r} \phi_i^*(\mathbf{r}) \phi_j(\mathbf{r}) = \delta_{ij} \quad \text{for } i, j = 1, 2, \tag{A.18}$$

$$V_H(\mathbf{r}) = \sum_j \int d\mathbf{r}' \frac{|\phi_j(\mathbf{r}')|^2}{|\mathbf{r} - \mathbf{r}'|}, \qquad V_X(\mathbf{r}, \mathbf{r}') = -\sum_j \frac{\phi_j^*(\mathbf{r}')\phi_j(\mathbf{r})}{|\mathbf{r} - \mathbf{r}'|}. \tag{A.19}$$

That is, formally they are identical to the corresponding expressions in terms of the functions ψ_1 and ψ_2. Equations A.16–A.18 are precisely the Hartree–Fock equations discussed in Section 2.8.

In the case of N electrons the derivation is identical to the above, the only differences being that we obtain N equations like eqn A.16, and the sums in eqn A.19 run over the N wavefunctions ϕ_i with the lowest-energy eigenvalues, ε_i.

The introduction of the spin (see Chapter 11) in the previous derivation can be achieved by considering the variables $\mathbf{r}_1$ and $\mathbf{r}_2$ as describing both the positions of the two electrons and the orientation of their spin with respect to an infinitesimal uniform magnetic field: $\mathbf{r}_1 \rightarrow (\mathbf{r}_1, \sigma_1)$ and $\mathbf{r}_2 \rightarrow (\mathbf{r}_2, \sigma_2)$, with $\sigma = \uparrow$ or $\downarrow$. Following this replacement each integral should be considered as an integration over the space variable and a sum over the spin variable: $\int d\mathbf{r} \rightarrow \sum_\sigma \int d\mathbf{r}$. This is a consequence of the fact that we need to take expectation values over spinors, as in eqn 11.38. After these replacements eqn A.16 remains almost unchanged. The only novelty is that only the wavefunctions ϕ_j with the same spin as ϕ_1 contribute to the sum defining the exchange potential of eqn A.19.

Appendix B
Derivation of the Kohn–Sham equations

In this appendix we go through the derivation of the Kohn–Sham equations of Section 3.2. The procedure is similar to what we have seen in Appendix A; therefore we address directly the case of N electrons.

The Hohenberg–Kohn theorem (Section 3.1.1) states that the total energy, E, in the electronic ground state is a functional of the electron charge density, $n(\mathbf{r})$, $E = F[n]$. In the Kohn–Sham approach the functional F is decomposed as in eqn 3.10. We report the decomposition below, denoting the wavefunctions by ψ_i for convenience:

$$F[n] = \int d\mathbf{r}\, n(\mathbf{r}) V_n(\mathbf{r}) - \sum_i \int d\mathbf{r}\, \psi_i^*(\mathbf{r}) \frac{\nabla^2}{2} \psi_i(\mathbf{r}) + \frac{1}{2} \iint d\mathbf{r} d\mathbf{r}' \frac{n(\mathbf{r})n(\mathbf{r}')}{|\mathbf{r}-\mathbf{r}'|} + E_{xc}[n].$$

(B.1)

Here the first term describes the interaction of the electrons with the nuclei, the second term is the kinetic energy, the third term is the Hartree energy and the last one is the exchange and correlation energy. The index i runs over N wavefunctions, as usual.

The density, $n(\mathbf{r})$, is expressed in terms of the (yet unknown) Kohn–Sham wavefunctions, $\psi_i(\mathbf{r})$, using eqn 2.44:

$$n(\mathbf{r}) = \sum_i |\psi_i(\mathbf{r})|^2.$$

(B.2)

The Hohenberg–Kohn theorem states that the total energy, E, reaches its minimum value in correspondence with the ground-state electron density. This property provides us with a recipe for determining the Kohn–Sham wavefunctions. In fact, the minimum principle can be expressed using a functional derivative, as in eqn 3.11:

$$\frac{\delta F}{\delta n} = 0.$$

(B.3)

This property is entirely analogous to the minimum principle used in Appendix A for deriving the Hartree–Fock equations. In order to make use of eqn B.3 we note that the chain rule of standard calculus also applies to the case of functional derivatives (the definition of functional derivative is given by eqn 3.24). If we apply the chain rule to the functional derivative of F with respect to any of the wavefunctions, ψ_i^*, we find:

$$\frac{\delta F}{\delta \psi_i^*} = \frac{\delta F}{\delta n} \frac{\delta n}{\delta \psi_i^*} = \frac{\delta F}{\delta n} \psi_i,$$

(B.4)

where the last equality follows from eqn B.2. As discussed in Appendix A, also in this case the derivatives with respect to ψ_i or ψ_i^* can be treated as if independent (i.e. the final result would be the same if we were correctly treating them as interdependent). By combining eqns B.3 and B.4 we see that the Kohn–Sham orbitals must satisfy the following condition:

$$\frac{\delta F}{\delta \psi_i^*} = 0, \tag{B.5}$$

which is reminiscent of what was found in Appendix A for the Hartree–Fock equations. At this point it is convenient to request that the Kohn–Sham wavefunctions satisfy the orthonormality constraints:

$$\langle \psi_i | \psi_j \rangle = \delta_{ij}, \tag{B.6}$$

which guarantees that the density in eqn B.2 is correctly normalized to N electrons. The constrained optimization problem defined by eqns B.5 and B.6 is formally identical to the one that was solved in Appendix A, and hence the same procedure can be followed. To this aim we introduce the Lagrange functional:

$$L = F - \sum_{ij} \lambda_{ij} \left[\langle \psi_i | \psi_j \rangle - \delta_{ij} \right], \tag{B.7}$$

with λ_{ij} the Lagrange multipliers, and we search for the extrema of this new functional. We obtain:

$$\frac{\delta L}{\delta \psi_i^*} = 0 \quad \longrightarrow \quad \frac{\delta F}{\delta \psi_i^*} = \sum_j \lambda_{ij} \, \psi_j. \tag{B.8}$$

The functional derivatives $\delta F / \delta \psi_i^*$ can be evaluated by inserting eqn B.1 inside eqn B.4:

$$-\frac{\nabla^2}{2} \psi_i(\mathbf{r}) + \frac{\delta}{\delta n} \left\{ \int d\mathbf{r}\, n(\mathbf{r}) V_n(\mathbf{r}) + \frac{1}{2} \iint d\mathbf{r} d\mathbf{r}' \frac{n(\mathbf{r}) n(\mathbf{r}')}{|\mathbf{r} - \mathbf{r}'|} + E_{xc}[n] \right\} \psi_i(\mathbf{r})$$
$$= \sum_j \lambda_{ij}\, \psi_j(\mathbf{r}). \tag{B.9}$$

By using the definition of functional derivative in eqn 3.24, this equation can be rewritten rapidly as:

$$\left[-\frac{\nabla^2}{2} + V_n(\mathbf{r}) + \int d\mathbf{r}' \frac{n(\mathbf{r}')}{|\mathbf{r} - \mathbf{r}'|} + \frac{\delta E_{xc}}{\delta n} \right] \psi_i(\mathbf{r}) = \sum_j \lambda_{ij}\, \psi_j(\mathbf{r}). \tag{B.10}$$

Here we can recognize the Hartree potential, $V_H(\mathbf{r})$, and the exchange and correlation potential, $V_{xc}(\mathbf{r})$, in the third and fourth terms inside the square brackets, respectively. As in the case of the Hartree–Fock equations in Appendix A, the matrix λ_{ij} of the Lagrange multipliers can be diagonalized (see eqn A.15). By introducing 'rotated' wavefunctions, ϕ_i, as in eqn A.16, we finally obtain the Kohn–Sham equations of Section 3.2:

$$\left[-\frac{1}{2} \nabla^2 + V_n(\mathbf{r}) + V_H(\mathbf{r}) + V_{xc}(\mathbf{r}) \right] \phi_i(\mathbf{r}) = \varepsilon_i \phi_i(\mathbf{r}). \tag{B.11}$$

In the case of spin-DFT the procedure for deriving the Kohn–Sham equations remains the same. The only difference is that we need to consider Kohn–Sham wavefunctions with a spin label, and start from the minimum principle expressed by eqn 11.69. By using the chain rule similarly to eqn B.4, we arrive at the spin-dependent Kohn–Sham formulation given by eqn 11.70.

Appendix C
Numerical solution of the Kohn–Sham equations

In this appendix we review some numerical procedures used for finding self-consistent solutions to the Kohn–Sham equations. Atomic Hartree units will be used throughout. For clarity we report below eqns 3.27–3.32. The Kohn–Sham equations read:

$$-\frac{1}{2}\nabla^2\phi_i(\mathbf{r}) + V_{tot}(\mathbf{r})\phi_i(\mathbf{r}) = \varepsilon_i\phi_i(\mathbf{r}), \tag{C.1}$$

where the total potential is the sum of the nuclear potential, V_n, the Hartree potential from the electrons, V_H, and the exchange and correlation potential, V_{xc}:

$$V_{tot}(\mathbf{r}) = V_n(\mathbf{r}) + V_H(\mathbf{r}) + V_{xc}(\mathbf{r}). \tag{C.2}$$

The Hartree potential is obtained by solving Poisson's equation using the electron charge density:

$$\nabla^2 V_H(\mathbf{r}) = -4\pi n(\mathbf{r}). \tag{C.3}$$

This operation requires the charge density, which is obtained from the Kohn–Sham wavefunctions using:

$$n(\mathbf{r}) = \sum_i |\phi_i(\mathbf{r})|^2. \tag{C.4}$$

In the case of N electrons, the sum runs over the N wavefunctions with the smallest eigenvalues, ε_i. The density is also used to calculate the exchange and correlation potential, as discussed in Section 3.3.3. The calculation of the nuclear potential, V_n, is reviewed separately in Appendix E, since it requires the introduction of the concept of 'pseudopotentials'. In the following it will be assumed that we already decided the elemental composition of the atomistic model, and we know the location of all the nuclei.

The starting point for solving eqn C.1 is to decide the spatial domain where the wavefunctions, ϕ_i, are defined. For simplicity here we consider a cubic box of dimension a, with $0 \leq x, y, z \leq a$, which we call the *computational cell*. Since eqn C.1 is a second-order partial differential equation, we need to specify two boundary conditions on the surfaces of the cell (for each Cartesian direction).

One possible set of boundary conditions corresponds to having $\phi_i = 0$ whenever x, y or z is either 0 or a. These conditions correspond to the 'infinite square well' of introductory quantum mechanics textbooks. These boundary conditions are

Water molecule

Amorphous SiO$_2$

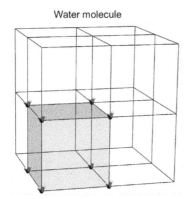
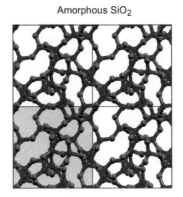

Fig. C.1   Schematic representation of computational cells for DFT calculations with periodic boundary conditions. The left panel shows the case of one isolated water molecule in a cubic computational cell of size 10 Å. The electronic wavefunctions are described only within the fundamental cell (highlighted), and periodic boundary conditions are imposed on the cell surfaces. In this example the periodic replicas of the computational cell define a simple-cubic lattice of water molecules. The cell needs to be large enough so that the electron charge density associated with the replicas do not overlap. The right panel shows a side view of a periodic model of amorphous SiO$_2$. The fundamental cell is highlighted, and we can easily identify the replicas of each atom in the neighbouring cells. The atomistic model of SiO$_2$ is from Pasquarello and Car (1997).

appropriate for studying confined systems, e.g. an isolated molecule; however they are inadequate when addressing solids, for which it is usually advantageous to exploit the regularity of the atomic structure. A choice which is more appropriate for solids corresponds to *periodic boundary conditions*. In this case we have:

$$\phi_i(x + a, y, z) = \phi_i(x, y, z), \qquad \nabla\phi_i(x + a, y, z) = \nabla\phi_i(x, y, z), \qquad \text{(C.5)}$$

and similar for y and z. Equation C.5 simply means that the wavefunction and its gradient repeat themselves as we cross the surfaces of the computational cell. A schematic representation of computational cells with periodic boundary conditions is shown in Figure C.1. For definiteness these boundary conditions will be assumed in the following.

Real-space representation

The most intuitive (although not necessarily the most efficient) way of setting up a numerical solution of eqn C.1 is to discretize the computational cell using a grid of $N_p \times N_p \times N_p$ equally spaced points:

$$\mathbf{r} = \frac{a}{N_p}(p\mathbf{u}_x + q\mathbf{u}_y + r\mathbf{u}_z),$$

with p, q and r integers taking the values $0, 1, \ldots, N_p - 1$. With this choice the solution of eqn C.1 will yield the values of the wavefunctions, $\phi_i(\mathbf{r})$, at the points of the chosen grid. For convenience, in order to denote the wavefunctions on the grid, we can use the notation:

$$\Phi_i(p, q, r) = \phi_i \left(\frac{p}{N_p} a \mathbf{u}_x + \frac{q}{N_p} a \mathbf{u}_y + \frac{r}{N_p} a \mathbf{u}_z \right). \tag{C.6}$$

With this choice the kinetic energy term in eqn C.1 can be approximated using finite-differences formulae, as was done in Exercise 8.1. For example, the second derivative with respect to x can be written as:

$$\frac{\partial^2}{\partial x^2} \phi_i(\mathbf{r}) \approx \left(\frac{N_p}{a} \right)^2 \left[\Phi_i(p+1, q, r) - 2\Phi_i(p, q, r) + \Phi_i(p-1, q, r) \right], \tag{C.7}$$

and similarly for y and z. This is really the simplest possible approximation, and more refined strategies are used in practice. The total potential, $V_{\text{tot}}(\mathbf{r})$, is also represented on a discrete grid, as in eqn C.6, therefore the potential term in eqn C.1 is simply given by:

$$V_{\text{tot}}(p, q, r)\, \Phi_i(p, q, r). \tag{C.8}$$

The periodic boundary conditions of eqn C.5 are imposed by requiring that:

$$\Phi_i(N_p, q, r) = \Phi_i(0, q, r) \qquad \text{(continuity of the wavefunction)}, \tag{C.9}$$

$$\Phi_i(-1, q, r) = \Phi_i(N_p - 1, q, r) \text{ (continuity of the gradient)}, \tag{C.10}$$

and similarly for the other variables, q and r. At this point we can assemble all the pieces together and rewrite eqn C.1 for the wavefunction on the grid points.

Since the final expression is somewhat cumbersome, it is convenient to transform the three-dimensional discrete set of values $\Phi_i(p, q, r)$ into a one-dimensional array. This can be achieved by mapping points (p, q, r) of the three-dimensional grid onto an array index, n, as follows: $(p, q, r) \rightarrow n = 1 + p + qN_p + rN_p^2$. This map allows us to replace $\Phi_i(p, q, r)$ by the one-dimensional array $\Phi_i(n)$, with $n = 1, 2, \ldots, N_p^3$. By combining eqn C.1 and eqns C.6–C.10 the Kohn–Sham equations become a standard problem of matrix diagonalization:

$$\mathbb{H} \begin{bmatrix} \Phi_i(1) \\ \Phi_i(2) \\ \cdots \\ \Phi_i(N_p^3) \end{bmatrix} = \varepsilon_i \begin{bmatrix} \Phi_i(1) \\ \Phi_i(2) \\ \cdots \\ \Phi_i(N_p^3) \end{bmatrix}, \tag{C.11}$$

where the matrix $\mathbb{H}$ contains $N_p^3 \times N_p^3$ elements, and is understood as the representation of the Kohn–Sham Hamiltonian in a real-space grid. For simplicity we will not give the explicit form of this matrix, but simply point out that it is a band matrix, as shown in Figure C.2.

In principle one could attempt the solution of the eigenvalue problem in eqn C.11 using standard linear algebra techniques (e.g. using 'Octave' as it was done in Exercise 8.1). However in realistic calculations it is necessary to use rather specialized techniques. In order to clarify this point let us consider the following example.

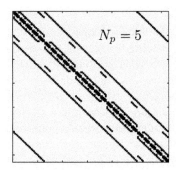 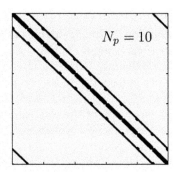

Fig. C.2 Schematic representations of the Kohn–Sham Hamiltonian, $\mathbb{H}$, of eqn C.11 in a real-space grid. The black points indicate the non-zero entries of the matrix, the grey regions indicate zeros. The matrix on the left, of size 125×125, corresponds to a cubic computational cell with $5\times5\times5$ grid points. The matrix on the right, of size $1{,}000\times1{,}000$, is for $10\times10\times10$ grid points. In typical calculations the number of grid points is much larger than in these examples; indeed it can be of the order of $100\times100\times100$ and beyond.

In a moderately accurate calculation the separation between nearest-neighbour grid points is of the order of $a/N_p \sim 0.1$ Å. If we consider a computational cell of side $a = 20$ Å, then the wavefunction arrays will contain $N_p^3 = 8 \cdot 10^6$ elements, and the Hamiltonian matrix, $\mathbb{H}$, will have $64 \cdot 10^{12}$ elements. Since complex numbers in double-precision floating-point arithmetics correspond to 16 bytes of information, each Kohn–Sham wavefunction will require 122 Mbytes of memory. On the other hand, storing all the elements of the matrix $\mathbb{H}$ is obviously impossible. Fortunately only a small fraction of the elements of $\mathbb{H}$ are non-zero (Figure C.2); therefore one can store the matrix in a highly compressed form where each non-zero element and its 'coordinates' in the matrix (i.e. row and column index) are retained. There exist specialized linear algebra techniques for solving this kind of eigenvalue problem involving compressed matrices, and the interested reader will find a review in Torsti *et al.* (2006).

After the arrays $\{\Phi_i(1),\ldots,\Phi_i(N_p^3)\}$ which are solutions of eqn C.11 have been determined, the electronic charge density can be calculated on the real-space grid using the discretized version of eqn C.4. From the charge density it is then straightforward to evaluate the exchange and correlation potential at every grid point. For the calculation of the Hartree potential the most common approach consists of solving eqn C.3 directly. This can be achieved by discretizing $\nabla^2 V_{\mathrm{H}}$ in the same way as for the kinetic energy in eqn C.7. This procedure transforms the Poisson equation into a linear system where the matrix of coefficients is highly sparse (similar to Figure C.2). This new linear system can be solved using the same techniques as for the wavefunctions. Once the updated potential, V_{tot}, is known on the real-space grid, one can repeat the entire procedure from eqn C.11. The loop described in Figure 3.3 is executed in this way until convergence is reached.

Planewaves representation

An alternative possibility for solving the Kohn–Sham equations consists of representing the wavefunctions using a Fourier series, and solving for the Fourier coefficients. This approach is particularly effective in combination with periodic boundary conditions, and represents indeed the most common strategy in the condensed matter physics literature.

The Fourier expansion of the electron wavefunctions proceeds in a similar way as for the electron charge density in Section 5.3. In the simplest case of a cubic computational cell we define the 'primitive vectors of the reciprocal lattice' as follows:

$$\mathbf{b_1} = \frac{2\pi}{a}\mathbf{u}_x, \qquad \mathbf{b_2} = \frac{2\pi}{a}\mathbf{u}_y, \qquad \mathbf{b_3} = \frac{2\pi}{a}\mathbf{u}_z, \tag{C.12}$$

and the 'reciprocal lattice vectors' $\mathbf{G}$ as in eqn 5.6:

$$\mathbf{G} = m_1\mathbf{b_1} + m_2\mathbf{b_2} + m_3\mathbf{b_3}, \text{ with } m_1, m_2, m_3 \text{ integers.} \tag{C.13}$$

It is immediately possible to verify that the complex exponentials

$$\exp(i\mathbf{G} \cdot \mathbf{r}) \tag{C.14}$$

automatically satisfy the boundary conditions defined by eqn C.5. For example we have:

$$\exp[i\mathbf{G} \cdot (\mathbf{r} + a\mathbf{u}_x)] = \exp(i\mathbf{G} \cdot \mathbf{r})\exp(ia\mathbf{G} \cdot \mathbf{u}_x) = \exp(i\mathbf{G} \cdot \mathbf{r})\exp(i2\pi m_1) = \exp(i\mathbf{G} \cdot \mathbf{r}),$$

and similarly for the gradient $\nabla\exp(i\mathbf{G} \cdot \mathbf{r})$. The functions defined by eqn C.14 are called *planewaves* since they can be interpreted as the time-independent part of a planar propagating wave, as in eqns 7.30 and 7.31.

Since planewaves satisfy periodic boundary conditions in the computational cell by construction, it seems sensible to write the Kohn–Sham wavefunctions as a linear combination of such waves:

$$\phi_i(\mathbf{r}) = \sum_{\mathbf{G}} c_i(\mathbf{G})\exp(i\mathbf{G} \cdot \mathbf{r}), \tag{C.15}$$

where the sum runs over the $\mathbf{G}$-vectors defined by all the integers, m_1, m_2, m_3, in eqn C.13. The relation between the 'planewave coefficients', $c_i(\mathbf{G})$, and the original wavefunctions is obtained by multiplying by $\exp(-i\mathbf{G}' \cdot \mathbf{r})$ both sides of the last equation, and then integrating over the computational cell. We find:

$$\int d\mathbf{r}\exp(-i\mathbf{G}' \cdot \mathbf{r})\phi_i(\mathbf{r}) = \sum_{\mathbf{G}} c_i(\mathbf{G})\int d\mathbf{r}\exp(-i\mathbf{G}' \cdot \mathbf{r})\exp(i\mathbf{G} \cdot \mathbf{r}) = a^3 c_i(\mathbf{G}'),$$

$$\tag{C.16}$$

where the last equality follows directly from an explicit evaluation of the integral. We can recognize in eqns C.15 and C.16 the standard prescriptions for the Fourier series of a periodic function. In the DFT literature it is common to call these relations a *planewaves representation* of the Kohn–Sham wavefunctions.

Although in principle the Fourier series in eqn C.15 should contain an infinite number of terms, for obvious reasons it is necessary to truncate the series and use a finite number of **G**-vectors. The largest **G**-vector included defines the level of detail with which we are describing the computational cell. In fact, if we truncate the series to the **G**-vectors defined by $m_1 = -N_{\max}, \ldots, N_{\max} - 1$ and similarly for m_2 and m_3, then the smallest length scale that we can describe corresponds to the wavelength $\lambda_{\min} = a/N_{\max}$. This is entirely analogous to the discretization of the computational cell discussed in the previous section. The set of all the planewaves included in the expansion of eqn C.15 is referred to as the *basis set*.

Traditionally the planewaves forming the basis set are specified by one single parameter called the *planewaves kinetic energy cutoff*. This parameter is defined as the quantum-mechanical kinetic energy associated with the planewave having the largest **G**-vector. If we denote such vector by $\mathbf{G}_{\max}$, the planewave cutoff is given by:

$$E_{\text{cut}} = \frac{|\mathbf{G}_{\max}|^2}{2},$$

and the basis set includes all and only the functions $\exp(i\mathbf{G} \cdot \mathbf{r})$ with $|\mathbf{G}|^2/2 \leq E_{\text{cut}}$. Typical planewaves cutoff values range between 10 and 50 Ha.

The Kohn–Sham equations can be rewritten in a planewaves representation by inserting eqn C.15 inside eqn C.1 and then proceeding as for eqn C.16. The result is:

$$\frac{|\mathbf{G}|^2}{2}c_i(\mathbf{G}) + \sum_{\mathbf{G}'} v_{\text{tot}}(\mathbf{G} - \mathbf{G}')c_i(\mathbf{G}') = \varepsilon_i\, c_i(\mathbf{G}), \qquad (\text{C.17})$$

where we define

$$v_{\text{tot}}(\mathbf{G}) = \frac{1}{a^3} \int d\mathbf{r} \exp(-i\mathbf{G} \cdot \mathbf{r})V_{\text{tot}}(\mathbf{r}).$$

We see that in this form the unknowns are the wavefunction coefficients, $c_i(\mathbf{G})$. Similarly to the case of real-space grids in the previous section, these coefficients define an array of complex numbers, with size $N_G = (2N_{\max})^3$. Therefore also in this case the solution of the Kohn–Sham equations corresponds to finding the eigenvalues and eigenvectors of a matrix.

In practical calculations the situation is slightly more complicated than it would appear from eqn C.17. In fact, the evaluation of the term involving the total potential requires one multiplication for each pair of **G** and **G**', i.e. N_G^2 floating-point operations. Since this procedure can become very time-consuming when N_G is large, it is more convenient to perform the multiplication of $V_{\text{tot}}(\mathbf{r})$ and $\phi_i(\mathbf{r})$ on a real-space grid (as in the previous section), and then calculate the Fourier transform of the result. Without going into the details, we just mention that this alternative strategy requires only $12N_G \log_2 N_G$ multiplications when using 'fast Fourier transform' algorithms (Johnson and Frigo, 2007). For instance, in the case of a calculation with a basis set of 50,000 planewaves, the latter strategy would be about 300 times faster.

The same considerations made for the term $V_{\text{tot}}(\mathbf{r})\phi_i(\mathbf{r})$ in the Kohn–Sham equations apply to the calculation of the charge density and the exchange and correlation potential. For example, the density is obtained from products like $\phi_i^*(\mathbf{r})\phi_i(\mathbf{r})$, which are most conveniently performed on a real-space grid and then Fourier-transformed.

On the other hand, the Hartree potential in eqn C.3 is calculated most efficiently in the planewaves representation. In fact, by defining:

$$v_{\mathrm{H}}(\mathbf{G}) = \frac{1}{a^3} \int d\mathbf{r} \exp(-i\mathbf{G} \cdot \mathbf{r}) V_{\mathrm{H}}(\mathbf{r}) \quad \text{and} \quad n(\mathbf{G}) = \frac{1}{a^3} \int d\mathbf{r} \exp(-i\mathbf{G} \cdot \mathbf{r}) n(\mathbf{r}),$$

we find immediately from eqn C.3:

$$v_{\mathrm{H}}(\mathbf{G}) = \frac{4\pi}{|\mathbf{G}|^2} n(\mathbf{G}). \tag{C.18}$$

Once we have determined the total potential starting from the solutions, $c_i(\mathbf{G})$, to eqn C.17, the self-consistent cycle of Figure 3.3 can be repeated until convergence is achieved.

Atomic orbitals representation

As a final example of representation of the Kohn–Sham wavefunctions we consider briefly the case of a basis of atomic orbitals. This choice allows us to establish a connection with Exercises 2.2, 4.4 and 4.5, and with Section 10.1. For simplicity we only consider a computational cell containing atoms of the same type.

Let us assume that we have been able to determine the solutions, ψ_{nlm}, of the Kohn–Sham equations for one isolated atom, i.e. the *atomic orbitals*:

$$-\frac{1}{2}\nabla^2 \psi_{nlm}(\mathbf{r}) + V_{\mathrm{tot}}^{\mathrm{atom}}(\mathbf{r}) \, \psi_{nlm}(\mathbf{r}) = \varepsilon_{nl} \, \psi_{nlm}(\mathbf{r}), \tag{C.19}$$

with n, l and m the principal quantum number, the angular quantum number (e.g. s, p, d, f) and the azimuthal quantum number, respectively.

If we now consider a computational cell containing M atoms of this type at the positions $\mathbf{R}_1, \ldots, \mathbf{R}_M$, it seems sensible to write the Kohn–Sham wavefunctions of this system by using *linear combinations of atomic orbitals*, or in short LCAO:

$$\phi_i(\mathbf{r}) = \sum_{I,nlm} c_{I,nlm} \psi_{nlm}(\mathbf{r} - \mathbf{R}_I). \tag{C.20}$$

In this expansion the sum runs over all the atoms $I = 1, \ldots, M$, and over some 'relevant' quantum numbers. For instance, if we were to describe the valence electrons of silicon we would include the orbitals ψ_{3s}, ψ_{3p_x}, ψ_{3p_y} and ψ_{3p_z}, and possibly the five Si-$3d$ orbitals.

In order to avoid confusion with the indices we can use the compact notation:

$$\varphi_\nu(\mathbf{r}) = \psi_{nlm}(\mathbf{r} - \mathbf{R}_I),$$

where the composite index ν identifies both the atomic position (I) and the orbital (nlm) centred on that position. Using this compact notation the wavefunctions are rewritten as:

$$\phi_i(\mathbf{r}) = \sum_\nu c_{i\nu} \, \varphi_\nu(\mathbf{r}).$$

As already done in the previous sections, we can replace this last expression inside eqn C.1 in order to obtain a matrix equation for the unknown coefficients, $c_{i\nu}$:

$$\sum_{\nu} c_{i\nu} \left[-\frac{1}{2}\nabla^2 \varphi_\nu(\mathbf{r}) + V_{\text{tot}}(\mathbf{r})\varphi_\nu(\mathbf{r}) - \varepsilon_i \varphi_\nu(\mathbf{r}) \right] = 0. \qquad (\text{C.21})$$

In order to extract the unknown coefficients, in eqn C.16 we multiplied both sides of the equation by $\exp(-i\mathbf{G}' \cdot \mathbf{r})$ and we integrated over the computational cell. Here we can proceed similarly by multiplying eqn C.21 by $\varphi_\mu^*(\mathbf{r})$ and performing the integrals. After defining the *Hamiltonian matrix in the atomic orbitals representation*:

$$(\mathbb{H})_{\mu\nu} = \int d\mathbf{r}\, \varphi_\mu^*(\mathbf{r}) \left[-\frac{1}{2}\nabla^2 + V_{\text{tot}}(\mathbf{r}) \right] \varphi_\nu(\mathbf{r}), \qquad (\text{C.22})$$

and the *overlap matrix*:

$$(\mathbb{S})_{\mu\nu} = \int d\mathbf{r}\, \varphi_\mu^*(\mathbf{r})\varphi_\nu(\mathbf{r}), \qquad (\text{C.23})$$

we obtain:

$$\sum_{\nu} (\mathbb{H} - \varepsilon_i \mathbb{S})_{\mu\nu}\, c_{i\nu} = 0. \qquad (\text{C.24})$$

This is a 'generalized eigenvalue problem', and differs from eqns C.11 and C.17 by the presence of the overlap matrix, $\mathbb{S}$. The appearance of an overlap matrix is a consequence of the fact that the atomic orbitals associated with different atoms are not orthogonal in general: $\int d\mathbf{r}\, \varphi_\mu^*(\mathbf{r})\varphi_\nu(\mathbf{r}) \neq 0$. Note that the concept of overlap integrals already appeared in Exercises 4.4 and 11.5.

If the atomic orbitals, $\varphi_\nu(\mathbf{r})$, are sufficiently localized around each atom, the Hamiltonian and the overlap matrices will be highly sparse. This situation is similar to what was discussed for real-space methods. Also in this case the numerical solution of eqn C.24 can be performed using very efficient linear algebra techniques for sparse eigenproblems.

Once we have determined the arrays of coefficients, $c_{i\nu}$, which are solutions of the Kohn–Sham equations, it is possible to construct the charge density and the total potential, and then proceed with the DFT self-consistent cycle until convergence. Here we refrain from going into more details on these aspects, since at this point the computational strategies become rather specialized.

As one might imagine, there exist many other possible *numerical representations* of the Kohn–Sham equations beyond the three cases discussed in this appendix. The interested reader will find a comprehensive and up-to-date review of the principal techniques in Chapters 12–17 of Martin (2004).

Appendix D
Reciprocal lattice and Brillouin zone

In order to keep this book self-contained, here we review briefly the concepts of direct lattice, reciprocal lattice and Brillouin zone. The reader who is not already familiar with these concepts is advised to consult the classic presentations by Kittel (1976) or Ashcroft and Mermin (1976).

The notions of direct and reciprocal lattices and Brillouin zone are introduced as a mathematical tool for describing the properties of materials which can be considered (to a certain approximation) as perfect crystals, where the atoms are arranged in patterns which repeat themselves periodically.

The smallest portion of a crystal from which we can generate the entire solid using rigid translations is called the *primitive unit cell*, and corresponds to a parallelepiped whose edges are denoted by $\mathbf{a}_1$, $\mathbf{a}_2$ and $\mathbf{a}_3$. These vectors are called the *primitive lattice vectors*. The cubic computational cell in Appendix C can be considered as a particular case of primitive unit cell, where the three primitive lattice vectors are orthogonal and of the same length.

The volume, Ω, of the primitive unit cell can be obtained from the primitive lattice vectors using the triple product:

$$\Omega = |\mathbf{a}_1 \cdot \mathbf{a}_2 \times \mathbf{a}_3|. \tag{D.1}$$

The entire crystal can be thought of as if it had been obtained by repeating the primitive unit cell everywhere in space through rigid translations corresponding to the vectors:

$$\mathbf{R} = n_1\mathbf{a}_1 + n_2\mathbf{a}_2 + n_3\mathbf{a}_3, \text{ with } n_1, n_2, n_2 \text{ integers.} \tag{D.2}$$

These vectors are said to define the *direct lattice*. An example of such a crystal and its primitive unit cell in two dimensions is shown in Figure D.1.

In order to describe the properties of the crystal by means of Fourier series and planewaves as in Appendix C, it is useful to introduce the *reciprocal lattice*. For this purpose we first construct the *primitive vectors of the reciprocal lattice*, $\mathbf{b}_1$, $\mathbf{b}_2$ and $\mathbf{b}_3$, using the prescription:

$$\mathbf{b}_1 = 2\pi \frac{\mathbf{a}_2 \times \mathbf{a}_3}{\mathbf{a}_1 \cdot \mathbf{a}_2 \times \mathbf{a}_3}, \tag{D.3}$$

and similarly for $\mathbf{b}_2$ and $\mathbf{b}_3$ after performing cyclic permutations of the indices. Then we define the reciprocal lattice as the set of all the vectors $\mathbf{G}$ obtained as:

$$\mathbf{G} = m_1\mathbf{b}_1 + m_2\mathbf{b}_2 + m_3\mathbf{b}_3, \text{ with } m_1, m_2, m_2 \text{ integers.} \tag{D.4}$$

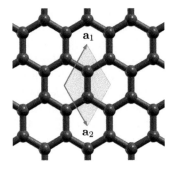
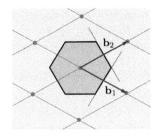

Fig. D.1 Graphene is a two-dimensional crystal consisting of carbon atoms arranged in a honeycomb lattice (Geim and Novoselov, 2007). The left panel shows a ball-and-stick representation of the ideal graphene structure. The C–C bond length is 1.42 Å. The primitive lattice vectors, $\mathbf{a}_1$ and $\mathbf{a}_2$, define the primitive unit cell (yellow), which contains two carbon atoms. We can generate the entire crystal by performing rigid translations of this cell using the direct-lattice vectors $\mathbf{R} = n_1\mathbf{a}_1 + n_2\mathbf{a}_2$ (with n_1 and n_2 integers). The right panel shows the primitive vectors of the reciprocal lattice, $\mathbf{b}_1$ and $\mathbf{b}_2$, as obtained from $\mathbf{a}_1$ and $\mathbf{a}_2$ using eqn D.5. The linear combinations of $\mathbf{b}_1$ and $\mathbf{b}_2$ with integer coefficients define the reciprocal lattice (grey dots). The light blue area indicates the first Brillouin zone. The bisectors of $\mathbf{b}_1$ and $\mathbf{b}_2$ are also indicated. The first Brillouin zone was constructed by taking the smallest region around the origin which is enclosed by the bisectors of all the reciprocal lattice vectors, $\mathbf{G}$. In this case we need to consider the lines bisecting $\pm\mathbf{b}_1$, $\pm\mathbf{b}_2$ and $\pm(\mathbf{b}_1 - \mathbf{b}_2)$.

The definition of the primitive vectors of the reciprocal lattice in eqn D.3 is chosen so as to obtain the following property:

$$\mathbf{a}_i \cdot \mathbf{b}_j = 2\pi \, \delta_{ij}, \tag{D.5}$$

with δ_{ij} the Kronecker delta. In the language of linear algebra this property corresponds to saying that the vectors $\mathbf{a}_1, \mathbf{a}_2, \mathbf{a}_3$ and $\mathbf{b}_1, \mathbf{b}_2, \mathbf{b}_3$ are 'bi-orthogonal', and that the direct and reciprocal lattices form 'dual' vector spaces. From the point of view of practical calculations eqn D.5 implies that, for any direct lattice vector $\mathbf{R}$ and reciprocal lattice vector $\mathbf{G}$, we have $\exp(i\mathbf{G} \cdot \mathbf{R}) = 1$, as can be verified by combining eqns D.2, D.4 and D.5. This result is very useful when calculating integrals involving planewaves (see for example Exercise 10.5).

In Section 9.2 it was shown that, in crystalline solids, any Kohn–Sham wavefunction can be written as the product of a planewave and a function which has the same periodicity as the direct lattice:

$$\phi_{i\mathbf{k}}(\mathbf{r}) = e^{i\mathbf{k}\cdot\mathbf{r}} u_{i\mathbf{k}}(\mathbf{r}) \quad \text{with} \quad u_{i\mathbf{k}}(\mathbf{r} + \mathbf{R}) = u_{i\mathbf{k}}(\mathbf{r}).$$

Furthermore, in Exercise 9.3 we have seen that the eigenvalues corresponding to wavevectors $\mathbf{k}$ and $\mathbf{k} + \mathbf{G}$ coincide; that is, the energy bands in a crystal are *periodic functions in reciprocal space*.

Owing to this property, it is natural to study the band structure only within one unit cell of the reciprocal lattice. For example, one might consider wavevectors $\mathbf{k}$ inside the primitive cell of the reciprocal lattice, i.e. the parallelepiped defined by $\mathbf{b}_1$, $\mathbf{b}_2$ and $\mathbf{b}_3$. In practice it is more useful to consider $\mathbf{k}$-vectors inside a volume in reciprocal space called the *first Brillouin zone*. The first Brillouin zone is formally defined as the volume containing $\mathbf{k}$ vectors whose distance from $\mathbf{G} = 0$ is smaller than the distance from any other $\mathbf{G}$-vector. In practice the first Brillouin zone is constructed by first determining the planes which bisect all the $\mathbf{G}$-vectors, and then taking the smallest volume centred at $\mathbf{G} = 0$ which is enclosed by these planes. An example of a reciprocal lattice and first Brillouin zone in two dimensions is shown in Figure D.1.

The volume of the first Brillouin zone is equal to the volume of a primitive cell of the reciprocal lattice. Therefore it is given by:

$$\Omega_{\text{BZ}} = |\mathbf{b}_1 \cdot \mathbf{b}_2 \times \mathbf{b}_3| = \frac{(2\pi)^3}{\Omega}, \tag{D.6}$$

where the last equality follows from eqns D.3 and D.1.

The use of the first Brillouin zone instead of the primitive cell of the reciprocal lattice is convenient for the simple reason that, by construction, the former is more symmetric; therefore it leads to a more intuitive and compact representation of band structures.

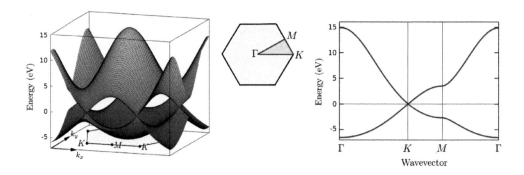

Fig. D.2 Comparison between a 'volumetric plot' and a 'band structure plot' of single-particle eigenvalues in a crystal. In order to make the visualization easier we consider graphene (see Figure D.1) as an example of a two-dimensional crystal. For the sake of clarity we represent the band structure of graphene using the simplest possible approximation, i.e. a tight-binding model (not *ab initio*). The model in this figure is taken from the book by Saito *et al.* (1998, page 27). In this case we have two eigenvalues for each wavevector; therefore the plots $\varepsilon_i(\mathbf{k})$ vs. $\mathbf{k}$ define two surfaces, as shown in the left panel. Already in this very simple example it is difficult to extract *quantitative* information from the surface plot; therefore it is convenient to introduce line-plots as shown in the right panel. In the line-plots (band structures) the wavevector, $\mathbf{k}$, is varied along the path $\Gamma \to K \to M \to \Gamma$, shown in the middle. In this case two bands are clearly visible in the plot. The zero of the energy axis corresponds to the top of the valence band. The *irreducible part* of the Brillouin zone is indicated by the yellow region in the middle panel. Starting from the irreducible zone we can generate the full zone by performing a reflection around the ΓM line and all the possible $60°$ rotations around Γ.

The energy bands of a crystal define volumetric plots $\varepsilon_i(\mathbf{k})$ vs. $\mathbf{k}$, where $\mathbf{k}$ identifies a point in the Brillouin zone. These volumetric plots are difficult to visualize and interpret, since they contain too much information.

In order to make the visualization clear and intuitive, it is customary to produce line-plots with $\mathbf{k}$ varying along straight segments inside the Brillouin zone. For example, the function:

$$E_i(\kappa) = \varepsilon_i[\mathbf{k}_1 + \kappa(\mathbf{k}_2 - \mathbf{k}_1)] \quad \text{with} \quad \kappa \in [0, 1],$$

describes how the energy, $\varepsilon_i(\mathbf{k})$, of band i changes when $\mathbf{k}$ sweeps the segment between $\mathbf{k}_1$ and $\mathbf{k}_2$ in the Brillouin zone.

The straight segments are typically chosen among the edges of the irreducible part of the Brillouin zone, i.e. the portion from which the entire zone can be generated by applying the symmetries of the crystal. Figure D.2 shows a comparison between a volumetric plot of $\varepsilon_i(\mathbf{k})$ vs. $\mathbf{k}$ and a band structure plot for the two-dimensional crystal of Figure D.1.

Appendix E
Pseudopotentials

As already mentioned in a few occasions (e.g. Sections 5.1 and 5.2, and Exercise 6.5), in many cases it is convenient to perform DFT calculations by describing explicitly the valence electrons only. This simplification can be achieved by introducing so-called 'pseudopotentials'. In this appendix we briefly discuss the principles underlying the use of pseudopotentials in DFT calculations.

Figure E.1 shows the Kohn–Sham electron wavefunctions of an isolated Si atom, as calculated using DFT/LDA. In this figure we see both the core states, $1s$, $2s$ and $2p$, and the valence states, $3s$ and $3p$. When we consider the electronic charge densities associated with all the core states or all the valence states (right panel), we realize that the core electrons are tightly bound to the Si nucleus, while the valence electrons tend to localize further away. In this example the valence electron density reaches a maximum near a radius corresponding to half the length of an Si–Si bond in silicon

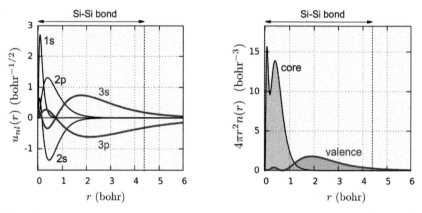

Fig. E.1 The left panel shows the radial wavefunctions, $u_{nl}(r)$, calculated for the Si atom using DFT/LDA. The complete wavefunctions are given by $\psi_{nl}(\mathbf{r}) = r^{-1}u_{nl}(r)Y_{lm}(\mathbf{r}/r)$, with Y_{lm} the spherical harmonic for the angular quantum number l and the azimuthal quantum number m. The nucleus is located at $r = 0$. For comparison we show the length of the Si–Si bond in silicon (diamond structure). The right panel shows the total (radial) charge densities of the core electrons and of the valence electrons. We see that the valence charge density peaks at a radius corresponding approximately to half the Si–Si bond length in silicon, while the core electrons are localized near the nucleus.

(diamond structure). In addition, the core density is negligible where the valence density is large and vice versa.

The spatial distribution of the electron densities in Figure E.1 suggests that valence electrons will be the most sensitive to changes in the chemical bonding environment, while core electrons (which are located near the Coulomb singularity) will be relatively immune to such changes. This intuitive observation is often expressed by stating that, to a certain approximation, only the valence electrons participate in chemical bonding, while core electrons are inert. Obviously this is a well established empirical fact, known since the early work of Lewis (1916), i.e. well before the development of DFT.

The separation of core states from valence states suggests that we should be able to perform DFT calculations on poly-atomic systems by keeping the core electrons as they appear in the isolated atom. This choice corresponds to making the *frozen core* approximation.

If the core is to be held 'frozen', then there is not much point in describing the Kohn–Sham wavefunctions of core states. The next step is therefore to completely remove the core electrons from our description. This choice leads to a substantial computational saving. For example, in the case of tungsten, we would only describe six valence electrons instead of 74 electrons in total.

How do we decide which wavefunctions should be considered 'core' and which ones 'valence' states? As a rule of thumb, in the context of DFT calculations the 'valence' corresponds to the outermost shell of the atom in the periodic table; for example, for tungsten we would have $6s^2 5d^4$. However, there are cases where one might need to include more electronic states in the set of 'valence electrons'. For example, in the case of bismuth it is important to describe on an equal footing both the nominal valence shell, $6s^2 6p^3$, and the 'semi-core' shell, $5d^{10}$. In practice the distinction between core and valence is not a strict one, and depends on the level of accuracy that one is trying to achieve. When in doubt, an inspection of the spatial extent of all the atomic wavefunctions as in Figure E.1 represents the first point of call for identifying core and valence states.

After having decided which electrons should be considered as the valence states, the procedure for eliminating the core electrons is not trivial and requires some care. For a start, near the nucleus the valence electronic wavefunctions must undergo a change of sign in order to be orthogonal to the core states. For example, in Figure E.1 we see that the 2s state changes sign (i.e. it exhibits a 'node') in the region where the 1s state is localized, and the 3s state changes sign twice (two nodes) so that the overlap integrals with the 1s or the 2s state vanish. If we simply ignored the core states, then the valence states would not exhibit the correct 'nodal structure' near the nucleus.

A second issue is that, even if we were able to obtain the correct oscillating features in the wavefunctions, then it would be very difficult to describe them using a real-space grid or a planewaves basis. In order to illustrate this aspect, Figure E.2 shows how the nodal structure of the 3s state of Si would appear if the function was sampled only at a small number of evenly spaced points. In this case the description of the oscillations would be very poor, thereby undermining the accuracy and the numerical stability of the calculations. Of course we may try to fix this issue by using a finer real-space grid, or a higher kinetic energy cutoff in the case of a planewaves basis (see Appendix C);

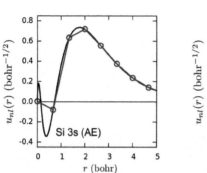 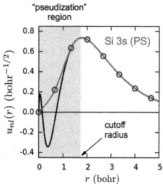

Fig. E.2 The left panel shows how the $3s$ wavefunction of Si from Figure E.1 appears when we use a real-space grid of points spaced by 0.3 Å (circles and blue line). The fine details of the wavefunction near the nucleus are completely missing, and this leads to inaccuracies in the calculations. The right panel shows the pseudization of the $3s$ all-electron wavefunction (AE) of Si. Inside a 'pseudization' region defined by the cutoff radius, r_c (set to 1.7 bohr in this example), the original wavefunction is replaced by a smooth function (red curve) which does not exhibit any fine structure near the nucleus. Using the same real-space grid of points as for the left panel, the pseudo-wavefunction (PS) can be described without difficulty. Note that for $r > r_c$ the all-electron wavefunction and the pseudo-wavefunction coincide, by construction. In this example the pseudo-wavefunction was obtained by using the procedure of Troullier and Martins (1991).

however, this would result into hopelessly time-consuming calculations.

The difficulties associated with the nodal structure of the valence wavefunctions can be overcome at once by replacing the oscillating part of the wavefunction by a *smooth and nodeless curve*, as shown in the right panel of Figure E.2. This operation can be performed in four steps:

○ Calculate the Kohn–Sham wavefunction in the presence of the core electrons (black curve in Figure E.2). This wavefunction is typically referred to as the 'all-electron wavefunction'.

○ Decide on a radial cutoff, r_c, that is the radius which sets the boundary of the region where the wavefunction is to be modified, as shown in Figure E.2. We will call the interval $0 < r < r_c$ the 'pseudization region'.

○ Inside this pseudization region replace the all-electron wavefunction by a smooth and nodeless function (e.g. a simple polynomial).

○ The new function is chosen so as to yield the same electron density as the all-electron function for $0 < r < r_c$, and to match its value and slope at $r = r_c$.

The result of this operation is shown by the red curve in Figure E.2, and is referred to as the 'pseudo-wavefunction'.

The introduction of a pseudo-wavefunction which is smooth in the pseudization region eliminates the problems connected with the nodal structure of the all-electron wavefunction. The remaining question now is how to obtain such a smooth pseudo-wavefunction directly by solving the Kohn–Sham equations.

The strategy underlying the pseudopotential method is to construct a modified nuclear potential which satisfies the following conditions: (i) outside the pseudization region the modified potential coincides with the original Kohn–Sham potential (obtained from a calculation including all the electrons); (ii) inside the pseudization region the potential is modified in such a way that the solution of the Kohn–Sham equation yields precisely the pseudo-wavefunction.

In order to clarify this procedure let us consider the Kohn–Sham equation for the $3s$ state of Si. In a spherical coordinate system centred on the Si nucleus we have:

$$\psi_{3s}^{\mathrm{AE}}(\mathbf{r}) = \frac{1}{r} u_{3s}^{\mathrm{AE}}(r)\, Y_{00}(\hat{\mathbf{r}}),$$

where Y_{00} denotes the $(l = 0, m = 0)$ spherical harmonic (which is simply a constant in the the case of $l = 0$), $\hat{\mathbf{r}} = \mathbf{r}/r$, and u_{3s}^{AE} is the solution of the radial Kohn–Sham equation:

$$-\frac{1}{2}\frac{d^2}{dr^2} u_{3s}^{\mathrm{AE}}(r) + V^{\mathrm{AE}}(r)\, u_{3s}^{\mathrm{AE}}(r) = E_{3s}\, u_{3s}^{\mathrm{AE}}(r). \tag{E.1}$$

In these expressions the superscript 'AE' indicates that the calculations are performed by including all the electrons, core and valence. The potential V^{AE} is the total potential experienced by the Kohn–Sham electrons, and includes the nuclear, Hartree, and exchange and correlations terms.

Once u_{3s}^{AE} has been calculated as in Figure E.1, we can proceed with the 'pseudization' procedure illustrated in Figure E.2. We denote the pseudo-wavefunction thus obtained u_{3s}^{PS}. At this point we can determine the 'pseudo-potential', V_{3s}^{PS}, which would give u_{3s}^{PS} as a solution, by requiring that the following equation be satisfied, in analogy with eqn E.1:

$$-\frac{1}{2}\frac{d^2}{dr^2} u_{3s}^{\mathrm{PS}}(r) + V_{3s}^{\mathrm{PS}}(r)\, u_{3s}^{\mathrm{PS}}(r) = E_{3s}\, u_{3s}^{\mathrm{PS}}(r). \tag{E.2}$$

In this case *the pseudo-wavefunction is known* (because it was determined by smoothing u_{3s}^{AE}), but *the pseudopotential is unknown*. Therefore this time we need to solve for the potential, and a straightforward inversion of eqn E.2 yields:

$$V_{3s}^{\mathrm{PS}} = E_{3s} + \frac{1}{2 u_{3s}^{\mathrm{PS}}}\frac{d^2 u_{3s}^{\mathrm{PS}}}{dr^2}. \tag{E.3}$$

If we use the pseudopotential, V^{PS}, obtained from this equation instead of V^{AE} in eqn E.1, then we obtain a pseudo-wavefunction which is smooth inside the core region, and has the *same Kohn–Sham eigenvalue* as the original all-electron wavefunction. In addition, by comparing eqns E.1 and E.2 for $r > r_c$, and remembering that in this region $u_{3s}^{\mathrm{PS}}(r) = u_{3s}^{\mathrm{AE}}(r)$, we see that $V_{3s}^{\mathrm{PS}}(r) = V_{3s}^{\mathrm{AE}}(r)$ for $r > r_c$.

The procedure described so far works for the $3s$ state of an isolated Si atom. However, in actual calculations, where the atom is forming chemical bonds with other atoms, we also need to deal with $3p$ states, and more generally with any electronic configuration which differs from that of the isolated atom. In other words, we need to make the pseudopotential *transferable* to other chemical environments.

Transferability can be achieved by removing from the pseudopotential the electronic contributions arising from the Hartree and the exchange and correlation potentials, which are specific to a particular electronic configuration, while keeping only the modified nuclear potential. This procedure is referred to as 'de-screening', meaning that the screening effect of the electrons is removed from the pseudopotential. The final potential thus obtained is called the *ionic pseudopotential*.

The ionic pseudopotential obtained for silicon using the procedure just described is shown in Figure E.3. We can see that, beyond the cutoff radius, the potential coincides with the original Coulomb potential of the nucleus, for a charge state given by the number, Z_v, of valence electrons: $V^{\mathrm{PS}}(r) = -Z_v/r$ for $r > r_c$. Inside the pseudization region the singularity at $r = 0$ is removed, and the pseudopotential resembles a potential well. If we use this ionic pseudopotential in the Kohn–Sham equations, we discover that the lowest-energy state is precisely u_{3s}^{PS}; that is, the core states, $1s$ and $2s$, have completely disappeared from this picture.

As a final consideration we mention that, in practical calculations on poly-atomic systems, we can no longer exploit the spherical symmetry of atomic calculations, and around the nuclei the wavefunctions will generally correspond to mixtures of s, p, d and f angular momentum components. In order to take this into account the pseudopotential is generalized as follows:

$$V^{\mathrm{PS}}(\mathbf{r}, \mathbf{r}') = \sum_{lm} V_l^{\mathrm{PS}}(r) \frac{\delta(r - r')}{r^2} Y_{lm}(\hat{\mathbf{r}}) Y_{lm}^*(\hat{\mathbf{r}}'), \qquad (\mathrm{E.4})$$

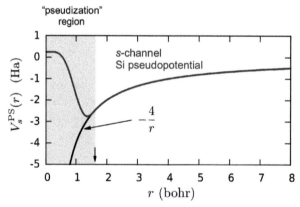

Fig. E.3 The blue curve represents the pseudopotential for the $3s$ pseudo-wavefunction of Si shown in Figure E.2, as obtained by using eqn E.3 and removing the electronic contributions (i.e. the Hartree and the exchange and correlation potentials). Beyond the cutoff radius, $r_c = 1.7$ bohr (arrow), the pseudopotential coincides with the nuclear Coulomb potential corresponding to the valence electronic charge, $-Z_v/r$ with $Z_v = 4$ for Si. The pseudopotential shown in this figure is for the $l = 0$ angular momentum component of the wavefunction, called the 's-channel' in jargon. Similar pseudopotentials need to be calculated for the p-channel, d-channel and possibly the f-channel. All the angular components $V_l^{\mathrm{PS}}(r)$ are then assembled in a non-local ionic pseudopotential, as in eqn E.4.

where V_l^{PS} refers to the angular momentum $l = 0$ (s), $l = 1$ (p), $l = 2$ (d) or $l = 3$ (f). This potential is applied to a wavefunction, $\psi^{\text{PS}}(\mathbf{r})$, by using an integral as follows:

$$\int d\mathbf{r}'\, V^{\text{PS}}(\mathbf{r}, \mathbf{r}')\, \psi^{\text{PS}}(\mathbf{r}').$$

For example, if we consider the $3s$ pseudo-wavefunction of Si, it can be verified that:

$$\int d\mathbf{r}' V^{\text{PS}}(\mathbf{r}, \mathbf{r}')\, \psi_{3s}^{\text{PS}}(\mathbf{r}') = V_{l=0}^{\text{PS}}(r)\, \psi_{3s}^{\text{PS}}(r).$$

In other words, for wavefunctions of well-defined angular momentum the generalized pseudopotential of eqn E.4 reduces to a standard local potential.

Equation E.4 defines a *non-local pseudopotential*, in the sense that $V^{\text{PS}}(\mathbf{r}, \mathbf{r}')$ depends both on $\mathbf{r}$ and $\mathbf{r}'$, and samples the values of the wavefunction at the points $\mathbf{r}'$ (while the original potential is a function of the position, $\mathbf{r}$, and samples the wavefunction at $\mathbf{r}$). The non-locality arises from the fact that we need to 'extract' the angular momentum components, s, p, d or f, of a given wavefunction, and act on each component separately using the appropriate pseudopotential, V_l^{PS}.

Summing up the preceding discussion we can say that, in the case of calculations using pseudopotentials, the nuclear potential experienced by the electrons

$$V_{\text{n}}(\mathbf{r}) = -\sum_I \frac{Z_I}{|\mathbf{r} - \mathbf{R}_I|},$$

must be replaced by a non-local ionic pseudopotential:

$$V_{\text{ion}}^{\text{PS}}(\mathbf{r}) = \sum_I V_I^{\text{PS}}(\mathbf{r} - \mathbf{R}_I, \mathbf{r}' - \mathbf{R}_I),$$

where each atom, I, carries its own angular momentum-dependent non-local pseudopotential, V_I^{PS}, as in eqn E.4. Obviously, after generating the pseudopotentials V_I^{PS} for each atom, it is necessary to perform extensive tests in order to ensure that the pseudization will not compromise the accuracy of the final DFT calculations.

The present discussion of pseudopotentials is somewhat oversimplified in order to keep the presentation as intuitive as possible. In reality the theory of pseudopotentials can be developed on very firm mathematical grounds using rigorous derivations. In fact, the pseudopotential method is among the most finely tuned components of DFT software packages, and the methods are so specialized that the pseudopotential theory can almost be regarded as a discipline in its own right.

As a final remark, we note that the success of the pseudopotential method has been instrumental to the introduction of DFT in materials modelling. In fact, without this method we would not be able to study complex systems involving many atoms.

The interested reader will find an authoritative historical perspective on the development of the pseudopotential method in Cohen (1984). Details on some of the popular pseudopotential methods which are currently in use in first-principles materials modelling can be found in Chapter 11 of Martin (2004), or directly in the original scientific literature (e.g. Hamann *et al.*, 1979; Troullier and Martins, 1991; Vanderbilt, 1990; Blöchl, 1994).

References

Abramowitz, M. and Stegun, I. A. (1965). *Handbook of mathematical functions.* Dover, New York.

Ackland, G. J., Warren, M. C., and Clark, S. J. (1997). Practical methods in ab initio lattice dynamics. *Journal of Physics: Condensed Matter*, **9**, 7861.

Adler, S. L. (1962). Quantum theory of the dielectric constant in real solids. *Physical Review*, **126**, 413.

Aebi, P., Osterwalder, J., Fasel, R., Naumović, D., and Schlapbach, L. (1994). Fermi surface mapping with photoelectrons at UV energies. *Surface Science*, **331**, 917.

Albrecht, S., Reining, L., Del Sole, R., and Onida, G. (1998). *Ab initio* calculation of excitonic effects in the optical spectra of semiconductors. *Physical Review Letters*, **80**, 4510.

Alfè, D., Gillan, M. J., and Price, G. D. (1999). The melting curve of iron at the pressures of the Earth's core from *ab initio* calculations. *Nature*, **401**, 462.

Allen, P. B. and Mitrovich, B. (1982). Theory of superconducting T_c. In *Solid State Physics*, **37**, ed. H. Ehrenreich, F. Seitz, D. Turnbull, Academic Press, New York, 1.

An, J. M. and Pickett, W. E. (2001). Superconductivity of MgB_2: Covalent bonds driven metallic. *Physical Review Letters*, **86**, 4366.

Andersen, H. C. (1980). Molecular dynamics simulations at constant pressure and/or temperature. *Journal of Chemical Physics*, **72**, 2384.

Anderson, P. W. (1950). Antiferromagnetism. Theory of superexchange interaction. *Physical Review*, **79**, 350.

Anderson, P. W. (1972). More is different. *Science*, **177**, 393.

Andersson, Y., Langreth, D. C., and Lundqvist, B. I. (1996). van der Waals interactions in density-functional theory. *Physical Review Letters*, **76**, 102.

Anisimov, V. I., Zaanen, J., and Andersen, O. K. (1991). Band theory and Mott insulators: Hubbard U instead of Stoner I. *Physical Review B*, **44**, 943.

Aryasetiawan, F. and Gunnarsson, O. (1998). The GW method. *Reports on Progress in Physics*, **61**, 237.

Ashcroft, N. W. and Mermin, N. D. (1976). *Solid state physics.* Harcourt College Publishers, New York.

Aspnes, D. E. and Studna, A. A. (1983). Dielectric functions and optical parameters of Si, Ge, GaP, GaAs, GaSb, InP, InAs, and InSb from 1.5 to 6.0 eV. *Physical Review B*, **27**, 985.

Atkins, P. and de Paula, J. (2006). *Atkins' physical chemistry* (8th edn). Oxford University Press, Oxford.

Bachelet, G. B., Hamann, D. R., and Schlüter, M. (1982). Pseudopotentials that work: From H to Pu. *Physical Review B*, **26**, 4199.

Bardeen, J. (1961). Tunnelling from a many-particle point of view. *Physical Review Letters*, **6**, 57.

Bardeen, J., Cooper, L. N., and Schrieffer, J. R. (1957). Theory of superconductivity. *Physical Review*, **108**, 1175.

Bardyszewski, W. and Hedin, L. (1985). A new approach to the theory of photoemission from solids. *Physica Scripta*, **32**, 439.

Baroni, S., Giannozzi, P., and Testa, A. (1987). Elastic constants of crystals from linear-response theory. *Physical Review Letters*, **59**, 2662.

Baroni, S., de Gironcoli, S., Dal Corso, A., and Giannozzi, P. (2001). Phonons and related crystal properties from density-functional perturbation theory. *Review of Modern Physics*, **73**, 515.

Baroni, S., Giannozzi, P., and Isaev, E. (2010). Density-functional perturbation theory for quasi-harmonic calculations. *Reviews in Mineralogy and Geochemistry*, **71**, 39.

Beamson, G., Briggs, D., Davies, S. F., Fletcher, I. W., Clark, D. T., Howard, J., Gelius, U., Wannberg, B., and Balzer, P. (1990). Performance and application of the Scienta ESCA300 spectrometer. *Surface and Interface Analysis*, **15**, 541.

Becke, A. D. (1988). Density-functional exchange-energy approximation with correct asymptotic behavior. *Physical Review A*, **38**, 3098.

Becke, A. D. (1993). Density-functional thermochemistry. III. The role of exact exchange. *Journal of Chemical Physics*, **98**, 5648.

Bendtsen, J. (1974). The rotational and rotation-vibrational Raman spectra of $^{14}N_2$, $^{14}N^{15}N$ and $^{15}N_2$. *Journal of Raman Spectroscopy*, **2**, 133.

Benedict, L. X., Shirley, E. L., and Bohn, R. B. (1998). Optical absorption of insulators and the electron–hole interaction: An *ab initio* calculation. *Physical Review Letters*, **80**, 4514.

Binnig, G., Rohrer, H., Gerber, Ch., and Weibel, E. (1982). Surface studies by scanning tunneling microscopy. *Physical Review Letters*, **49**, 57.

Bitzek, E., Koskinen, P., Gähler, F., Moseler, M., and Gumbsch, P. (2006). Structural relaxation made simple. *Physical Review Letters*, **97**, 170201.

Bloch, F. (1928). Über die quantenmechanik der elektronen in kristallgittern. *Zeitschrift für Physik*, **52**, 555.

Blöchl, P. E. (1994). Projector augmented-wave method. *Physical Review B*, **50**, 17953.

Blundell, S. (2001). *Magnetism in condensed matter*. Oxford University Press, Oxford.

Boehler, R. (1993). Temperatures in the Earth's core from melting-point measurements of iron at high static pressures. *Nature*, **363**, 534.

Bohr, N. (1913). On the constitution of atoms and molecules. *Philosophical Magazine*, **26**, 1.

Bonhomme, C., Gervais, C., Babonneau, F., Coelho, C., Pourpoint, F., Azaïs, T., Ashbrook, S. E., Griffin, J. M., Yates, J. R., Mauri, F., and Pickard, C. J. (2012). First-principles calculation of NMR parameters using the gauge including projector augmented wave method: A chemist's point of view. *Chemical Reviews*, **112**, 5733.

Born, M. and Huang, K. (1954). *Dynamical theory of crystal lattices*. Oxford University Press, Oxford.

Born, M. and Oppenheimer, R. (1927). Zur quantentheorie der molekeln. *Annalen der Physik*, **389**, 457.

Born, M. and Wolf, E. (1999). *Principles of optics* (7th edn). Cambridge University Press, Cambridge.

Bouquet, F., Fisher, R. A., Phillips, N. E., Hinks, D. G., and Jorgensen, J. D. (2001). Specific heat of $Mg^{11}B_2$: Evidence for a second energy gap. *Physical Review Letters*, **87**, 047001.

Bransden, B. H. and Joachain, C. J. (1983). *Physics of atoms and molecules*. Longman, London.

Breit, G. (1929). The effect of retardation on the interaction of two electrons. *Physical Review*, **34**, 553.

Brockhouse, B. N. and Stewart, A. T. (1958). Normal modes of aluminum by neutron spectrometry. *Review of Modern Physics*, **30**, 236.

Brown, J. M. and McQueen, R. G. (1986). Phase transitions, Grüneisen parameter, and elasticity for shocked iron between 77 GPa and 400 GPa. *Journal of Geophysical Research: Solid Earth*, **91**, 7485.

Brüesch, P. (1986). *Phonons: theory and experiments II*. Springer, Berlin.

Buchenau, U., Prager, M., Nücker, N., Dianoux, A. J., Ahmad, N., and Phillips, W. A. (1986). Low-frequency modes in vitreous silica. *Physical Review B*, **34**, 5665.

Burke, K. (2012). Perspective on density functional theory. *Journal of Chemical Physics*, **136**, 150901.

Burkel, E. (2000). Phonon spectroscopy by inelastic X-ray scattering. *Reports on Progress in Physics*, **63**, 171.

Burkel, E., Peisl, J., and Dorner, B. (1987). Observation of inelastic X-ray scattering from phonons. *Europhysics Letters*, **3**, 957.

Cahangirov, S., Topsakal, M., and Ciraci, S. (2010). Long-range interactions in carbon atomic chains. *Physical Review B*, **82**, 195444.

Callen, H. B. (1985). *Thermodynamics and an introduction to thermostatistics* (2nd edn). Wiley, New York.

Car, R. and Parrinello, M. (1985). Unified approach for molecular dynamics and density-functional theory. *Physical Review Letters*, **55**, 2471.

Carpenter, J. M. and Price, D. L. (1985). Correlated motions in glasses studied by coherent inelastic neutron scattering. *Physical Review Letters*, **54**, 441.

Carrier, P., Wentzcovitch, R., and Tsuchiya, J. (2007). First-principles prediction of crystal structures at high temperatures using the quasiharmonic approximation. *Physical Review B*, **76**, 064116.

Ceder, G. (2010). Opportunities and challenges for first-principles materials design and applications to Li battery materials. *MRS Bulletin*, **35**, 693.

Ceperley, D. M. and Alder, B. J. (1980). Ground state of the electron gas by a stochastic method. *Physical Review Letters*, **45**, 566.

Chandler-Horowitz, D. and Amirtharaj, P. M. (2005). High-accuracy, midinfrared (450 cm$^{-1} \leq \omega \leq 4000$ cm^{-1}) refractive index values of silicon. *Journal of Applied Physics*, **97**, 123526.

Chandrasekhar, S. (1944). Some remarks on the negative hydrogen ion and its absorption coefficient. *Astrophysical Journal*, **100**, 176.

Cheetham, A. K. and Hope, D. A. O. (1983). Magnetic ordering and exchange effects in the antiferromagnetic solid solutions $Mn_xNi_{1-x}O$. *Physical Review B*, **27**, 6964.

Choi, H. J., Roundy, D., Sun, H., Cohen, M. L., and Louie, S. G. (2002). The origin of the anomalous superconducting properties of MgB_2. *Nature*, **418**, 758.

Chow, T. L. (2000). *Mathematical methods for physicists: A concise introduction.* Cambridge University Press, Cambridge.

Chupp, E. L., Dotchin, L. W., and Pegg, D. J. (1968). Radiative mean-life measurements of some atomic-hydrogen excited states using beam-foil excitation. *Physical Review*, **175**, 44.

Clark, C. D., Dean, P. J., and Harris, P. V. (1964). Intrinsic edge absorption in diamond. *Proceedings of the Royal Society of London A*, **277**, 312.

Cohen, M. L. (1984). Application of the pseudopotential model to solids. *Annual Review of Materials Science*, **14**, 119.

Cohen, M. L. (2000). The theory of real materials. *Annual Review of Materials Science, ***30**, 1.

Cohen, M. L. and Bergstresser, T. K. (1966). Band structures and pseudopotential form factors for fourteen semiconductors of the diamond and zinc-blende structures. *Physical Review*, **141**, 789.

Cohen-Tannoudji, C., Diu, B., and Laloë, F. (1977). *Quantum mechanics, Volume II*. Wiley.

Condon, E. U. (1930). The theory of complex spectra. *Physical Review*, **36**, 1121.

Courths, R. and Hüfner, S. (1984). Photoemission experiments on copper. *Physics Reports*, **112**, 53.

Cowley, R. A. (1968). Anharmonic crystals. *Reports on Progress in Physics*, **31**, 123.

Croat, J. J., Herbst, J. F., Lee, R. W., and Pinkerton, F. E. (1984). High-energy product Nd-Fe-B permanent magnets. *Applied Physics Letters*, **44**, 148.

Csányi, G., Albaret, T., Payne, M. C., and De Vita, A. (2004). 'Learn on the fly': A hybrid classical and quantum-mechanical molecular dynamics simulation. *Physical Review Letters*, **93**, 175503.

Dal Corso, A., Baroni, S., and Resta, R. (1994). Density-functional theory of the dielectric constant: Gradient-corrected calculation for silicon. *Physical Review B*, **49**, 5323.

Dal Corso, A. and Resta, R. (1994). Density-functional theory of macroscopic stress: Gradient-corrected calculations for crystalline Se. *Physical Review B*, **50**, 4327.

Dalgarno, A. and Lewis, J. T. (1955). The exact calculation of long-range forces between atoms by perturbation theory. *Proceedings of the Royal Society of London A*, **233**, 70.

Damascelli, A., Hussain, Z., and Shen, Z.-X. (2003). Angle-resolved photoemission studies of the cuprate superconductors. *Review of Modern Physics*, **75**, 473.

Darling, B. T. and Dennison, D. M. (1940). The water vapor molecule. *Physical Review*, **57**, 128.

Dederichs, P. H., Blügel, S., Zeller, R., and Akai, H. (1984). Ground states of constrained systems: Application to Cerium impurities. *Physical Review Letters*, **53**, 2512.

Dieter, G. E. (1986). *Mechanical metallurgy* (3rd edn). McGraw-Hill, New York.

Dion, M., Rydberg, H., Schröder, E., Langreth, D. C., and Lundqvist, B. I. (2004). van der Waals density functional for general geometries. *Physical Review Letters*, **92**, 246401.

Dirac, P. A. M. (1927). The quantum theory of the emission and absorption of radiation. *Proceedings of the Royal Society of London A*, **114**, 243.

Dirac, P. A. M. (1928). The quantum theory of the electron. *Proceedings of the Royal Society of London A*, **117**, 610.

Dirac, P. A. M. (1930). Note on exchange phenomena in the Thomas atom. *Mathematical Proceedings of the Cambridge Philosophical Society*, **26**, 376.

Dresselhaus, G., Kip, A. F., and Kittel, C. (1955). Cyclotron Resonance of electrons and holes in silicon and germanium crystals. *Physical Review*, **98**, 368.

Dubrovinsky, L., Dubrovinskaia, N., Prakapenka, V. B., and Abakumov, A. M. (2012). Implementation of micro-ball nanodiamond anvils for high-pressure studies above 6 Mbar. *Nature Communications*, **3**, 1163.

Dufek, P., Blaha, P., Sliwko, V., and Schwarz, K. (1994). Generalized-gradient-approximation description of band splittings in transition-metal oxides and fluorides. *Physical Review B*, **49**, 10170.

Dyall, K. G. and Faegri, K. (2007). *Introduction to relativistic quantum chemistry*. Oxford University Press, Oxford.

Dziewonski, A. M. and Anderson, D. L. (1981). Preliminary reference Earth model. *Physics of the Earth and Planetary Interiors*, **25**, 297.

Eerenstein, W., Mathur, N. D., and Scott, J. F. (2006). Multiferroic and magneto-electric materials. *Nature*, **442**, 759.

Ehrenreich, H. and Cohen, M. H. (1959). Self-consistent field approach to the many-electron problem. *Physical Review*, **115**, 786.

Einstein, A. (1905*a*). Über einen die erzeugung und verwandlung des lichtes betreffenden heuristischen gesichtspunkt. *Annalen der Physik*, **322**, 132.

Einstein, A. (1905*b*). Zur elektrodynamik bewegter körper. *Annalen der Physik*, **322**, 891.

Eisenschitz, R. and London, F. (1930). Über das verhältnis der van der Waalsschen kräfte zu den homöopolaren bindungskräften. *Zeitschrift für Physik*, **60**, 491.

Eliashberg, G. M. (1960). Interactions between electrons and lattice vibrations in a superconductor. *Soviet Physics Journal of Experimental and Theoretical Physics*, **11**, 696.

Engel, G. E. and Farid, B. (1993). Generalized plasmon-pole model and plasmon band structures of crystals. *Physical Review B*, **47**, 15931.

Eremets, M. I. and Troyan, I. A. (2011). Conductive dense hydrogen. *Nature Materials*, **10**, 927.

Errandonea, D., Boehler, R., and Ross, M. (2001). Melting of the alkaline-earth metals to 80 GPa. *Physical Review B*, **65**, 012108.

Farid, B. and Godby, R. W. (1991). Cohesive energies of crystals. *Physical Review B*, **43**, 14248.

Favot, F. and Dal Corso, A. (1999). Phonon dispersions: Performance of the generalized gradient approximation. *Physical Review B*, **60**, 11427.

Featherston, F. H. and Neighbours, J. R. (1963). Elastic constants of tantalum, tungsten, and molybdenum. *Physical Review*, **130**, 1324.

Fermi, E. (1928). Eine statistische methode zur bestimmung einiger eigenschaften des atoms und ihre anwendung auf die theorie des periodischen systems der elemente. *Zeitschrift für Physik*, **48**, 73.

Fermi, E. (1932). Quantum theory of radiation. *Review of Modern Physics*, **4**, 87.

Fermi, E. (1956). *Thermodynamics*. Dover, New York.

Fermi, E. (1966). *Molecules, crystals, and quantum statistics*. Benjamin. Translated from 'Molecole e cristalli' by M. Ferro-Luzzi, ed. L. Motz.

Ferrari, A. C. and Robertson, J. (2004). Raman spectroscopy of amorphous, nano-structured, diamondlike carbon, and nanodiamond. *Philosophical Transactions of the Royal Society of London A*, **362**, 2477.

Ferraro, J. R., Nakamoto, K., and Brown, C. W. (2003). *Introductory Raman spectroscopy*. Academic Press, New York.

Ferrenberg, A. M. and Swendsen, R. H. (1988). New Monte Carlo technique for studying phase transitions. *Physical Review Letters*, **61**, 2635.

Fetter, A. L. and Walecka, J. D. (2003). *Quantum theory of many-particle systems*. Dover, New York.

Feynman, R. P. (1939). Forces in molecules. *Physical Review*, **56**, 340.

Floris, A., Profeta, G., Lathiotakis, N. N., Lüders, M., Marques, M. A. L., Franchini, C., Gross, E. K. U., Continenza, A., and Massidda, S. (2005). Superconducting properties of MgB_2 from first principles. *Physical Review Letters*, **94**, 037004.

Fock, V. (1930a). Bemerkung zum virialsatz. *Zeitschrift für Physik*, **63**, 855.

Fock, V. (1930b). Näherungsmethode zur lösung des quantenmechanischen mehr-körperproblems. *Zeitschrift für Physik*, **61**, 126.

Fuchs, M. and Scheffler, M. (1999). Ab initio pseudopotentials for electronic structure calculations of poly-atomic systems using density-functional theory. *Computer Physics Communications*, **119**, 67.

Fujiwara, H. (2007). *Spectroscopic ellipsometry: Principles and applications*. Wiley, New York.

Furche, F. (2001). Molecular tests of the random phase approximation to the exchange-correlation energy functional. *Physical Review B*, **64**, 195120.

Geim, A. K. (2011). Random walk to graphene. *Angewandte Chemie International Edition*, **50**, 6966.

Geim, A. K. and Novoselov, K. S. (2007). The rise of graphene. *Nature Materials*, **6**, 183.

Georges, A., Kotliar, G., Krauth, W., and Rozenberg, M. J. (1996). Dynamical mean-field theory of strongly correlated fermion systems and the limit of infinite dimensions. *Review of Modern Physics*, **68**, 13.

Giacomazzi, L., Umari, P., and Pasquarello, A. (2009). Medium-range structure of vitreous SiO_2 obtained through first-principles investigation of vibrational spectra. *Physical Review B*, **79**, 064202.

Giannozzi, P., Baroni, S., Bonini, N., Calandra, M., Car, R., Cavazzoni, C., Ceresoli, D., Chiarotti, G. L., Cococcioni, M., Dabo, I., Dal Corso, A., de Gironcoli, S., Fabris, S., Fratesi, G., Gebauer, R., Gerstmann, U., Gougoussis, C., Kokalj, A., Lazzeri, M., Martin-Samos, L., Marzari, N., Mauri, F., Mazzarello, R., Paolini, S.,

Pasquarello, A., Paulatto, L., Sbraccia, C., Scandolo, S., Sclauzero, G., Seitsonen, A. P., Smogunov, A., Umari, P., and Wentzcovitch, R. M. (2009). QUANTUM ESPRESSO: A modular and open-source software project for quantum simulations of materials. *Journal of Physics: Condensed Matter*, **21**, 395502.

Gillan, M. J., Alfè, D., Brodholt, J., Vočadlo, L., and Price, G. D. (2006). First-principles modelling of earth and planetary materials at high pressures and temperatures. *Reports on Progress in Physics*, **69**, 2365.

Giustino, F. and Pasquarello, A. (2006). Mixed Wannier–Bloch functions for electrons and phonons in periodic systems. *Physical Review Letters*, **96**, 216403.

Godby, R. W., Schlüter, M., and Sham, L. J. (1986). Accurate exchange-correlation potential for silicon and its discontinuity on addition of an electron. *Physical Review Letters*, **56**, 2415.

Goldstein, H. (1950). *Classical mechanics*. Addison-Wesley, Reading.

Gonze, X. (1995). Perturbation expansion of variational principles at arbitrary order. *Physical Review A*, **52**, 1086.

Gonze, X. and Lee, C. (1997). Dynamical matrices, Born effective charges, dielectric permittivity tensors, and interatomic force constants from density-functional perturbation theory. *Physical Review B*, **55**, 10355.

Graves, J. L. and Parr, R. G. (1985). Possible universal scaling properties of potential-energy curves for diatomic molecules. *Physical Review A*, **31**, 1.

Greeley, J., Jaramillo, T. F., Bonde, J., Chorkendorff, I. B., and Nørskov, J. K. (2006). Computational high-throughput screening of electrocatalytic materials for hydrogen evolution. *Nature Materials*, **5**, 909.

Green, M. A. and Keevers, M. J. (1995). Optical properties of intrinsic silicon at 300 K. *Progress in Photovoltaics: Research and Applications*, **3**, 189.

Gregory, R. D. (2006). *Classical mechanics: An undergraduate text*. Cambridge University Press, Cambridge.

Grimvall, G. (1981). *The electron–phonon interaction in metals*. North Holland, Amsterdam.

Grimvall, G. (1999). *Thermophysical properties of materials*. North Holland, Amsterdam.

Grossman, J. C., Schwegler, E., Draeger, E. W., Gygi, F., and Galli, G. (2004). Towards an assessment of the accuracy of density functional theory for first principles simulations of water. *Journal of Chemical Physics*, **120**, 300.

Gunnarsson, O. (1976). Band model for magnetism of transition metals in the spin-density-functional formalism. *Journal of Physics F: Metal Physics*, **6**, 587.

Gunnarsson, O. and Lundqvist, B. I. (1976). Exchange and correlation in atoms, molecules, and solids by the spin-density-functional formalism. *Physical Review B*, **13**, 4274.

Hafner, J., Wolverton, C., and Ceder, G. (2006). Toward computational materials design: The impact of density functional theory on materials research. *MRS Bulletin*, **31**, 659.

Hall, J. J. (1967). Electronic effects in the elastic constants of n-type silicon. *Physical Review*, **161**, 756.

Halliday, D., Resnick, R., and Walker, J. (1997). *Fundamentals of physics* (5th edn). Wiley, New York.

Hamann, D. R., Schlüter, M., and Chiang, C. (1979). Norm-conserving pseudo-potentials. *Physical Review Letters*, **43**, 1494.

Hanke, W. and Sham, L. J. (1980). Many-particle effects in the optical spectrum of a semiconductor. *Physical Review B*, **21**, 4656.

Hartree, D. R. (1928). The wave mechanics of an atom with a non-Coulomb central field. Part II. Some results and discussion. *Mathematical Proceedings of the Cambridge Philosophical Society*, **24**, 111.

Hasan, M. Z. and Kane, C. L. (2010). Topological insulators. *Review of Modern Physics*, **82**, 3045.

Heath, J. R., Zhang, Q., O'Brien, S. C., Curl, R. F., Kroto, H. W., and Smalley, R. E. (1987). The formation of long carbon chain molecules during laser vaporization of graphite. *Journal of the American Chemical Society*, **109**, 359.

Hedin, L. (1965). New method for calculating the one-particle Green's function with application to the electron-gas problem. *Physical Review*, **139**, A796.

Hedin, L. and Lundqvist, S. (1969). Effects of electron–electron and electron–phonon interactions on the one-electron states of solids. In *Solid State Physics, 23*, ed. H. Ehrenreich, F. Seitz, D. Turnbull, Academic Press, New York, 1.

Heitler, W. H. and London, F. (1927). Wechselwirkung neutraler atome und homöopolare bindung nach der quantenmechanik. *Zeitschrift für Physik*, **44**, 455.

Hellmann, H. (1933). Zur rolle der kinetischen elektronenenergie für die zwischenatomaren kräfte. *Zeitschrift für Physik*, **85**, 180.

Herbst, J. F., Croat, J. J., Pinkerton, F. E., and Yelon, W. B. (1984). Relationships between crystal structure and magnetic properties in $Nd_2Fe_{14}B$. *Physical Review B*, **29**, 4176.

Hertz, H. (1887). Über einen einfluss des ultravioletten lichtes auf die electrische entladung. *Annalen der Physik*, **267**, 983.

Hill, R. (1952). The elastic behaviour of a crystalline aggregate. *Proceedings of the Physical Society, Section A*, **65**, 349.

Himpsel, F. J. (1983). Angle-resolved measurements of the photoemission of electrons in the study of solids. *Advances in Physics*, **32**, 1.

Himpsel, F. J. (1990). Inverse photoemission from semiconductors. *Surface Science Reports*, **12**, 3.

Hofer, W. A. (2003). Challenges and errors: Interpreting high resolution images in scanning tunneling microscopy. *Progress in Surface Science*, **71**, 147.

Hohenberg, P. and Kohn, W. (1964). Inhomogeneous electron gas. *Physical Review*, **136**, B864.

Hopcroft, M. A., Nix, W. D., and Kenny, T. W. (2010). What is the Young's modulus of silicon? *Journal of Microelectromechanical Systems*, **19**, 229.

Hu, C. C. (2009). *Modern semiconductor devices for integrated circuits*. Prentice Hall, New York.

Hu, J. Z., Merkle, L. D., Menoni, C. S., and Spain, I. L. (1986). Crystal data for high-pressure phases of silicon. *Physical Review B*, **34**, 4679.

Humphrey, W., Dalke, A., and Schulten, K. (1996). VMD – Visual Molecular Dynamics. *Journal of Molecular Graphics*, **14**, 33.

Hybertsen, M. S. and Louie, S. G. (1986). Electron correlation in semiconductors and insulators: Band gaps and quasiparticle energies. *Physical Review B*, **34**, 5390.

Ibach, H. (1970). Optical surface phonons in zinc oxide detected by slow-electron spectroscopy. *Physical Review Letters*, **24**, 1416.

Iitaka, T., Hirose, K., Kawamura, K., and Murakami, M. (2004). The elasticity of the MgSiO$_3$ post-perovskite phase in the Earth's lowermost mantle. *Nature*, **430**, 442.

Ismail-Beigi, S., Chang, E. K., and Louie, S. G. (2001). Coupling of nonlocal potentials to electromagnetic fields. *Physical Review Letters*, **87**, 087402.

Itskov, M. (2007). *Tensor algebra and tensor analysis for engineers: With applications to continuum mechanics.* Springer, Berlin.

Jackson, J. D. (1998). *Classical electrodynamics* (3rd edn). Wiley, New York.

Jaklevic, R. C. and Lambe, J. (1966). Molecular vibration spectra by electron tunneling. *Physical Review Letters*, **17**, 1139.

Janak, J. F. (1978). Proof that $\partial E / \partial n_i = \epsilon_i$ in density-functional theory. *Physical Review B*, **18**, 7165.

Jayaraman, A. (1983). Diamond anvil cell and high-pressure physical investigations. *Review of Modern Physics*, **55**, 65.

Johnson, D. L. (1974). Local field effects and the dielectric response matrix of insulators: a model. *Physical Review B*, **9**, 4475.

Johnson, D. L. (1975). Local-field effects, X-ray diffraction, and the possibility of observing the optical Borrmann effect: Solutions to Maxwell's equations in perfect crystals. *Physical Review B*, **12**, 3428.

Johnson, S. G. and Frigo, M. (2007). A modified split-radix FFT with fewer arithmetic operations. *IEEE Transactions on Signal Processing*, **55**, 111.

Kang, K., Meng, Y. S., Bréger, J., Grey, C. P., and Ceder, G. (2006). Electrodes with high power and high capacity for rechargeable lithium batteries. *Science*, **311**, 977.

Kaxiras, E. (2003). *Atomic and electronic structure of solids.* Cambridge University Press, Cambridge.

Kermode, J. R., Albaret, T., Sherman, D., Bernstein, N., Gumbsch, P., Payne, M. C., Csányi, G., and De Vita, A. (2008). Low-speed fracture instabilities in a brittle crystal. *Nature*, **455**, 1224.

Kim, Y.-S., Marsman, M., Kresse, G., Tran, F., and Blaha, P. (2010). Towards efficient band structure and effective mass calculations for III-V direct band-gap semiconductors. *Physical Review B*, **82**, 205212.

King-Smith, R. D. and Vanderbilt, D. (1993). Theory of polarization of crystalline solids. *Physical Review B*, **47**, 1651.

Kioupakis, E., Zhang, P., Cohen, M. L., and Louie, S. G. (2008). GW quasiparticle corrections to the LDA+U/GGA+U electronic structure of bcc hydrogen. *Physical Review B*, **77**, 155114.

Kitaura, R., Nakanishi, R., Saito, T., Yoshikawa, H., Awaga, K., and Shinohara, H. (2009). High-yield synthesis of ultrathin metal nanowires in carbon nanotubes. *Angewandte Chemie International Edition*, **48**, 8298.

Kittel, C. (1958). *Elementary statistical physics.* Wiley, New York.

Kittel, C. (1976). *Introduction to solid state physics* (5th edn). Wiley, New York.

Kiusalaas, J. (2005). *Numerical methods in engineering with Matlab*. Cambridge University Press, Cambridge.

Kleinman, L. and Bylander, D. M. (1982). Efficacious form for model pseudopotentials. *Physical Review Letters*, **48**, 1425.

Koch, W. and Holthausen, M. C. (2001). *A chemist's guide to density functional theory* (2nd edn). Wiley, New York.

Kohn, W. (1999). Nobel Lecture. Electronic structure of matter: Wave functions and density functionals. *Review of Modern Physics*, **71**, 1253.

Kohn, W. and Sham, L. J. (1965). Self-consistent equations including exchange and correlation effects. *Physical Review*, **140**, A1133.

Kokalj, A. (2003). Computer graphics and graphical user interfaces as tools in simulations of matter at the atomic scale. *Computational Materials Science*, **28**, 155.

Kong, Y., Dolgov, O. V., Jepsen, O., and Andersen, O. K. (2001). Electron–phonon interaction in the normal and superconducting states of MgB_2. *Physical Review B*, **64**, 020501.

Kortus, J., Mazin, I. I., Belashchenko, K. D., Antropov, V. P., and Boyer, L. L. (2001). Superconductivity of metallic boron in MgB_2. *Physical Review Letters*, **86**, 4656.

Kotliar, G., Savrasov, S. Y., Haule, K., Oudovenko, V. S., Parcollet, O., and Marianetti, C. A. (2006). Electronic structure calculations with dynamical mean-field theory. *Review of Modern Physics*, **78**, 865.

Kramers, H. A. (1934). L'interaction entre les atomes magnétogènes dans un cristal paramagnétique. *Physica*, **1**, 182.

Kresse, G. and Furthmüller, J. (1996). Efficient iterative schemes for *ab initio* total-energy calculations using a plane-wave basis set. *Physical Review B*, **54**, 11169.

Kübler, J. (2000). *Theory of itinerant electron magnetism*. Oxford University Press, Oxford.

Kübler, J., Hock, K.-H., Sticht, J., and Williams, A. R. (1988). Density functional theory of non-collinear magnetism. *Journal of Physics F: Metal Physics*, **18**, 469.

Kunc, K. and Syassen, K. (2010). P(V) equations of state of solids: Density functional theory calculations and LDA versus GGA scaling. *Physical Review B*, **81**, 134102.

Laio, A., Bernard, S., Chiarotti, G. L., Scandolo, S., and Tosatti, E. (2000). Physics of iron at Earth's core conditions. *Science*, **287**, 1027.

Landau, L. D. and Lifshitz, E. M. (1960). *Electrodynamics of continuous media*. Pergamon, Oxford.

Landau, L. D. and Lifshitz, E. M. (1965). *Quantum mechanics, non-relativistic theory* (2nd edn). Pergamon, Oxford.

Landau, L. D. and Lifshitz, E. M. (1969). *Statistical physics* (2nd edn). Pergamon, Oxford.

Langreth, D. C. and Mehl, M. J. (1983). Beyond the local-density approximation in calculations of ground-state electronic properties. *Physical Review B*, **28**, 1809.

Lassner, E. and Schubert, W. D. (1999). *Tungsten: Properties, chemistry, technology of the element, alloys, and chemical compounds*. Kluwer, New York.

Lazzeri, M., Calandra, M., and Mauri, F. (2003). Anharmonic phonon frequency shift in MgB_2. *Physical Review B*, **68**, 220509.

Lee, A. M. and Handy, N. C. (1993). Dissociation of hydrogen and nitrogen molecules studied using density functional theory. *Journal of the Chemical Society, Faraday Transactions*, **89**, 3999.

Lee, C., Yang, W., and Parr, R. G. (1988). Development of the Colle–Salvetti correlation-energy formula into a functional of the electron density. *Physical Review B*, **37**, 785.

Lennard-Jones, J. E. (1929). The electronic structure of some diatomic molecules. *Transactions of the Faraday Society*, **25**, 668.

Lewis, G. N. (1916). The atom and the molecule. *Journal of the American Chemical Society*, **38**, 762.

Liu, G., Wang, G., Zhu, Y., Zhang, H., Zhang, G., Wang, X., Zhou, Y., Zhang, W., Liu, H., Zhao, L., Meng, J., Dong, X., Chen, C., Xu, Z., and Zhou, X. J. (2008). Development of a vacuum ultraviolet laser-based angle-resolved photoemission system with a superhigh energy resolution better than 1 meV. *Review of Scientific Instruments*, **79**, 023105.

Lu, D., Vishik, I. M., Yi, M., Chen, Y., Moore, R. G., and Shen, Z.-X. (2012). Angle-resolved photoemission studies of quantum materials. *Annual Review of Condensed Matter Physics*, **3**, 129.

Lu, Z. W., Zunger, A., and Deutsch, M. (1993). Electronic charge distribution in crystalline diamond, silicon, and germanium. *Physical Review B*, **47**, 9385.

Lüders, M., Marques, M. A. L., Lathiotakis, N. N., Floris, A., Profeta, G., Fast, L., Continenza, A., Massidda, S., and Gross, E. K. U. (2005). *Ab initio* theory of superconductivity. I. Density functional formalism and approximate functionals. *Physical Review B*, **72**, 024545.

Macfarlane, G. G., McLean, T. P., Quarrington, J. E., and Roberts, V. (1958). Fine structure in the absorption-edge spectrum of Si. *Physical Review*, **111**, 1245.

Manzano, C., Soe, W.-H., Kawai, H., Saeys, M., and Joachim, C. (2011). Origin of the apparent (2×1) topography of the $Si(100)-c(4\times2)$ surface observed in low-temperature STM images. *Physical Review B*, **83**, 201302.

Marini, A. (2008). *Ab initio* finite-temperature excitons. *Physical Review Letters*, **101**, 106405.

Marini, A., Onida, G., and Del Sole, R. (2001). Plane-wave DFT-LDA calculation of the electronic structure and absorption spectrum of copper. *Physical Review B*, **64**, 195125.

Marlow, W. C. and Bershader, D. (1964). Shock-tube measurement of the polarizability of atomic hydrogen. *Physical Review*, **133**, A629.

Marques, M. A. L. and Gross, E. K. U. (2004). Time-dependent density functional theory. *Annual Reviews of Physical Chemistry*, **55**, 427.

Marques, M. A. L., Lüders, M., Lathiotakis, N. N., Profeta, G., Floris, A., Fast, L., Continenza, A., Gross, E. K. U., and Massidda, S. (2005). *Ab initio* theory of superconductivity. II. Application to elemental metals. *Physical Review B*, **72**, 024546.

Martin, R. M. (2004). *Electronic structure: Basic theory and practical methods.* Cambridge University Press, Cambridge.

Mattuck, R. D. (1976). *A guide to Feynman diagrams in the many-body problem* (2nd edn). McGraw-Hill, New York.

Mauri, F., Galli, G., and Car, R. (1993). Orbital formulation for electronic-structure calculations with linear system-size scaling. *Physical Review B*, **47**, 9973.

Mauri, F. and Louie, S. G. (1996). Magnetic susceptibility of insulators from first principles. *Physical Review Letters*, **76**, 4246.

McSkimin, H. J. and Andreatch, P. Jr. (1964). Elastic moduli of silicon vs hydrostatic pressure at 25.0°C and -195.8°C. *Journal of Applied Physics*, **35**, 2161.

Mehl, M. J., Osburn, J. E., Papaconstantopoulos, D. A., and Klein, B. M. (1990). Structural properties of ordered high-melting-temperature intermetallic alloys from first-principles total-energy calculations. *Physical Review B*, **41**, 10311.

Mehta, S., Price, G. D., and Alfè, D. (2006). *Ab initio* thermodynamics and phase diagram of solid magnesium: A comparison of the LDA and GGA. *Journal of Chemical Physics*, **125**, 194507.

Merzbacher, E. (1998). *Quantum mechanics* (3rd edn). Wiley, New York.

Messiah, A. (1965). *Quantum mechanics. Volume II*. North Holland, Amsterdam.

Migdal, A. B. (1958). Interaction between electrons and lattice vibrations in a normal metal. *Soviet Physics Journal of Experimental and Theoretical Physics*, **7**, 996.

Mochalin, V. N., Shenderova, O., Ho, D., and Gogotsi, Y. (2011). The properties and applications of nanodiamonds. *Nature Nanotechnology*, **7**, 11.

Moll, N., Bockstedte, M., Fuchs, M., Pehlke, E., and Scheffler, M. (1995). Application of generalized gradient approximations: The diamond-β-tin phase transition in Si and Ge. *Physical Review B*, **52**, 2550.

Momma, K. and Izumi, F. (2008). VESTA: A three-dimensional visualization system for electronic and structural analysis. *Journal of Applied Crystallography*, **41**, 653.

Monserrat, B., Drummond, N. D., and Needs, R. J. (2013). Anharmonic vibrational properties in periodic systems: Energy, electron–phonon coupling, and stress. *Physical Review B*, **87**, 144302.

Moon, C.-Y., Kim, Y.-H., and Chang, K. J. (2004). Dielectric-screening properties and Coulomb pseudopotential μ^* for MgB_2. *Physical Review B*, **70**, 104522.

Moriarty, J. A. and Althoff, J. D. (1995). First-principles temperature–pressure phase diagram of magnesium. *Physical Review B*, **51**, 5609.

Morrone, J. A. and Car, R. (2008). Nuclear quantum effects in water. *Physical Review Letters*, **101**, 017801.

Morse, P. M. (1929). Diatomic molecules according to the wave mechanics. II. Vibrational levels. *Physical Review*, **34**, 57.

Moruzzi, V. L., Marcus, P. M., Schwarz, K., and Mohn, P. (1986). Ferromagnetic phases of bcc and fcc Fe, Co, and Ni. *Physical Review B*, **34**, 1784.

Murakami, M., Hirose, K., Kawamura, K., Sata, N., and Ohishi, Y. (2004). Post-perovskite phase transition in $MgSiO_3$. *Science*, **304**, 855.

Muramatsu, H., Hayashi, T., Kim, Y. A., Shimamoto, D., Endo, M., Terrones, M., and Dresselhaus, M. S. (2008). Synthesis and isolation of molybdenum atomic wires. *Nano Letters*, **8**, 237.

Murphy, W. F. (1977). The ro-vibrational Raman spectrum of water vapour ν_2 and $2\nu_2$. *Molecular Physics*, **33**, 1701.

Nagamatsu, J., Nakagawa, N., Muranaka, T., Zenitani, Y., and Akimitsu, J. (2001). Superconductivity at 39 K in magnesium diboride. *Nature*, **410**, 63.

Nielsen, M. A. and Chuang, I. L. (2010). *Quantum computation and quantum information: 10th anniversary edition*. Cambridge University Press, Cambridge.

Nielsen, O. H. and Martin, R. M. (1983). First-principles calculation of stress. *Physical Review Letters*, **50**, 697.

Nielsen, O. H. and Martin, R. M. (1985). Quantum-mechanical theory of stress and force. *Physical Review B*, **32**, 3780.

Nocedal, J. and Wright, S. J. (1999). *Numerical optimization*. Springer, Berlin.

Noffsinger, J., Kioupakis, E., Van de Walle, C. G., Louie, S. G., and Cohen, M. L. (2012). Phonon-assisted optical absorption in silicon from first principles. *Physical Review Letters*, **108**, 167402.

Nørskov, J. K., Bligaard, T., Logadottir, A., Kitchin, J. R., Chen, J. G., Pandelov, S., and Stimming, U. (2005). Trends in the exchange current for hydrogen evolution. *Journal of the Electrochemical Society*, **152**, J23.

Novoselov, K. S., Jiang, D., Schedin, F., Booth, T. J., Khotkevich, V. V., Morozov, S. V., and Geim, A. K. (2005). Two-dimensional atomic crystals. *Proceedings of the National Academy of Sciences of the United States of America*, **102**, 10451.

Novotny, L. and Hecht, B. (2006). *Principles of nano-optics*. Cambridge University Press, Cambridge.

Nye, J. F. (1985). *Physical properties of crystals: Their representation by tensors and matrices*. Oxford University Press, Oxford.

Oganov, A. R. and Ono, S. (2004). Theoretical and experimental evidence for a post-perovskite phase of $MgSiO_3$. *Nature*, **430**, 445.

Oganov, A. R., Price, G. D., and Scandolo, S. (2005). *Ab initio* theory of planetary materials. *Zeitschrift für Kristallographie*, **220**, 531.

Olijnyk, H. and Holzapfel, W. B. (1985). High-pressure structural phase transition in Mg. *Physical Review B*, **31**, 4682.

Onida, G., Reining, L., Godby, R. W., Del Sole, R., and Andreoni, W. (1995). *Ab initio* calculations of the quasiparticle and absorption spectra of clusters: The sodium tetramer. *Physical Review Letters*, **75**, 818.

Onida, G., Reining, L., and Rubio, A. (2002). Electronic excitations: Density-functional versus many-body Green's-function approaches. *Review of Modern Physics*, **74**, 601.

Ordejón, P., Drabold, D. A., Grumbach, M. P., and Martin, R. M. (1993). Unconstrained minimization approach for electronic computations that scales linearly with system size. *Physical Review B*, **48**, 14646.

Oshikiri, M., Aryasetiawan, F., Imanaka, Y., and Kido, G. (2002). Quasiparticle effective-mass theory in semiconductors. *Physical Review B*, **66**, 125204.

Parlinski, K., Li, Z. Q., and Kawazoe, Y. (1997). First-principles determination of the soft mode in cubic ZrO_2. *Physical Review Letters*, **78**, 4063.

Parr, R. G. and Yang, W. (1989). *Density-functional theory of atoms and molecules*. Oxford University Press, Oxford.

Parrinello, M. and Rahman, A. (1980). Crystal structure and pair potentials: A molecular-dynamics study. *Physical Review Letters*, **45**, 1196.

Parsons, R. (1958). The rate of electrolytic hydrogen evolution and the heat of adsorption of hydrogen. *Transactions of the Faraday Society*, **54**, 1053.

Pask, J. E., Singh, D. J., Mazin, I. I., Hellberg, C. S., and Kortus, J. (2001). Structural, electronic, and magnetic properties of MnO. *Physical Review B*, **64**, 024403.

Pasquarello, A. and Car, R. (1997). Dynamical charge tensors and infrared spectrum of amorphous SiO_2. *Physical Review Letters*, **79**, 1766.

Pasquarello, A., Sarnthein, J., and Car, R. (1998). Dynamic structure factor of vitreous silica from first principles: Comparison to neutron-inelastic-scattering experiments. *Physical Review B*, **57**, 14133.

Patrick, C. E. and Giustino, F. (2013). Quantum nuclear effects in the photophysics of diamondoids. *Nature Communications*, **4**, 2006.

Pauli, W. (1927). Zur quantenmechanik des magnetischen elektrons. *Zeitschrift für Physik*, **43**, 601.

Pauli, W. (1940). The connection between spin and statistics. *Physical Review*, **58**, 716.

Pauling, L. (1928). The application of the quantum mechanics to the structure of the hydrogen molecule and hydrogen molecule-ion and to related problems. *Chemical Reviews*, **5**, 173.

Pauling, L. and Beach, J. Y. (1935). The van der Waals interaction of hydrogen atoms. *Physical Review*, **47**, 686.

Pavone, P., Karch, K., Schütt, O., Strauch, D., Windl, W., Giannozzi, P., and Baroni, S. (1993). Ab-initio lattice dynamics of diamond. *Physical Review B*, **48**, 3156.

Peierls, R. E. (1955). *Quantum theory of solids*. Clarendon Press, Oxford.

Pekeris, C. L. (1958). Ground state of two-electron atoms. *Physical Review*, **112**, 1649.

Perdew, J. P. and Zunger, A. (1981). Self-interaction correction to density-functional approximations for many-electron systems. *Physical Review B*, **23**, 5048.

Perdew, J. P., Parr, R. G., Levy, M., and Balduz, J. L. (1982). Density-functional theory for fractional particle number: Derivative discontinuities of the energy. *Physical Review Letters*, **49**, 1691.

Perdew, J. P. and Levy, M. (1983). Physical content of the exact Kohn–Sham orbital energies: Band gaps and derivative discontinuities. *Physical Review Letters*, **51**, 1884.

Perdew, J. P., Chevary, J. A., Vosko, S. H., Jackson, K. A., Pederson, M. R., Singh, D. J., and Fiolhais, C. (1992). Atoms, molecules, solids, and surfaces: Applications of the generalized gradient approximation for exchange and correlation. *Physical Review B*, **46**, 6671.

Perdew, J. P., Burke, K., and Ernzerhof, M. (1996). Generalized gradient approximation made simple. *Physical Review Letters*, **77**, 3865.

Perdigão, L., Deresmes, D., Grandidier, B., Dubois, M., Delerue, C., Allan, G., and Stiévenard, D. (2004). Semiconducting surface reconstructions of p-type Si(100) substrates at 5 K. *Physical Review Letters*, **92**, 216101.

Petersen, K. E. (1982). Silicon as a mechanical material. *Proceedings of the IEEE*, **70**, 420.

Pettifor, D. G. (1995). *Bonding and structure of molecules and solids*. Oxford University Press, Oxford.

Phani, M. K., Lebowitz, Joel L., and Kalos, M. H. (1980). Monte Carlo studies of an fcc Ising antiferromagnet with nearest- and next-nearest-neighbor interactions. *Physical Review B*, **21**, 4027.

Phillips, J. C. and Kleinman, L. (1959). New method for calculating wave functions in crystals and molecules. *Physical Review*, **116**, 287.

Pick, R. M., Cohen, M. H., and Martin, R. M. (1970). Microscopic theory of force constants in the adiabatic approximation. *Physical Review B*, **1**, 910.

Pickard, C. J. and Mauri, F. (2001). All-electron magnetic response with pseudo-potentials: NMR chemical shifts. *Physical Review B*, **63**, 245101.

Pickard, C. J. and Mauri, F. (2003). Nonlocal pseudopotentials and magnetic fields. *Physical Review Letters*, **91**, 196401.

Pickett, W. E. (1989). Electronic structure of the high-temperature oxide super-conductors. *Review of Modern Physics*, **61**, 433.

Prawer, S. and Nemanich, R. J. (2004). Raman spectroscopy of diamond and doped diamond. *Philosophical Transactions of the Royal Society of London A*, **362**, 2537.

Prendergast, D. and Galli, G. (2006). X-ray absorption spectra of water from first principles calculations. *Physical Review Letters*, **96**, 215502.

Price, D. L. and Carpenter, J. M. (1987). Scattering function of vitreous silica. *Journal of Non-Crystalline Solids*, **92**, 153.

Purcell, E. M. (1938). The focusing of charged particles by a spherical condenser. *Physical Review*, **54**, 818.

Rajagopal, A. K. and Callaway, J. (1973). Inhomogeneous electron gas. *Physical Review B*, **7**, 1912.

Raman, C. V. (1928). A new radiation. *Indian Journal of Physics*, **2**, 387.

Raman, C. V. and Krishnan, K. S. (1928). A new type of secondary radiation. *Nature*, **121**(3048), 501.

Ramesh, R. and Spaldin, N. A (2007). Multiferroics: Progress and prospects in thin films. *Nature Materials*, **6**, 21.

Rastogi, V. K. and Girvin, M. E. (1999). Structural changes linked to proton translocation by subunit c of the ATP synthase. *Nature*, **402**, 263.

Raty, J.-Y. and Galli, G. (2003). Ultradispersity of diamond at the nanoscale. *Nature Materials*, **2**, 792.

Raty, J.-Y., Galli, G., Bostedt, C., van Buuren, T. W., and Terminello, L. J. (2003). Quantum confinement and fullerenelike surface reconstructions in nanodiamonds. *Physical Review Letters*, **90**, 037401.

Read, A. J. and Needs, R. J. (1991). Calculation of optical matrix elements with nonlocal pseudopotentials. *Physical Review B*, **44**, 13071.

Resta, R. (1994). Macroscopic polarization in crystalline dielectrics: The geometric phase approach. *Review of Modern Physics*, **66**, 899.

Resta, R. (2000). Manifestations of Berry's phase in molecules and condensed matter. *Journal of Physics: Condensed Matter*, **12**, R107.

Roaf, D. J. (1962). The Fermi surfaces of copper, silver and gold II. Calculation of the Fermi surfaces. *Philosophical Transactions of the Royal Society of London A*, **255**, 135.

Rohlfing, M. and Louie, S. G. (1998). Electron–hole excitations in semiconductors and insulators. *Physical Review Letters*, **81**, 2312.

Rose, J. H., Ferrante, J., and Smith, J. R. (1981). Universal binding energy curves for metals and bimetallic interfaces. *Physical Review Letters*, **47**, 675.

Rostgaard, C., Jacobsen, K. W., and Thygesen, K. S. (2010). Fully self-consistent GW calculations for molecules. *Physical Review B*, **81**, 085103.

Roundy, D., Krenn, C. R., Cohen, M. L., and Morris, J. W. (2001). The ideal strength of tungsten. *Philosophical Magazine A*, **81**, 1725.

Runge, E. and Gross, E. K. U. (1984). Density-functional theory for time-dependent systems. *Physical Review Letters*, **52**, 997.

Saito, R., Dresselhaus, G., and Dresselhaus, M. S. (1998). *Physical Properties of Carbon Nanotubes*. Imperial College Press, London.

Sakurai, J. J. and Napolitano, J. (2011). *Modern quantum mechanics*. Addison-Wesley, San Francisco.

Sarnthein, J., Pasquarello, A., and Car, R. (1995). Structural and electronic properties of liquid and amorphous SiO_2: An *ab initio* molecular dynamics study. *Physical Review Letters*, **74**, 4682.

Sarnthein, J., Pasquarello, A., and Car, R. (1997). Origin of the high-frequency doublet in the vibrational spectrum of vitreous SiO_2. *Science*, **275**, 1925.

Savrasov, S. Y. (1996). Linear-response theory and lattice dynamics: A muffin-tin-orbital approach. *Physical Review B*, **54**, 16470.

Sayle, R. A. and Milner-White, E. J. (1995). RASMOL: Biomolecular graphics for all. *Trends in Biochemical Sciences*, **20**, 374.

Schaad, L. J. and Hicks, W. V. (1970). Equilibrium bond length in H_2^+. *Journal of Chemical Physics*, **53**, 851.

Schilling, A., Cantoni, M., Guo, J. D., and Ott, H. R. (1993). Superconductivity above 130 K in the Hg-Ba-Ca-Cu-O system. *Nature*, **363**(6424), 56.

Schreckenbach, G. and Ziegler, T. (1998). Density functional calculations of NMR chemical shifts and ESR g-tensors. *Theoretical Chemistry Accounts*, **99**, 71.

Schrödinger, E. (1926). Quantisierung als eigenwertproblem. *Annalen der Physik*, **385**, 437.

Schwoerer-Böhning, M., Macrander, A. T., and Arms, D. A. (1998). Phonon dispersion of diamond measured by inelastic X-ray scattering. *Physical Review Letters*, **80**, 5572.

Senn, H. M. and Thiel, W. (2009). QM/MM methods for biomolecular systems. *Angewandte Chemie International Edition*, **48**, 1198.

Sham, L. J. and Rice, T. M. (1966). Many-particle derivation of the effective-mass equation for the Wannier exciton. *Physical Review*, **144**, 708.

Sham, L. J. and Schlüter, M. (1983). Density-functional theory of the energy gap. *Physical Review Letters*, **51**, 1888.

Shen, G., Mao, H.-K., Hemley, R. J., Duffy, T. S., and Rivers, M. L. (1998). Melting and crystal structure of iron at high pressures and temperatures. *Geophysical Research Letters*, **25**, 373.

Shoenberg, D. (1962). The Fermi surfaces of copper, silver and gold. I. The de Haas–van Alphen effect. *Philosophical Transactions of the Royal Society of London A*, **255**, 85.

Sidorin, I., Gurnis, M., and Helmberger, D. V. (1999). Evidence for a ubiquitous seismic discontinuity at the base of the mantle. *Science*, **286**, 1326.

Siegbahn, K. (1982). Electron spectroscopy for atoms, molecules, and condensed matter. *Review of Modern Physics*, **54**, 709.

Sinnokrot, M. O. and Sherrill, C. D. (2001). Density functional theory predictions of anharmonicity and spectroscopic constants for diatomic molecules. *Journal of Chemical Physics*, **115**, 2439.

Škoro, G. P., Bennett, J. R. J., Edgecock, T. R., Gray, S. A., McFarland, A. J., Booth, C. N., Rodgers, K. J., and Back, J. J. (2011). Dynamic Young's moduli of tungsten and tantalum at high temperature and stress. *Journal of Nuclear Materials*, **409**, 40.

Skylaris, C.-K., Haynes, P. D., Mostofi, A. A., and Payne, M. C. (2005). Introducing ONETEP: Linear-scaling density functional simulations on parallel computers. *Journal of Chemical Physics*, **122**, 084119.

Slater, J. C. (1929). The theory of complex spectra. *Physical Review*, **34**, 1293.

Slater, J. C. (1951). A simplification of the Hartree–Fock method. *Physical Review*, **81**, 385.

Slater, J. C. and Johnson, K. H. (1972). Self-consistent-field $X\alpha$ cluster method for polyatomic molecules and solids. *Physical Review B*, **5**, 844.

Solin, S. A. and Ramdas, A. K. (1970). Raman spectrum of diamond. *Physical Review B*, **1**, 1687.

Souza, I., Íñiguez, J., and Vanderbilt, D. (2002). First-principles approach to insulators in finite electric fields. *Physical Review Letters*, **89**, 117602.

Spiegel, M., Lipschutz, S., Schiller, J., and Spellman, D. (2009). *Schaum's outline of complex variables* (2nd edn). McGraw-Hill, New York.

Squires, G. L. (1978). *Introduction to the theory of thermal neutron scattering*. Cambridge University Press, Cambridge.

Stampfl, A. P. J., Con Foo, J. A., Leckey, R. C. G., Riley, J. D., Denecke, R., and Ley, L. (1995). Mapping the Fermi surface of Cu using ARUPS. *Surface Science*, **331**, 1272.

Starace, A. F. (1971). Length and velocity formulas in approximate oscillator-strength calculations. *Physical Review A*, **3**, 1242.

Steeb, W. H. and Shi, T. K. (1997). *Matrix calculus and Kronecker product with applications and C++ programs*. World Scientific, Singapore.

Stipe, B. C., Rezaei, M. A., and Ho, W. (1998). Single-molecule vibrational spectroscopy and microscopy. *Science*, **280**, 1732.

Stoner, E. C. (1938). Collective electron ferromagnetism. *Proceedings of the Royal Society of London A*, **165**, 372.

Strinati, G. (1984). Effects of dynamical screening on resonances at inner-shell thresholds in semiconductors. *Physical Review B*, **29**, 5718.

Strinati, G., Mattausch, H. J., and Hanke, W. (1982). Dynamical aspects of correlation corrections in a covalent crystal. *Physical Review B*, **25**, 2867.

Stuart, B. (2004). *Infrared spectroscopy*. Wiley, New York.

Sze, S. M. (1981). *Physics of semiconductor devices* (2nd edn). Wiley, New York.

Tersoff, J. and Hamann, D. R. (1985). Theory of the scanning tunneling microscope. *Physical Review B*, **31**, 805.

Teter, M. P., Payne, M. C., and Allan, D. C. (1989). Solution of Schrödinger's equation for large systems. *Physical Review B*, **40**, 12255.

Thaddeus, P. and McCarthy, M. C. (2001). Carbon chains and rings in the laboratory and in space. *Spectrochimica Acta Part A*, **57**, 757.

Thomas, L. H. (1927). The calculation of atomic fields. *Mathematical Proceedings of the Cambridge Philosophical Society*, **23**, 542.

Tiedje, T., Yablonovitch, E., Cody, G. D., and Brooks, B. G. (1984). Limiting efficiency of silicon solar cells. *IEEE Transactions on Electron Devices*, **31**, 711.

Timoshenko, S. (1951). *Theory of elasticity*. McGraw-Hill, New York.

Torsti, T., Eirola, T., Enkovaara, J., Hakala, T., Havu, P., Havu, V., Höynälänmaa, T., Ignatius, J., Lyly, M., Makkonen, I., Rantala, T. T., Ruokolainen, J., Ruotsalainen, K., Räsänen, E., Saarikoski, H., and Puska, M. J. (2006). Three real-space discretization techniques in electronic structure calculations. *Physica Status Solidi B*, **243**, 1016.

Tran, F. and Blaha, P. (2009). Accurate band gaps of semiconductors and insulators with a semilocal exchange-correlation potential. *Physical Review Letters*, **102**, 226401.

Troullier, N. and Martins, J. L. (1991). Efficient pseudopotentials for plane-wave calculations. *Physical Review B*, **43**, 1993.

Tsai, D. H. (1979). The virial theorem and stress calculation in molecular dynamics. *Journal of Chemical Physics*, **70**, 1375.

Umari, P. and Pasquarello, A. (2002). *Ab initio* molecular dynamics in a finite homogeneous electric field. *Physical Review Letters*, **89**, 157602.

Umari, P., Gonze, X., and Pasquarello, A. (2003). Concentration of small ring structures in vitreous silica from a first-principles analysis of the Raman spectrum. *Physical Review Letters*, **90**, 027401.

Umrigar, C. J. and Gonze, X. (1994). Accurate exchange-correlation potentials and total-energy components for the helium isoelectronic series. *Physical Review A*, **50**, 3827.

Van Hove, L. (1954). Correlations in space and time and Born approximation scattering in systems of interacting particles. *Physical Review*, **95**, 249.

Vanderbilt, D. (1990). Soft self-consistent pseudopotentials in a generalized eigenvalue formalism. *Physical Review B*, **41**, 7892.

Verlet, L. (1967). Computer 'experiments' on classical fluids. I. Thermodynamical properties of Lennard-Jones molecules. *Physical Review*, **159**, 98.

Vignale, G. and Rasolt, M. (1988). Current- and spin-density-functional theory for inhomogeneous electronic systems in strong magnetic fields. *Physical Review B*, **37**, 10685.

von Ballmoos, C., Wiedenmann, A., and Dimroth, P. (2009). Essentials for ATP synthesis by F1F0 ATP synthases. *Annual Review of Biochemistry*, **78**, 649.

von Barth, U. and Hedin, L. (1972). A local exchange-correlation potential for the spin polarized case. I. *Journal of Physics C: Solid State Physics*, **5**, 1629.

Vosko, S. H., Wilk, L., and Nusair, M. (1980). Accurate spin-dependent electron liquid correlation energies for local spin density calculations: A critical analysis. *Canadian Journal of Physics*, **58**, 1200.

Wallace, D. C. (1998). *Thermodynamics of Crystals*. Dover, New York.

Wang, J., Neaton, J. B., Zheng, H., Nagarajan, V., Ogale, S. B., Liu, B., Viehland, D., Vaithyanathan, V., Schlom, D. G., Waghmare, U. V., Spaldin, N. A., Rabe, K. M., Wuttig, M., and Ramesh, R. (2003). Epitaxial $BiFeO_3$ multiferroic thin film heterostructures. *Science*, **299**, 1719.

Wang, Y., Plackowski, T., and Junod, A. (2001). Specific heat in the superconducting and normal state (2-300 K, 0-16 T), and magnetic susceptibility of the 38 K superconductor MgB_2: Evidence for a multicomponent gap. *Physica C: Superconductivity*, **355**, 179.

Warren, J. L., Yarnell, J. L., Dolling, G., and Cowley, R. A. (1967). Lattice dynamics of diamond. *Physical Review*, **158**, 805.

Warshel, A. and Levitt, M. (1976). Theoretical studies of enzymic reactions: Dielectric, electrostatic and steric stabilization of the carbonium ion in the reaction of lysozyme. *Journal of Molecular Biology*, **103**, 227.

Wentzcovitch, R. M. and Stixrude, L. (2010). Theoretical and computational methods in mineral physics: Geophysical applications. *Reviews in Mineralogy and Geochemistry*, **71**, 1.

Wiesendanger, R. (1994). *Scanning probe microscopy and spectroscopy: Methods and applications*. Cambridge University Press, Cambridge.

Wigner, E. and Huntington, H. B. (1935). On the possibility of a metallic modification of hydrogen. *Journal of Chemical Physics*, **3**, 764.

Williams, Q., Jeanloz, R., Bass, J., Svendsen, B., and Ahrens, T. J. (1987). The melting curve of iron to 250 gigapascals: A constraint on the temperature at Earth's center. *Science*, **236**, 181.

Wiser, N. (1963). Dielectric constant with local field effects included. *Physical Review*, **129**, 62.

Yang, H. D., Lin, J.-Y., Li, H. H., Hsu, F. H., Liu, C. J., Li, S.-C., Yu, R.-C., and Jin, C.-Q. (2001). Order parameter of MgB_2: A fully gapped superconductor. *Physical Review Letters*, **87**, 167003.

Yang, W. (1991). Direct calculation of electron density in density-functional theory. *Physical Review Letters*, **66**, 1438.

Yin, M. T. and Cohen, M. L. (1982a). Theory of lattice-dynamical properties of solids: Application to Si and Ge. *Physical Review B*, **26**, 3259.

Yin, M. T. and Cohen, M. L. (1982b). Theory of static structural properties, crystal stability, and phase transformations: Application to Si and Ge. *Physical Review B*, **26**, 5668.

Yokoya, T., Nakamura, T., Matushita, T., Muro, T., Okazaki, H., Arita, M., Shimada, K., Namatame, H., Taniguchi, M., Takano, Y., Nagao, M., Takenouchi, T., Kawarada, H., and Oguchi, T. (2006). Soft X-ray angle-resolved photoemission spectroscopy of heavily boron-doped superconducting diamond films. *Science and Technology of Advanced Materials*, **7**, Supplement 1, S12.

Yokoyama, T. and Takayanagi, K. (2000). Anomalous flipping motions of buckled dimers on the Si(001) surface at 5 K. *Physical Review B*, **61**, R5078.

Yoo, C. S., Holmes, N. C., Ross, M., Webb, D. J., and Pike, C. (1993). Shock temperatures and melting of iron at Earth core conditions. *Physical Review Letters*, **70**, 3931.

Yu, P. Y. and Cardona, M. (2010). *Fundamentals of semiconductors*. Springer, Berlin.

Zeller, R. (2006). Spin-polarized DFT calculations and magnetism. *Computational nanoscience: Do it yourself*, **31**, 419.

Ziman, J. M. (1960). *Electrons and phonons: the theory of transport phenomena in solids*. Clarendon Press, Oxford.

Ziman, J. M. (1972). *Principles of the theory of solids*. Cambridge University Press, Cambridge.

Ziman, J. M. (1975). *Elements of advanced quantum theory*. Cambridge University Press, Cambridge.

Index

ab initio, 1

absorption
  coefficient, 192
  cross-section, 187
  direct, 198
  phonon-assisted, 201
anharmonic effects, 134
approximation
  adiabatic, 53
  Born–Oppenheimer, 53
  clamped nuclei, 25
  frozen core, 259
  generalized gradient, 14, 69
  harmonic, 105, 108, 133
  independent electrons, 27
  local density, 13, 40, 45
  mean-field, 31
  quasiharmonic, 146

band
  conduction, 171
  gap, 168, 173
  structure, 159, 166, 170, 255
  valence, 171
basis set, 251
Bethe–Salpeter approach, 202
Bloch sphere, 217
Bohr magneton, 211
Born–von Karman force constants, 107
bosons, 145
Brillouin zone, 117, 157, 256
  first, 256
brittle fracture, 10
bulk modulus, 97

calculation
  non-self-consistent, 159
  self-consistent, 47
catalyst, 11
collective displacements, 110
computational cell, 246
copper, 159
correlation, 35, 57
  LDA, 43
cumulene, 116

decay rate, 184
density functional
  perturbation theory, 14, 121

theory, 36
  limitations, 49
  time-dependent, 13, 205
density matrix, 229
density of states
  electrons, 83, 160
  phonon, 141, 145
descriptor, 11
diamond, 121, 127, 166
  nano, 4
dielectric
  constant, 199, 200, 205
  function, 188, 205
  matrix, 204
dipole
  electric, 185
    transitions, 186, 198
  magnetic, 208
Dirac
  delta function, 74
  equation, 209
  notation, 32
dispersion relations, 118
distribution
  Bose–Einstein, 144
  Fermi–Dirac, 82
dynamical matrix, 109, 120

effective mass, 171
  longitudinal, 173
  transverse, 173
Einstein convention, 91
elastic regime, 87
electron
  affinity, 174
  core, 67, 258
  density, 21, 215, 229
  dynamics, 182
  excited states, 193
  gas, 40, 44, 234
  valence, 67, 258
electron–hole interaction, 201
electronic structure theory, 26
energy
  band structure, 153
  binding, 66, 165
  cohesive, 69
  correlation, 42
  dissociation, 67
  exchange, 41